THE ANTHOCYANIN PIGMENTS OF PLANTS

THE ANTHOCYANIN PIGMENTS OF PLANTS

BY

MURIEL WHELDALE ONSLOW, M.A.

FORMERLY FELLOW OF NEWNHAM COLLEGE, CAMBRIDGE, AND
RESEARCH STUDENT AT THE JOHN INNES HORTICULTURAL
INSTITUTION, MERTON, SURREY

SECOND EDITION

CAMBRIDGE
AT THE UNIVERSITY PRESS
1925

CAMBRIDGE
UNIVERSITY PRESS

University Printing House, Cambridge CB2 8BS, United Kingdom

Published in the United States of America by Cambridge University Press, New York

Cambridge University Press is part of the University of Cambridge.

It furthers the University's mission by disseminating knowledge in the pursuit of education, learning and research at the highest international levels of excellence.

www.cambridge.org
Information on this title: www.cambridge.org/9781107630901

First edition 1916
Second edition 1925
First published 1925
First paperback edition 2014

A catalogue record for this publication is available from the British Library

ISBN 978-1-107-63090-1 Paperback

PREFACE

OF the various investigations which have been made upon the anthocyanin pigments along botanical, chemical and genetical lines, no complete account has yet been written. It is the object of this book to provide such an account of the work which has been done. Although it is only within recent years that any very notable researches have been made upon these pigments, I feel that consideration is due to the many workers who, in the course of the last century, have paved the way for their successors. This I offer as my excuse for dwelling in the following pages upon some researches which are now almost entirely superseded.

I do not pretend to claim that anthocyanins will ever have a great significance from the strictly botanical point of view. Even when the obscurity which surrounds their physiological functions is elucidated, it can scarcely be expected that they will have a significance in the least comparable, for instance, to that of chlorophyll. From the strictly chemical standpoint, as chemical compounds, they have a certain interest. But I believe it to be in connection with problems of inheritance that they will provide a great and interesting field for research.

We have now ample evidence that the development in plants of many and various anthocyanin pigments affords an almost unlimited supply of material for the study of inheritance. It must also be patent to those who have been working on the subject of Genetics that a proper conception of the inter-relationships and inheritance of the manifold characters of animals and plants will be greatly facilitated by a knowledge of the chemical substances and reactions of which these characters are largely the outward expression. Herein lies the interest connected with anthocyanin pigments. For we have now, on the one hand, satisfactory methods for the isolation, analyses and determination of the constitutional formulae of these pigments. On the other hand, we have the

Mendelian methods for determining the laws of their inheritance. By a combination of the two methods, we are within reasonable distance of being able to express some of the phenomena of inheritance in terms of chemical composition and structure. There can be little doubt that exact information of this kind must be at least helpful for the true understanding of the vital and important subject of Heredity.

In the preparation of this book I gratefully acknowledge the help afforded to me by many of my friends, and I am especially indebted to Mrs E. A. Newell Arber for kindly correcting my proofs.

To Professor Bateson, F.R.S., my sincerest thanks are due for the great interest he has taken in much of my work which is included in this volume, and for his many valuable suggestions and criticisms. I wish also to record my thanks to Dr F. F. Blackman, F.R.S., for criticisms and assistance with the manuscript.

I regret that some of the most recent and important work on the subject has not been altogether successfully incorporated in the book, owing to the difficulty I have experienced in learning, at the earliest opportunity, of the results obtained by scientists in other countries during this and the preceding year.

M. W.

CAMBRIDGE,
May, 1916.

PREFACE TO SECOND EDITION

SINCE the appearance of the first edition the publications of greatest value on the subject of anthocyanin pigments have been in connection with the chemistry and biochemistry of these substances. This later work has now been included, and the present state of our knowledge of the significance of the pigments in relation to plant metabolism has, as far as possible, been indicated.

The consideration of the more recent investigations on Genetics in which the anthocyanin pigments are involved has presented difficulties, in view of the large number of papers published. An attempt has been made to record the majority of these publications, but it has been impossible, within the size and scope of this book, to give detailed accounts of the researches on the inheritance of soluble pigments. Such curtailment, however, should not be serious, for many of the results indicated have not greatly increased our special knowledge, either of the inter-relationships of the pigments, or of the bearing of their chemical constitution on the inheritance of colour.

Recent work on Genetics, nevertheless, has considerably broadened our general conceptions of Mendelian factors and has prepared the way towards that elucidation which, unquestionably, in the course of time, will be found of the relations between the chemistry and biochemistry of the anthocyanin pigments and the factors for colour-inheritance.

M. W. O.

CAMBRIDGE,
April, 1925.

CONTENTS

PART I

GENERAL ACCOUNT

PART II

ANTHOCYANINS AND GENETICS

PART I

GENERAL ACCOUNT

Idea 27. And *first*, their *Colours*; where, with respect to several *Plants* and *Parts*, they are more *Changeable*; as Red, in *Flowers*; or *Constant*, as Green in *Leaves*. Which, with respect to several Ages of one *Part*, are more *fading*, as Green in *Fruits*; or *durable*, as Yellow in *Flowers*. In what *Parts* more *Single*, as always in the *Seed*; or more *Compounded*, as in the *Flower*; and in what *Plants* more especially, as in *Pancy*. Which proper to *Plants* that have such a *Taste* or *Smell*, as both, in *White Flowers*, are usually less strong. To *Plants* that flower in such a *Season*, as a *Yellow Flower*, I think, chiefly, to *Spring Plants*. And to *Plants* that are natural to such a *Soil* or *Seat*, as to *Water-plants*, more usually, a *white Flower*. What, amongst all *Colours*, more Common to *Plants*, as *Green*; or more Rare, as *Black*. And what all these Varieties of *Colours* are upon *Cultivation*, but chiefly, in their natural *Soil*. To observe also with their superficial *Colours*, those within: so the *Roots* of *Docks*, are *Yellow*; of *Bistort*, *Red*; of *Avens*, Purple; but of most, White. Where the Inward, and Superficial *Colours* agree; as in *Leaves*; or vary, as in the other Parts frequently. And in what manner they are *Situated*; some universally spreading, others running only along with the *Vessels*, as in the *Leaves* of Red *Dock*, and the *Flowers* of *Wood-Sorrel*.

From Nehemiah Grew's *Anatomy of Plants, with an Idea of a Philosophical History of Plants*. 1682.

PART I
GENERAL ACCOUNT

CHAPTER I
INTRODUCTORY

BY comparison with other products the anthocyanin pigments of plants have now received considerable attention. Such colouring matters have sometimes been spoken of as soluble pigments, since they are in a state of solution in the cell-sap, as contrasted with those which are in some way bound up with the structure of organised protoplasmic bodies known as plastids. The innumerable shades of blue, purple, violet, mauve and magenta, and nearly all the reds, which appear in flowers, fruits, leaves and stems are due to anthocyanin pigments. On the other hand, green, and the large majority of orange and yellow colours, are in the form of plastid pigments. Other colours, again, notably some scarlets and orange-reds, browns and even black, are produced by the presence of both plastid and anthocyanin pigments together in the tissues.

For almost a hundred years botanists have employed one term to denote the soluble pigments—anthocyanin[1] (ἄνθος, κύανος), a word first coined by Marquart (5)[2] for these substances in 1835 and retained in the same sense to the present day; other rival terms, now obsolete, such as erythrophyll, cyanophyll and cyanin have also been used from time to time.

Vegetable pigments served as matter for investigation at a very early date, and the property which first attracted attention was their behaviour towards acids and bases, that is, the reddening by acids and the formation of green coloured products with a base. Thus, in 1664, in the *Experiments and Considerations touching Colours* of Robert Boyle (121), we find the following directions: "Take good Syrrup of Violets, Imprægnated with the Tincture of the flowers, drop a little of it upon a

[1] In the present volume anthocyanin is largely used in the collective sense, that is as a term which we know to include a number of substances; sometimes however, when the context demands it, the plural form is used.

[2] Numbers in brackets refer to papers, etc., in the Bibliography at the end of the book, and such references are either entirely concerned with anthocyanins, or have some direct bearing upon them. References, on the contrary, in the foot-notes are not directly concerned with anthocyanins.

White Paper...and on this Liquor let fall two or three drops of Spirit either of Salt or Vinegar, or almost any other eminently Acid Liquor, & upon the Mixture of these you shall find the Syrrup immediately turn'd Red,...But to improve the Experiment, let me add what has not...been hitherto observ'd,...namely, that if instead of Spirit of Salt, or that of Vinegar, you drop upon the Syrrup of Violets a little Oyl of Tartar *per Deliquium*, or the like quantity of Solution of Potashes, & rubb them together with your finger, you shall find the Blew Colour of the Syrrup turn'd in a moment into a perfect Green."

And again in *The Anatomy of Plants* of 1682, Nehemiah Grew (1) tells us that "...*Spirit* of *Harts Horn* droped upon a *Tincture* of the *Flower* of *Lark-heel* & *Borage* turn them to a *verdegreese Green.*"

"*Spirit* of *Sulphur* on a *Tincture* of *Violets* turns it from *Blew* to a true *Lacke*, or midle *Crimson.*"

"*Spirit* of *Sulphur* upon a *Tincture* of *Clove-July-Flowers* makes a bright blood *Red*. Into the like *Colour*, it hightens a *Tincture* of *Red Roses.*"

It is not clear from the writings of Boyle and Nehemiah Grew whether these authors considered the great variety of shades to be the expression of one or of many substances, though it is clear that Nehemiah Grew differentiated between red, blue and green pigments as regards their solubilities. For he says, "...The *Liquors* I made use of for this purpose, were three, *sc. Oyl* of *Olives, Water,* & *Spirit* of *Wine*. The *Water* I used was from the *Thames*, because I could not procure any clear *Rain Water*, & had not leasure at present to distill any. But next to this, that yields as little *Salt*, as any." The oil he found dissolved the green colour, "...But there is no *Vegetable* yet known which gives a true *Red* to *Oyl*, except *Alkanet Root.*" This root "which immediately tinctures *Oyl* with a deeper *Red*, will not colour *Water* in the least. Next it is observable, That *Water* will take all the *Colours* of *Plants* in *Infusion* except a *Green*. So that as no *Plant* will by *Infusion* give a perfect *Blew* to *Oyl*; so their is none, that I know of, which, by *Infusion* will give a perfect *Green* to *Water*[1]." Boyle, on the other hand, was evidently impressed by the uniformity of the acid and alkali reactions: "...it is somewhat surprizing to see,...how Differingly-colour'd Flowers, or Blossoms,...how remote soever their Colours be from Green, would in a moment pass into a deep

[1] This is entirely in accordance with our present-day knowledge of the pigments. Chlorophyll, like all plastid pigments, is insoluble in water, but soluble in oil. The red, blue and purple anthocyanin pigments are soluble in water, but insoluble in oil. The red pigment of Alkanet root is not an anthocyanin and has quite different solubilities.

Degree of that Colour, upon the Touch of an Alcalizate Liquor." The gradual evolution of the idea of the multiplicity of pigments included under anthocyanin can only be realised in a comprehensive survey of the subject.

After the preliminary and somewhat diffuse observations of the earliest scientists, there follows a period during which certain definite lines of investigation emerge, and it is practically to these lines that the chapters in this book correspond. It may not be out of place, however, to give a general account of the different phases and kinds of investigation showing how they have developed, and how they are related to each other in the complete history.

The above lines of investigation may be enumerated thus: the morphological and histological distribution of the pigments, the factors controlling their formation, their function, their chemical composition, their mode of origin, and lastly, the part they have played in heredity.

First let us consider their distribution. Contributions to this portion of the subject naturally formed a great part, though by no means the whole, of the earlier work on pigments, since it merely involved general observation and microscopical examination of petals and leaves. Various writers published full accounts of the pigments of flowers, fruits, leaves, etc., showing how some colours are due to plastid, others to soluble pigments, and others again are the result of the combination of both in the cells. As would be expected, this form of investigation has tended to diminish in later years for, the histological basis once laid down, the physiological, chemical and biochemical aspects have come to the front. Mention must be made, however, of the publications of Buscalioni & Pollacci (17) in 1903 on the histological distribution of anthocyanin, and of a paper on similar lines read by Parkin (77) at the meeting of the British Association of the same year. Still later work of this kind is that of Gertz (19) which appeared in 1906; this author made a most thorough and systematic investigation of the occurrence of anthocyanin in representative genera of all natural orders, but the publication of his work in Swedish unfortunately restricts the circulation of his results.

The question of factors controlling anthocyanin formation is the next line of enquiry most convenient for consideration. The main factors concerned are light, temperature and nutrition.

The matter of fundamental significance in all questions of factor control, and one which it will be well to understand before turning to the factors in detail, is the dependence of anthocyanin formation upon the supply of an initial product or chromogen, as it is called, from which the pigment arises. This substance, like all others in plant metabolism, ultimately depends for its existence on a supply of carbohydrates (sugars), the first products of synthesis in the plant. The relationship between anthocyanin formation and the presence of sugars in the tissues has provided an important subject for research. Overton (420, 421), in 1899, first drew attention to its significance[1]; this botanist had noticed, while carrying out some experiments on osmosis, that culture of *Hydrocharis Morsus-ranae* in sugar solutions leads to greater production of anthocyanin in the leaves. Further experiments showed the phenomenon to be constant for quite a number of species when isolated leaves and twigs were fed on solutions of cane sugar, dextrose, laevulose and maltose. Repetition of experiments on these lines at later dates by Katić (441), Gertz (473) and others gave full confirmation to Overton's results. This discovery led Overton to the fairly obvious inference that possibly, in the normal plant, reddening of leaves, etc., is correlated with excess of sugar in the tissues, and he states that by tests upon red autumnal leaves he could detect more sugar in red than in green leaves.

More elaborate and conclusive work in this direction was commenced by Combes in 1909 and carried on in the following years. Combes (461, 472) had noted that decortication in some plants brings about a considerable development of anthocyanin in the leaves above the point of decortication. Analyses were made by him, not only of leaves reddened through this cause, but also of autumnal and other red leaves. In all cases he (222) claims to have shown that the red leaves contain greater quantities of sugars and glucosides than green ones from the same plant. From other, more general, phenomena, in addition to Combes's results, it may be safely inferred that an accumulation of such synthetic products as are manufactured in the leaves leads to production of anthocyanin. For example, we frequently find abnormal reddening of a single leaf on a plant otherwise in full vigour, and investigation almost invariably shows the reddening to be accompanied by injury, and the injury, whether it be due to mechanical cutting or breaking, or to the attacks of insects, will be found to affect those tissues which conduct away the synthetic products of the leaf.

There is little doubt that the chromogen of anthocyanin, in the form of a glucoside, is also manufactured in the leaves, and the whole trend of results goes to show that an enforced accumulation of this chromogen,

[1] The connection between pigmentation and the presence of sugars was also pointed out in 1899 by Mirande (419) as a result of his investigations on the genus *Cuscuta*.

together with sugars, by stoppage of the translocation current, leads to formation of anthocyanin in the leaf; and a similar result may arise from artificial feeding with sugars. To the more precise relationships between chromogen, anthocyanin and sugars attention will be given in a later chapter.

Thus it will be seen that in all problems connected with effects of temperature, light, etc., on anthocyanin formation, we are confronted with two distinct questions, i.e. the direct effect of light and temperature on the actual production of pigment, and the indirect effect of these factors on the supply of organic compounds from which the chromogen of anthocyanin is synthesised.

The relationship between pigment formation and light constitutes a problem to which there is no very satisfactory solution. Sachs (356, 358), Askenasy (369) and others have tried the obvious methods of growing plants in the dark with controls in the light, of darkening leaves while leaving inflorescences uncovered and so forth. The outcome of these researches, as well as of several others, has been to show that in many cases, for example, in flowers of *Tulipa*, *Hyacinthus*, *Iris* and *Crocus*, anthocyanin develops equally well in the dark; in other cases, such as *Pulmonaria*, *Antirrhinum*, and *Prunella*, the development is feeble or absent. A general survey of anthocyanin distribution leaves us in no doubt that, as far as organs where anthocyanin may be expected to develop are concerned, the greatest production takes place in the most illuminated parts. But we have on the other hand not a few examples, of which the root of *Beta* is a good illustration, of development of pigment in total darkness. In the absence of fuller evidence, the most reasonable point of view is that the actual process of pigment formation may in itself be entirely independent of light, should the tissues contain sufficient reserve materials to supply the chromogen. But if there is a shortage of reserve materials, such as would arise from diminished photosynthesis, the anthocyanin may fail to appear from lack of chromogen.

The problem of the effect of temperature offers similar difficulties. Does the temperature influence the actual formation of pigment, or is it again an indirect cause, making itself felt only through its effects upon the supply of materials from which the pigment is synthesised? That low temperature favours pigment formation would seem to be demonstrated by autumnal coloration, and the winter reddening of leaves of *Hedera*, *Ligustrum*, *Mahonia* and other evergreens. Conversely, Overton (420) found in *Hydrocharis*, the higher the temperature, the less anthocyanin. Klebs (447) also notes that flowers of *Campanula* and *Primula* may be almost white in a hot-house, but the same individuals kept in the cold will bear coloured flowers. The consideration of temperature is perhaps more difficult than that of light; for low temperature, on the one hand, retards photosynthesis by which sugars are formed, but, on the other hand, it also retards growth, starch formation and probably translocation, thereby tending to raise the sugar contents of the tissues. High temperature, on the contrary, accelerates growth and respiration, and consequently tends to prevent the accumulation of any excess of synthetic products.

An interesting application of these views upon light and temperature effects can be made in the case of Alpine flower coloration. The subject has been extensively studied by Gaston Bonnier (375, 376, 381, 394, 415), Flahault (375, 376, 377, 379) and others, and has had a great vogue with the writers on flower coloration in connection with insect pollination. The special features of the case are intensity of flower-colour and the formation of anthocyanin in the vegetative parts. Gaston Bonnier & Flahault have compared individuals grown at heights of 2300 metres with individuals grown in the plains, and have found that the latter produce paler flowers and less anthocyanin in the leaves and stems. It seems most reasonable to suppose that these phenomena form a natural demonstration of some of the relationships which we have just been considering between colour and factors. High Alpine plants are stunted in growth, i.e. little material is expended vegetatively; they are exposed by day to intense insolation while the night temperature is low. One may therefore suppose photosynthesis to be very active, whereas starch formation, translocation and growth are retarded, these being conditions which favour high sugar and chromogen concentrations in the tissues and resultant abundance of pigment.

From considerations of the above nature in greater detail in a later chapter, it will be seen that practically all the conditions which favour anthocyanin production also result in high sugar concentration. The latter leads to increased respiration and oxygen uptake—also a phenomenon accompanying pigment formation. The suggestion will be brought forward, though as yet there is little if any experimental evidence in support, that the acid products of respiration may play a part in augmenting the coloration.

The so-called functions of anthocyanin have provided material for another main line of research. Two essentially different types of function are readily distinguished, the biological and the physiological. The biological function is a subject which, in itself, needs extensive treatment, and does not lie within the scope of this book. It is solely connected with the attractive value of the coloured floral organs for pollination by insects, and the subsidiary question of the attractive value of ripe pigmented fruits for dispersal by birds. The relationship between flower-colour and entomophily has received great attention from botanists, and the whole matter is dealt with most thoroughly in Knuth's *Handbook of Flower Pollination* (527), which includes an excellent bibliography.

The physiological function is a very difficult and far less satisfactory matter. Several different functions of a physiological nature have been attributed to anthocyanin. One of the most famous is the screen theory, the idea of which was first based on work published in 1880 by Pringsheim, who showed that chlorophyll was bleached by intense light, but not if protected artificially by a red screen. Thus the view arose that anthocyanin might be protective in function, but experimental evidence does not altogether favour this hypothesis. For, in 1885, Reinke pointed out that it is those rays absorbed by chlorophyll which have the greatest destructive effect on chlorophyll, and Engelmann (494), in 1887, demonstrated that the absorption spectrum of anthocyanin is on the whole complementary to that of chlorophyll. Hence anthocyanin absorbs those rays which are least harmful to chlorophyll, and cannot therefore be said to provide an effective screen. A second suggestion, brought forward by Stahl (505) in 1896, and largely supported by him, is that anthocyanin absorbs certain of the sun's rays, and by converting them into heat, raises the temperature of the leaf, and this may serve to accelerate transpiration in difficult circumstances, as in damp regions of the tropics, or may protect leaves from low temperature as in Alpine regions. The chief points in favour of this hypothesis are the distribution of anthocyanin in leaves of shade-loving plants, and the fact, observed also by Stahl, and confirmed in 1909 by Smith (520), that the internal temperature of red leaves is greater than that of green.

The next line of investigation, the chemical composition of the pigments, is also difficult, and though spasmodically attacked from time to time, met with no very serious consideration till 1906. So intimately connected with its chemical composition that it can scarcely be considered separately, is the question of the mode of formation of anthocyanin, that is, the chemical reactions involved in the process. Closely connected also, though in a lesser degree, is the part played by anthocyanin pigments in heredity. It is proposed therefore to deal with these three lines of research more or less together.

As already pointed out, the reactions with acids and alkalies are the most obvious and striking chemical properties of anthocyanins, and they have helped to draw the attention of chemists to the subject; for much of the earlier chemical work on anthocyanin, notably of a group of French chemists (1800–1825), Braconnot (124), Payen & Chevallier (127, 128, 129) and Roux (130), centred round these reactions, especially in some cases round their rôle as indicators. But the idea of anthocyanin as an indicator was fully conceived long before 1807 by Robert Boyle (121). "When," he says, "we have a mind to examine, whether or no the Salt predominant in a Liquor or other Body, wherein 'tis Loose and Abundant, belong to the Tribe of *Acid* Salts or not...if such a Body turn the Syrrup of a Red or Reddish Purple Colour, it does for the most part argue the Body (especially if it be a distill'd Liquor) to abound with Acid Salt. But if the Syrrup be made Green, that argues the Predominant Salt to be of a Nature repugnant to that of the Tribe of Acids." After the reactions of anthocyanin with acids and alkalies, other reactions were noticed with iron salts and various reagents, many of which modify the colour as it is modified in nature. These reactions gave rise to views among some chemists, Fremy & Cloëz (140) and Wigand (150), that natural blue, purple and red pigments are modifications of the same substance, brought about by the presence of other compounds in the cell-sap. But as analyses and investigations proceeded, the view of a certain multiplicity of pigments gained the ascendency.

One of the first actual analyses of anthocyanin was carried out in 1849 by Morot (136) who isolated the blue pigment of the Cornflower, and found it to contain carbon, hydrogen and oxygen, with nitrogen as impurity. Ten years later Glénard (143, 144) isolated the pigment of wine, found it also to contain carbon, hydrogen and oxygen, and gave it a percentage formula.

An early suggestion as to the chemical nature of anthocyanin and its mode of formation was that of Wigand (150) in 1862. This author suggested that anthocyanin arises by the oxidation of a colourless tannin-like chromogen, a substance widely distributed in plants and giving a green reaction with iron salts and a yellow reaction with alkalies. The same substance obviously had been noted at an earlier date by Filhol (139, 146) who observed it to be widely distributed, and maintained that the green coloration of anthocyanin with alkalies was due to a mixture of a blue anthocyanin reaction plus the yellow reaction of these accompanying substances; a view also held by Wiesner (149). The idea of the formation of anthocyanin from a tannin-like chromogen by oxidation was, from this time onward, generally accepted by botanists and chemists. It was, for instance, again (1878) suggested by Gautier (164) on the strength of isolation and analyses of the pigments of various wines. Later, further analyses of Vine pigments were made by Gautier (190) and other investigators, notably Heise (182), but their results were in no way concordant except in so far as they agreed that the pigment contained carbon, hydrogen and oxygen only. These were followed again by analyses of the pigment of Bilberry fruits by Heise (193).

In 1895 Weigert (194) made the first attempt at a classification of anthocyanins based upon their chemical reactions, and he differentiated two groups, the 'Weinroth' and the 'Rübenroth,' according to the behaviour of the pigment to acids, alkalies and lead acetate solution. The Rübenroth group is however a small group of anthocyanin pigments of the orders Phytolaccaceae, Amarantaceae and Portulacaceae of the Centrospermae; the Weinroth group, including the greater number of anthocyanins, still remained undifferentiated.

In 1906 Grafe (212) published the results of some careful analyses of the pigment of the Hollyhock (*Althaea rosea*), followed by further analyses (237) of the pigment of scarlet *Pelargonium* flowers in 1911. In this publication Grafe clearly expresses the view that anthocyanin must be regarded as a general term for a number of substances, having similar properties and reactions, but differing somewhat from each other in chemical constitution.

Meanwhile from the year 1900 and onwards, the study of anthocyanin pigments gained a new impetus through their connection with the inheritance of colour in plants. For many experiments on the cross-breeding of colour-varieties were commenced about this time by Bateson, Punnett, Saunders (578, 590) and others working on the lines of Mendelian inheritance. Two of the first plants to be used for this purpose were the Stock (*Matthiola*) and the Sweet Pea (*Lathyrus*). Both these plants have a number of varieties, and although it is doubtful whether we know the character of the original type in *Matthiola*, it is fairly certain that the original wild Sweet Pea resembled the form known as 'Purple Invincible' which has a chocolate standard and purplish-blue alae. We now know from the teachings of heredity that all the coloured varieties of Sweet Pea, and probably many of those of Stocks, have arisen through the loss of certain factors from the type, and by recrossing selected varieties, the type can be obtained again, the process being known as the phenomenon of 'reversion to type.' One of the most striking results (599) of the experiments on Sweet Peas and Stocks was the phenomenon of the production of a plant bearing purple flowers (in the Sweet Pea the original type) by crossing two white-flowered strains. From such results the hypothesis arose, that red colour (anthocyanin) is due to the presence of two factors, *C* and *R*, in the plant, the loss of either factor resulting in an albino. A third factor, *B*, by its presence modifies the red colour to a blue or purple, but unless *C* and *R* are also both present in the plant, *B* is without effect. Hence white plants of known parentage can be selected which are carrying *C*, *R*, or *B* only, or *CB*, or *RB*.

Such a discovery appeared not only highly important in its connection with heredity, but it also provided well-defined material for the solution of the problem as to what chemical processes are involved in anthocyanin formation. Conversely, these processes once discovered, we should also have a chemical interpretation of the Mendelian factors for flower-colour.

The problem was first attacked by the author in 1909. The plant selected for investigation was the Snapdragon, *Antirrhinum majus*, since the inheritance of the flower-colour of this species had already been worked out (617, 638) during the years 1903–9, and confirmed independently by Baur (639). *Antirrhinum majus* presents a case of singular interest. In this species (as in *Matthiola* and *Lathyrus*) two white varieties, or more strictly two varieties which are albinos as regards anthocyanin, when crossed together, produce the magenta pigment (anthocyanin) characteristic of the type; but, whereas in *Lathyrus* and *Matthiola* the two whites producing a purple are identical in appearance, the two albinos in *Antirrhinum* can be distinguished at sight. One, which is known as ivory, is ivory-white in colour, the other, the true white, is dead white. These two varieties must obviously, between them, contain the chemical substances, or the power to produce the chemical substances, essential to the formation of anthocyanin.

The first clue towards an interpretation was provided by the difference in the action of ammonia vapour on the two flowers. The ivory exposed to ammonia turns bright yellow whereas the white is practically unaltered. It has been mentioned previously that Filhol (139), and in fact several of the earlier workers on plant pigments, had noted the occurrence of colourless substances in the plant which gave a bright yellow reaction with alkalies. Moreover, a tentative suggestion was made by Bidgood (18) some years ago (1905) in the *Journal* of the Royal Horticultural Society, that these substances were the colouring matters known as flavone and flavonol pigments. Reference to the work of A. G. Perkin on these pigments provided the information that they have a general distribution in nature, and had attracted the notice of chemists on account of their value as vegetable dyes. Some flavones and flavonols have been found in many species, others in but one, or a few, though this appearance of restriction may be due to lack of information. All members of the group are yellow substances differing slightly from each other in their chemical composition. In the plant they frequently occur as glucosides, in which condition their colour is much paler, and their solubility greater, than in the free state. All give intense yellow or orange-yellow coloration with alkalies, and green or brown coloration with iron salts. This information led the author to conclude that the widely distributed substances, occurring in all plant tissues which give the yellow reaction to ammonia vapour, are flavones or flavonols, and that the pigment of the ivory variety of *Antirrhinum* is of the same nature. The white variety of *Antirrhinum* is without the yellow pigment, and must contain some substance capable of action on the flavone or flavonol to form anthocyanin.

At about this period a series .of papers were published by Palladin (218, 225) setting forth what is known as the theory of 'Atmungspigmente.' Broadly speaking, Palladin supposes plants to contain chromogens which are capable of being oxidised by enzymes to pigments which in turn pass on oxygen to respirable substances in the plant. Among the 'Atmungspigmente' Palladin includes anthocyanin.

Palladin's conception, together with the evidence provided by *Antirrhinum*, led the author (227, 232, 241) to bring forward the hypothesis that anthocyanin is formed from a flavone by the action of an oxidising enzyme or oxidase. Thus the formation of magenta pigment in *Antirrhinum* would be brought about by the action of an oxidase produced by the white variety upon the flavone present in the ivory. Likewise in every plant anthocyanin would be formed by the action of an oxidase upon a flavone, and just as the flavones are a class of substances having similar properties, though showing some variety in structure, the anthocyanins also would have properties in common as a class but would, many of them, differ somewhat in chemical structure.

The oxidase hypothesis of anthocyanin formation was also taken up by Keeble & Armstrong (245, 246). These authors invented a microchemical method for testing for oxidases and peroxidases in plant tissues, and applied the method chiefly to colour varieties of the Chinese Primrose (*Primula sinensis*), a plant largely used for Mendelian experiments. By means of their test they discovered that peroxidases in *Primula* are chiefly confined to the epidermis and the bundle sheaths, both regions where anthocyanin also is localised. A further point of interest was discovered in connection with certain varieties of *Primula* known as dominant whites. These varieties have pigment (anthocyanin) in stems and leaves, though the flowers are white, and they are regarded, from their behaviour in crossing, as coloured forms in which the flower-colour is inhibited by some factor present in the plant. Keeble & Armstrong were able to show that flowers of these varieties gave ordinarily no peroxidase reaction, but after treatment with certain reagents, the inhibitor was removed, and the peroxidase reaction appeared in the petals. On the other hand, they found that the true albinos in *Primula* gave the peroxidase reactions as well as the coloured varieties. Hence to make the hypothesis fit the case of *Primula*, they were obliged to assume the lack of colour in the true albino to be due to lack of chromogen, an assumption which has never been verified.

Meanwhile the author's investigations on *Antirrhinum* proceeded on the lines of isolation and analysis (259) of the pigments of the colour-varieties. These pigments were prepared on as large a scale as possible by making water extracts of the flowers, precipitating the pigment as a lead salt by lead acetate, and again decomposing the salt with sulphuric acid. Filtered from lead sulphate, the pigment was obtained in dilute acid solution. It had been previously ascertained that anthocyanins are largely present in the plant in combination with sugar as glucosides. On boiling with acid, the glucoside is hydrolysed and the pigment, which is less soluble than the glucoside, separates out. This method was adopted for obtaining the pigment from the acid solution. The pigment of the ivory variety of *Antirrhinum* was identified by Bassett and the author with apigenin, a flavone of known constitution occurring in Parsley (*Apium Petroselinum*). Of the anthocyanin-containing varieties of *Antirrhinum* there are two, red and magenta, analogous to the red and purple varieties of *Lathyrus* and *Matthiola*. Analyses were made (270) of the red and magenta pigments prepared separately from several different varieties, and the results were concordant. In the case of both pigments the percentage of oxygen was greater than that in the flavone apigenin. Determination of the molecular weights of the red and magenta pigments, though not obtained to a high degree of accuracy, indicated that the molecules of anthocyanin are considerably larger than those of apigenin.

The conclusions drawn by Bassett and the author (270) were that in *Antirrhinum* the anthocyanins are derived from the flavone apigenin by oxidation, accompanied by condensation, possibly of two flavone molecules, or possibly of a flavone molecule with other aromatic substances present in the plant.

The problem of anthocyanin formation next received a most important series of contributions by Willstätter and a number of his co-workers. In 1913 Willstätter (260) first published some results on anthocyanins in general and the anthocyanin of the Cornflower (*Centaurea Cyanus*) in particular. Willstätter states that all natural anthocyanins are present in the plant in the condition of glucosides, and that many of them, moreover, including the anthocyanin of *Centaurea*, are very unstable in water solution, and readily change in these circumstances to a colourless isomer. The change can be prevented by adding certain neutral salts, and also acids, to the pigment solution. The explanation, according to Willstätter, of these phenomena lies in the fact that anthocyanin is an oxonium compound having a quinonoid structure and containing tetravalent oxygen. The quinonoid structure is rendered stable by the formation of oxonium salts with acids or with neutral salts, such as sodium nitrate and sodium chloride. The pigment itself, in the neutral state, is purple in colour, and has the structure of an inner oxonium salt. With acids it forms red oxonium salts, and with alkalies blue salts, the position of the metal being undetermined. More recently Willstätter (273, 274, 286–294, 300–307) has published the results of the isolation and analyses of a number of other anthocyanins, in addition to that of *Centaurea*, i.e. from the flowers of the Larkspur, Scarlet Geranium, Mallow, Hollyhock, Rose, Chrysanthemum, Viola, Petunia, Poppy, etc.; from Grapes, also, and the fruits of the Bilberry, Cranberry, Cherry, etc. The outcome of these analyses shows that all the above colouring matters are derivatives from three anthocyanin pigments, pelargonidin, cyanidin and delphinidin. In each of the above plants, one of these pigments is present either as a mono- or diglucoside, the sugar varying in different plants. In some cases one or more hydroxyl groups of cyanidin or delphinidin are replaced by the methoxy group OCH_3, that is the pigment is a mono- or dimethyl ester of these compounds. Cyanidin contains one more atom of oxygen than pelargonidin; delphinidin one more than cyanidin. Pelargonidin differs from the flavonol, kaempferol, mainly in the substitution of an oxygen atom by hydrogen; in the same way, cyanidin and delphinidin differ respectively from the flavonols, quercetin and myricetin. Thus, all Willstätter's anthocyanins are related to flavonols and not flavones. Willstätter, has, moreover, synthesised pelargonidin (275), so that the constitution of the anthocyanins is undoubtedly on a sound basis.

In June 1913 an account of certain phenomena first observed by Keeble & Armstrong (255) was published. These observations were concerned with the effect of reducing plant extracts with nascent hydrogen. A number of flowers, among which were the yellow Wallflower, Daffodil, *Crocus* and *Polyanthus*, were extracted with alcohol, and the extracts acidified and treated with zinc dust. Filtered from the zinc, and exposed to air, the solution develops a red colour which changes to green on addition of alkali. Thus Keeble & Armstrong were led to believe that they had made artificial anthocyanin, and that preliminary reduction and subsequent oxidation are the essential processes of anthocyanin formation. In November of the same year, a note was published by Combes (249) in the *Comptes Rendus*, giving a record of practically the same observation though in a much more complete form. By treating with sodium amalgam the acid alcohol solution of a yellowish crystalline substance which he had extracted from *Ampelopsis* leaves, he obtained a fine purple pigment which crystallised in needles. This pigment, he claims, on the basis of chemical and physical properties, to be identical with crystals of natural anthocyanin from the same plant; but no analyses are included. In addition, he found (250) that natural anthocyanin, after treatment with hydrogen peroxide, gives a yellow product identical with the natural yellow substance from which he started.

Combes believes that these results entirely revolutionise our ideas on anthocyanin formation. Thus he says: "*La production expérimentale d'une anthocyane peut donc être considérée comme réalisée.* Ce résultat permet d'entrevoir comme très proche la solution du problème de la formation des pigments anthocyaniques posé depuis plus de 120 ans et qui fut abordé par de nombreux physiologistes. On sait que dans toutes les hypothèses relatives à cette question, formulées depuis 1825, la pigmentation a toujours été considérée comme un phénomène d'oxydation; cette opinion ne peut plus être soutenue, puisqu'il apparaît que l'anthocyane des feuilles rouges prend naissance lorsque le composé correspondant contenu dans les feuilles vertes est soumis à l'hydrogène naissant, c'est-à-dire dans un milieu qui est au contraire réducteur."

The essential difference between Keeble & Armstrong's results and those of Combes can be explained in the following way. The purple pigment, as indeed the natural anthocyanin, will become colourless on powerful reduction, though it regains its colour in air. Thus Keeble & Armstrong's reduction was carried beyond the preliminary formation of pigment to the colourless stage, the colour returning again on exposure to air, whereas Combes's product was not reduced so far.

The production of a purple colour on reduction of flavone with sodium amalgam had been known to chemists some time before the work of Keeble & Armstrong. The almost universal distribution of flavones naturally explains the appearance of the colour on reduction of many plant extracts. The crystalline yellow product used by Combes is undoubtedly a flavone also, as will be seen on referring to his original communications. The question at issue is whether these purple pigments are anthocyanins.

The problem of the artificial formation of anthocyanin was attacked by Willstätter (274) also in 1914. By treating quercetin, according to Combes's method, Willstätter in the same way obtains a purple pigment, the bulk of which he finds is unstable in acid solution on boiling, though a certain amount of colour remains after this process. The product unstable to heat is, in Willstätter's opinion, not a true anthocyanin, but he maintains that the small amount of stable product left after heating is, when isolated, identical with the anthocyanin (cyanidin) of the Cornflower. Thus Willstätter regards this reaction as giving further confirmation of his suggested constitution of anthocyanin. To quote from his paper: "Die Bildung von Cyanidin aus Quercetin hat zweifache Bedeutung. Es ist dadurch eine Synthese von Cyanidin ausgeführt, da das Quercetin selbst vor zehn Jahren von St. von Kostanecki...synthetisch dargestellt worden ist. Ferner wird durch diese Umwandlung die Konstitutionsformel des Cyanidins bewiesen." Willstätter, however, draws no conclusion as to the origin of anthocyanin in the plant.

The author has been so far unable to bring the results provided by the case of *Antirrhinum* into line with Willstätter's views. In *Antirrhinum* the flavone is, without doubt, apigenin; when treated with nascent hydrogen this flavone produces a purple pigment, yet the latter does not resemble the natural anthocyanin of *Antirrhinum*, either in composition or properties (272).

In 1920 and 1921, the author[1] published a series of papers on oxidising enzymes. In these it was shown that about 60 % of the Angiosperms contain direct oxidases, i.e. the tissues blue guaiacum tincture. The remainder contain peroxidases, that is the tissues do not blue guaiacum unless hydrogen peroxide is added. An oxidase, moreover, is a system consisting of two enzymes, an oxygenase and a peroxidase, and a substance containing the dihydroxy grouping of catechol. The oxygenase catalyses the autoxidation of the 'catechol substance' with formation

[1] See pp. 108, 109.

of peroxide. The peroxidase decomposes the peroxide with liberation of active oxygen which may be accepted by guaiacum or other acceptor.

That the direct oxidase system is concerned with the formation of anthocyanin seems now very unlikely, as a considerable percentage of anthocyanin-forming plants are without the system and only contain peroxidase. *Primula* and *Lathyrus* are not oxidase plants. *Antirrhinum* contains oxidase, but there is apparently no connection between oxidase and factorial constitution.

There is, however, still some evidence (Keeble & Armstrong, 246) of a connection between the presence of anthocyanin and that of peroxidase in tissues. It is possible that the connection is merely indirect, and that conditions which favour the formation of anthocyanin are favourable to the development of the peroxidase reaction.

The artificial preparation of cyanidin from quercetin by Willstätter led to the wide-spread conclusion that *in the plant* anthocyanins are formed from flavones by simple reduction. Coupled with this were the observations already mentioned of Keeble & Armstrong and Combes and those further amplified by Everest (265), that extracts of the naturally-occurring flavones will give, on reduction with nascent hydrogen, red pigments said, on the basis of qualitative reactions, to be anthocyanins.

Certain difficulties, however, lie in the way of acceptance of this view. For the anthocyanins isolated by Willstätter must, as we have already seen, be derived by reduction from kaempferol, quercetin or myricetin. It is reasonable to expect that the anthocyanin of any plant should be accompanied by the flavone from which it is derived. In two plants only, *Delphinium Ajacis* and *Viola tricolor*, is there evidence based on sound experimental work. In *Delphinium Ajacis*, the anthocyanin is delphinidin, the flavone kaempferol; in *Viola tricolor*, the flavone is quercetin, the anthocyanin again a derivative of delphinidin. Hence the difficulty still remains. In no plant has both flavone and anthocyanin been isolated and analysed, and, at the same time, the artificial anthocyanin prepared by reduction from the extracted flavone, analysed and found to be identical, on accurate data, with the natural anthocyanin.

The chemical work on anthocyanin has not given the solution as to its mode of formation. The constitution of anthocyanins and flavones is known, but it is not clear whether, *in the plant*, the anthocyanins are derived from the flavones or formed independently.

Quite recently, the subject has been reopened. Various investigators, Noack (315), Jonesco (331, 341), and Kozlowski (335) have extracted, with various solvents, colourless or yellow compounds from plants, which, on heating with acids, either with or without oxidising agents, or by treatment with oxidase, have formed red pigments having the qualitative relations of anthocyanins. Some of these authors as a result have reverted to the view that anthocyanins are derived from flavones by oxidation, and also may exist in the plant in the form of their colourless isomers. The situation, as pointed out by Combes (336, 337, 349), is complicated by the fact that phlobatannins are frequently present in the plants used, and these, on treatment with acids, give rise to phlobaphenes having qualitative reactions very similar to anthocyanins. In the absence of isolation and complete analyses, no conclusions can be drawn from these results.

From the point of view of Genetics it is helpful to know from Willstätter's work the nature and constitution of many anthocyanins, and also that the same pigment may be common to different genera and species. That, moreover, the pigment of the 'true red' group, for example the pink variety of *Centaurea*, may be different from that of the 'purple-red' or 'blue' groups, for instance the purple and blue *Centaurea*.

If the colourless isomer of anthocyanin is present in plants, it is possible that the intensification of colour by some factors may be due to increased production of acid. There is, however, no experimental support of this suggestion at present.

Genetics, on the other hand, has given us no knowledge as to the origin of anthocyanin. The white and certain ivory varieties of *Antirrhinum* contain between them the components of anthocyanin; so also do the *C* and *R* whites of *Lathyrus*. The white of *Antirrhinum* is without flavone, but there are now no grounds for believing that the oxidase it contains acts on the flavone provided by the ivory to form anthocyanin. The nature of the reddening factor is still unknown.

The genetics of flower-colour in the Maize, Sweet Pea and Primula has reached a stage of considerable, though ordered, complexity, in which the many shades have their factorial counterparts, and even the location of some of the colour factors in the chromosomes is defined. These results have thrown no light on the system producing anthocyanin.

It is more than likely that the complexes of chemical substances, which produce the different pigments and different amounts of the same pigment, are the outcome of the activity of different "patterns of protoplasm," ultimately, in part at least, traceable back to chromosomes.

CHAPTER II

THE MORPHOLOGICAL DISTRIBUTION OF ANTHOCYANINS

It is a striking fact that anthocyanins are almost universally distributed. This is not fully realised until a systematic investigation is made, and then it is seen how few plants are entirely without these pigments. Many plants may appear to the casual observer to be free from anthocyanin, but a closer examination generally reveals its presence, often only in minute quantity, in such places as the base of the stem, the petioles, bud-scales, bracts or even in some unexpected organ, as for instance, the anthers and stigma. Again, an abnormal condition of drought, or fungal attack, may cause its appearance when otherwise the healthy plant is green. A number of plants, moreover, only form anthocyanin in their young leaves at the beginning, or in their old leaves at the close, of the vegetative season, or pigment may develop shortly before the death of the whole plant. Reviewing the flowering plants as a whole, two orders present an anomaly with respect to the distribution of anthocyanin. The power to form anthocyanin seems to be absent, first, from the Cucurbitaceae, as far as it has been possible to ascertain, and secondly, from certain genera of the Amaryllidaceae (*Leucojum*, *Galanthus*, *Ornithogalum* and *Narcissus*). This observation is confirmed by Gertz (19), who also adds to the list, *Herniaria* (Caryophyllaceae) and *Chrysosplenium* (Saxifragaceae), some species of *Potamogeton*, *Eleocharis* (Cyperaceae), *Kniphofia*, *Aloe*, *Reseda*, i.e. *R. odorata* and *R. lutea*, and *Buxus*. To the statement as regards *R. odorata* and *R. lutea* exception must be taken, since both these species form anthocyanin in their leaves under adverse conditions, such as are brought about by drought or injury.

The distribution of anthocyanin in the lower groups of plants has been investigated by Gertz (19). It appears in the young fronds of *Osmunda regalis*, of many species of *Adiantum*, of species of *Doodia*, *Pellaea* and *Davallia*. Gertz also notes it in *Blechnum* and *Azolla*, and among the Gymnosperms, in the cones of *Picea*, *Pinus* and *Larix*. Nagai (83), moreover, investigated anthocyanin in *Marchantia*.

As regards the Angiosperms, the present chapter contains an enumeration of the organs in which anthocyanin is formed in the varying circumstances of plant life. Some authors have attempted to classify the appearance of anthocyanin into permanent and transitory or periodical, normal and abnormal, etc., etc., but the variety and complexity of the underlying physiological causes in each case render such classifications of little value. No distinctions of this kind are attempted in the following classification.

1. The leaves of the majority of plants are green; nevertheless, in many genera and species, anthocyanin is always present *as a normal development*, and a complete series might be found ranging from plants whose leaves show a faint trace of anthocyanin to those whose leaves are more or less heavily pigmented. Some writers, notably Kerner (498), have pointed out that plants in which there is a considerable development of anthocyanin have often a special habitat. It would seem on the whole that this view is justified, though in many instances red pigmentation appears to be a specific character bearing no special relation to environment, as for example in *Geranium Robertianum*.

Kerner's views can be best expressed in his own words: "...anthocyanin frequently occurs only on the under side of foliage-leaves. This is observed especially among plants in the depths of shady forests, which, although belonging to widely-differing families, agree in a remarkable manner in this one point. One group of these plants has thick, almost leathery, evergreen leaves lying on the ground, which arise from subterranean tubers, or root-stocks, or from procumbent stems. The widely-distributed *Cyclamen europæum* may serve as a type of this group....Amongst other species belonging to this group may be mentioned *Cyclamen repandum* and *C. hederifolium*, *Cardamine trifolia*, *Soldanella montana*, *Hepatica triloba* and *Saxifraga Geum* and *cuneifolia*. Growing in habitats similar to these are to be met biennial, occasionally perennial, plants which in autumn form a rosette of leaves on their erect stems which survive the winter; these are always coloured violet on the side turned towards the ground, while the leaves which develop in the following warm summer on the elongated flower-stalks usually appear green below. To this group belong, especially, numerous Cruciferae (e.g. *Peltaria alliacea*, *Turritis glabra*, *Arabis brassicæformis*); species of spurge (e.g. *Euphorbia amygdaloides*), bell-flowers (e.g. *Campanula persicifolia*), and hawkweeds (e.g. *Hieracium tenuifolium*). Finally, deciduous shrubs are to be found in the depths and on the margins of forests whose leaves do not survive the winter, but which produce on the stems developing in the summer flat leaves whose under sides contain abundant anthocyanin, as, for example, *Senecio nemorensis* and *nebrodensis*, *Valeriana montana* and *tripteris*, *Epilobium montanum*,

Lactuca muralis, and many others." And again, "That which occurs in plants of the forest shade occurs similarly in those marsh plants whose leaf-like stems or flat, disc-like leaves float on the surface of the water. The green discs of duckweeds (e.g. *Lemna polyrrhiza*), of the Frogbit (*Hydrocharis morsus-ranæ*), of the Villarsia (*Villarsia nymphoides*), of water lilies (*Nymphæa Lotus* and *thermalis*), and of the magnificent *Victoria regia*, are strikingly bi-coloured, being light-green above and deep violet below." Occasionally anthocyanin is limited to definite areas in the leaf, and in this form is the cause of the dark brown or black spots on leaves (*Orchis maculata*, *Arum maculatum*[1], *Medicago maculata*, *Polygonum Persicaria*[2]), the dark colour being the result of the mixture of purple pigment and green chloroplastids.

Most of the above examples have been taken from flora of the temperate regions, but when we consider the tropical and sub-tropical regions, cases of red pigmentation are not only more numerous but are also much more striking. It is almost impossible to enumerate the species which have wonderfully variegated leaves, and constitute the class of 'beautiful-leaved plants' of cultivation. Examples are quoted by Hassack (493) in a paper on variegated leaves, and good plates have been published in a book by Lowe & Howard (29). A few of the most remarkable are as follows: *Calathea*, *Maranta* (Marantaceae), *Alocasia*, *Caladium*, *Xanthosoma violaceum* (Araceae), *Tradescantia discolor* (Commelinaceae), many Bromeliaceae (*Vriesia splendens*, *Nidularium* spp., *Billbergia gigantea*, *Tillandsia*), *Dracaena* spp. (Liliaceae), species of *Croton*, *Acalypha* and *Codiaeum* (Euphorbiaceae), many Gesneriaceae (*Sinningia atropurpurea*, *Episcia* sp., *Aeschynanthus* sp. and *Gesneria cinaberina*), *Coleus Verschaffeltii* (Labiatae), *Iresine* sp., *Alternanthera versicolor* (Amarantaceae), *Fittonia* sp. (Acanthaceae) and numerous Begonias. Hassack points out in a most interesting way how the colour in such leaves may vary from purple or crimson, through brown-reds to browns and bronze-greens, according to the amount of anthocyanin present, and its situation in the leaf, whether epidermal or deep-seated.

Stahl (505) lays great emphasis on the fact that many of these varie-

[1] In the case of *Arum maculatum* the leaves are not always spotted; in fact, according to the evidence of Colgan (73) and of Pethybridge (78), the spotted form is much less common, both in Great Britain and Ireland, than the unspotted. These two authors give some interesting data with regard to the distribution of both forms, as well as observations upon the structure and inheritance of the spot.

[2] Garjeanne (74) mentions the discovery in *Polygonum Persicaria* of a variegated form in which the chlorophyll is absent from large areas of the leaf, and when this is the case, the anthocyanin patch shows bright red.

gated-leaved plants, and those with purple and red under-surfaces, inhabit in the tropics, as in the temperate zones, the damp and shady forest regions; he quotes examples of such in support of his views upon the physiological significance of anthocyanin, to which we shall return in Chapter VIII. He notes that the orders specially represented are Araceae, Marantaceae, Piperaceae, Begoniaceae, Melastomaceae, Orchidaceae (*Pogonia crispa*) and Vitaceae (*Cissus discolor*); these probably include many of the plants already mentioned above. Some of these shade-loving plants are, in addition, characterised by a velvety surface due to epidermal papillae, which emphasises the richness of the variegation colour. Spotting and flecking with anthocyanin are also common in tropical plants, and occur on a much larger scale than in the temperate zones; instances are given by Hassack (493) and Stahl (505), such as *Gesneria cinaberina*, *Tradescantia zebrina*, *Musa zebrina Van Houtte* and *Costus zebrinus*.

Although the observations, which led Stahl to conclude that anthocyanin development is a characteristic of many shade-loving plants, are undoubtedly correct, yet it is also true that when a plant is capable of forming anthocyanin, those individuals which inhabit dry and sunny situations develop more anthocyanin than other individuals in moist and shady positions. That is, a plant may be green-leaved in the shade, and more or less red-leaved when exposed to the sun. This fact is at once patent to any one who gives attention to the matter, as the observations of F. Grace Smith (512) have shown. This author examined plants from regions in Massachusetts and Maine, and of 285 plants showing anthocyanin, 150 were from dry sunny places, 61 from dry shady places, 40 from wet sunny places and 34 from wet shady places. It seems likely that the phenomenon of pigmentation in shade, especially in the tropics, is a special adaptation, and of a different nature from the more general reddening of vegetation in sunny situations, and that some explanation may be found for the apparent contradiction between the two phenomena. Though F. G. Smith's observations do not give sufficient information to be conclusive, there is one significant fact among them, namely, that in wet shady places, the percentage of cases of red colour in the leaves is higher than that in the stems, whereas among the plants as a whole, anthocyanin was more frequently found in stems and petioles than in leaves. This may indicate that reddening in the shade is a peculiar function of the lamina, and is not connected with general reddening of the plant.

In petioles, anthocyanin is more frequent than in leaves, and in stems one might say that it is almost universal; the stem may be entirely red, or only red at the basal internodes: or the nodes may be red and the internodes green, or *vice versa.* A spotted effect produced by the local distribution of anthocyanin is seen in some stems, for example in *Chaerophyllum temulum* and *Conium maculatum* (Umbelliferae).

Red pigment is also very characteristic of the bud-scales of trees and shrubs; in some cases these scales represent leaf-bases (*Aesculus Hippocastanum, Acer pseudoplatanus, A. campestris*), in other cases, stipules (*Ulmus campestris, Tilia europaea*). In addition the stipules of herbaceous plants are often red, and the same statement holds good for enlarged leaf-bases (Umbelliferae).

2. Anthocyanin is characteristic, though by no means universally so, of young developing leaves and shoots. It is interesting to note how in some species one finds both reddening of young leaves and autumnal colouring, whereas in others this is not the case. There are, in fact, four possibilities:

(*a*) Both anthocyanin in young leaves and autumnal reddening. Species of *Acer, Rosa, Crataegus* and *Rubus.*

(*b*) Anthocyanin in young leaves but no autumnal reddening. *Corylus Avellana, Juglans regia, Fraxinus excelsior* and *Quercus Robur.*

(*c*) No anthocyanin in young leaves but some autumnal reddening. *Aesculus Hippocastanum.* This is a much rarer combination than (*a*) or (*b*).

(*d*) No anthocyanin in either young or old leaves. *Fagus sylvatica*[1].

Anthocyanin is especially abundant in the leaves of certain tropical trees (*Cinnamomum, Haematoxylon,* etc.); in addition to being coloured red, the young leaves often hang vertically downwards and straighten up again when mature (*Bauhinia, Dryobalanops, Cinnamomum*)[2]. In other species, especially among the Caesalpineae group of the Leguminosae, the whole young shoot has a vertical position, and in this respect, together with its intense red colour, forms a striking contrast to the older parts of the tree with green leaves in the normal horizontal position. As instances, certain species are mentioned by Stahl (505) and Keeble (503): *Amherstia nobilis* with young leaves of a bright red tint; a variety of *Cynometra cauliflora,* bright rose-colour; *Jonesia reclinata,* golden-green to bright red; *Brownea hybrida* and *B. grandiceps,* green spotted with red. Keeble further remarks that the coloration of young foliage in low latitudes is of such general occurrence that at the time of leaf-renewal, a tropical forest rivals in its tints the autumnal forests of the

[1] Pick (490) reports the presence of anthocyanin in young Beech leaves.

[2] Willis, J. C., *The Flowering Plants and Ferns,* Cambridge, 1904, p. 157.

temperate regions; he also quotes in confirmation the following extract from a paper by Johow (491), writing of the Lesser Antilles: "...all at once a red tint, due to the young foliage of the trees, appears in the landscape." Further details of such cases are given by Stahl (62, 505), Ewart (506), Keeble (503), Smith (520) and Weevers (521), and the physiological significance of anthocyanin formation in young leaves is discussed in Chapter VIII.

3. The older leaves of many herbaceous plants often acquire a considerable formation of anthocyanin by the end of the vegetative (not necessarily autumnal) season (*Lilium candidum*, *Digitalis purpurea*). Sometimes the whole plant becomes red in a striking way (*Chaerophyllum sylvestre*, *C. temulum*, *Galium Aparine*).

4. The autumnal coloration characteristic of temperate climates is one of the most pronounced cases of anthocyanin development. With the onset of the cooler weather of autumn, anthocyanin appears in quantity in the leaves of a number of trees, shrubs and climbing plants (species *Acer*, *Rhus*, *Euonymus*, *Crataegus*, *Viburnum*, *Cornus*, *Vitis*, *Ampelopsis*). In the *Pflanzenleben*, descriptions are given by Kerner (498) of the beauty of the autumnal foliage in the forests along the Rhine and the Danube, on the shores of the Canadian lakes and among the vegetation of the Alpine slopes. Kerner's descriptions give such vivid pictures of autumnal colouring that this opportunity is taken of quoting one of the passages from his account. "The heights along the middle course of the Danube, for example, the region known as the Wachan, below the town of Melk, show wide expanses of forest, in which beeches, hornbeams, evergreen oaks, common and Norway maples, birches, wild cherries and pears, mountain ashes and wild service-trees, aspens, limes, spruces, pines and firs take a share in the greatest variety. Bushes of Barberry (*Berberis vulgaris*), Dogwood (*Cornus sanguinea*), Cornel (*Cornus mas*), Spindle Tree (*Euonymus Europæus* and *verrucosus*), Dwarf cherry (*Prunus Chamæcerasus*), Sloe (*Prunus spinosa*), Juniper (*Juniperus communis*), and many other low shrubs arise as undergrowth, and spring up on the margins of the forests. The mountain slopes abutting on the valleys are planted with vines, and near by grow peach and apricot trees in great abundance. In the meadows on the shore, and on the islands of the Danube, rise huge abeles and black poplars, elms, willows, alders, and also an abundant sprinkling of trees of the bird cherry (*Prunus Padus*). ...The first frosts are the signal for the beginning of the vintage; all is busy in the vine-planted districts, and the call of the vine-dresser resounds from hill to hill. But it is also the signal for the forests on the

mountain slopes and in the meadows to change their hues. What an abundance of colour is then unfolded! The crowns of the pines bluish-green, the slender summits of the firs dark green, the foliage of hornbeams, maples, and white-stemmed birches pale yellow, the oaks brownish-yellow, the broad tracts of forest stocked with beeches in all gradations from yellowish to brownish-red, the mountain ashes, cherries and barberry bushes scarlet, the bird cherry and wild service-trees purple, the cornel and spindle-tree violet, aspens orange, abeles and silver willows white and grey, and alders a dull brownish-green. And all these colours are distributed in the most varied and charming manner. Here are dark patches traversed by broad light bands and narrow twisted stripes; there the forest is symmetrically patterned; there again the Chinese fire of an isolated cherry-tree or the summit of a single birch, with its lustrous gold springing up among the pines, illuminates the green background. To be sure this splendour of colour lasts but a short time. At the end of October the first frosts set in, and when the north wind rages over the mountain tops, all the red, violet, yellow, and brown foliage is shaken from the branches, tossed in a gay whirl to the ground, and drifted together along the banks and hedges. After a few days the mantle of foliage on the ground takes on a uniform brown tint, and in a few more days is buried under the winter coat of snow." The extreme beauty of the Alpine vegetation in autumn is also described by Overton (420), and led him to investigate the cause of the coloration.

Detailed accounts of the physiological processes producing autumnal coloration will be given in Chapter VI.

As in the case of reddening of young developing leaves, autumnal coloration is by no means universal, since, as we have seen, there are many trees and shrubs of which the leaves turn either yellow or brown and show no trace of pigment (*Fagus sylvatica*, *Carpinus Betulus*, *Quercus Robur*, *Betula alba*, *Alnus glutinosa*, *Populus alba*, *P. nigra*, *P. tremula*). There is some indication of a general tendency among genera of certain orders to develop red autumnal coloration, whereas genera of other orders are without it. Thus, for instance, the trees and shrubs of the Rosaceae (species of *Pyrus*, *Prunus*, *Rosa*, *Crataegus*) very readily produce anthocyanin. On the other hand, the majority of the trees in the Amentales do not form red pigment in the autumn (species of *Salix*, *Populus*, *Betula*, *Alnus*, *Juglans*, *Morus*, *Carpinus*, *Corylus*, *Castanea*, *Fagus* and *Quercus*). To the above, *Quercus rubra* and *Q. coccinea* form an exception. On the whole, however, little emphasis can be laid on the connection between autumnal coloration and relationship, as is illustrated by the Viburnums

—*V. Opulus* and *V. Lantana*; the former rapidly reddens in autumn whereas the latter forms little or no anthocyanin under similar conditions.

5. In a number of species, mechanical injury frequently leads to formation of anthocyanin in the portion of leaf, stem, or shoot, distal to the point of injury. Partial severing of a portion of the leaf blade may induce reddening in the severed portion. Crushing, or partial breaking of a stem or petiole, leads to the same result, as was shown by Gautier (190) in the Vine, or ringing of a stem will cause the leaves above to turn red, as Combes (461, 472) has demonstrated in *Spiraea*. The same kind of injury, with similar results, may be brought about by the attacks of insects. Under this heading must be included, as a result of injury, the copious formation of anthocyanin in galls, such as those which occur on species of *Rosa*, *Salix*, *Quercus* and many other plants. Also the local appearance of pigment in tissues infected with Fungi: as an example one may quote the purple blotches produced on leaves of *Tussilago Farfara* and the purple streaks on Wheat stems when these plants are infected with *Puccinia*. The physiological causes involved in the production of pigment on injury are dealt with in Chapter VI. Some plants show a greater tendency to redden on injury than others (*Oenothera*, *Rumex*, *Rheum*); also the reddening on injury is more rapid and frequent in autumn when the vegetative season is about at an end.

6. Low temperature may induce anthocyanin formation apart from autumnal coloration, for, in the development of the latter, factors other than temperature play an important part. Many evergreen shrubs show a reddening[1] in their leaves during winter (*Ligustrum vulgare*, *Mahonia* spp.). The same statement holds good for certain herbaceous plants which retain their leaves in winter (*Saxifraga umbrosa*).

7. Plants exposed to drought often develop anthocyanin. This may be seen sometimes in individuals in pots (*Pelargonium*) which have been insufficiently watered. Miyoshi (462) has observed that leaves of trees in the East Indies, Ceylon and Java redden during the dry period in the same way as autumnal leaves in the temperate regions.

8. In High Alpine plants there is undoubtedly, on the whole, a much stronger development of anthocyanin in all the parts—stem, leaves, bracts and flowers, than in plants growing in lowland regions. Gaston Bonnier (381, 394) has made observations upon this point, and has found

[1] This is of course quite different from the reddening shown by some evergreens in winter, notably species of the Gymnosperms, and which is due to the formation of a red product from chlorophyll in the chloroplastids.

that the flowers of many species are more highly coloured, the greater the altitude. Kerner (498) also remarks: "The leaflets and stem of the Alpine *Sedum atratum*, those of *Bartsia alpina*, and, above all, numerous species of *Pedicularis* (e.g. *Pedicularis incarnata, rostrata, recutita*) are coloured wholly purple or dark violet....It is also a very striking phenomenon that widely-distributed grasses (e.g. *Aira cæspitosa, Briza media, Festuca nigrescens, Milium effusum, Poa annua* and *nemoralis*), which in the valley possess pale green glumes, develop anthocyanin in them on lofty mountains, so that there the spikes and panicles exhibit a deep violet tint, and on this account the regions in which grasses of this kind grow in great quantities receive a peculiar dark colouring....The same occurs in the numerous sedges and rushes growing in the Alps, which have dark-violet, almost black, scales covering the flowers (e.g. *Carex nigra, atrata, aterrima, Juncus Jacquinii, trifidus, castaneus*)....It is known that the floral-leaves of many plants growing on lofty mountains, and in the far north, are coloured blue or red by anthocyanin, whilst in the same species, growing in warm lowlands and in southern districts, they appear white. Particularly noticeable in this respect are the Gypsophyllas (*Gypsophylla repens*), the Carline Thistle (*Carlina acaulis*), the large-flowered Bitter-cress (*Cardamine amara*), the Milfoil (*Achillea Millefolium*), and many of those Umbelliferae which have a very wide distribution, and occur all the way from the lowlands up to a height of 2500 metres in the Alps, such as *Pimpinella magna, Libanotis montana, Chærophyllum Cicutaria*, and *Laserpitium latifolium*." And again: "The flowers of species grown in the Alpine garden on the Blaser at a height of 2195 metres above the sea exhibited, as a rule, brilliant floral tints, and some were decidedly darker than the flowers grown in the Vienna Botanic Garden. *Agrostemma Githago, Campanula pusilla, Dianthus inodorus* (*sylvestris*), *Gypsophila repens, Lotus corniculatus, Saponaria ocymoides, Satureja hortensis, Taraxacum officinale, Vicia Cracca*, and *Vicia sepium* are good examples of this. Several species, which produced pure white petals in the Vienna gardens, e.g. *Libanotis montana*, had petals coloured reddish-violet by anthocyanin on their under sides in the Alpine garden. The glumes of all the Grasses which were green, or only just tinged with violet at a low level became a dark brownish-violet in the Alpine garden. The abundant formation of anthocyanin in the green tissue of the foliage-leaves and sepals, and in the stem, was particularly apparent. The leaves of the Stonecrops, *Sedum acre, album*, and *sexangulare* became purple-red, those of *Dracocephalum Ruyschianum* and *Leucanthemum vulgare* violet, those of *Lychnis Viscaria* and *Satureja hortensis* a brownish-red, and the foliage-leaves of *Bergenia crassifolia* and *Potentilla Tiroliensis*, even in August, had the scarlet-red colour which they usually assume in sunny spots in the valley in late autumn."

As regards Arctic vegetation, Gertz (19), on the authority of Wulff (71), maintains that abundant production of anthocyanin is a distinguishing feature; and so much so, that the periodicity of autumnal coloration is very little marked. The causes of the intense colour in Alpine and Arctic vegetation are dealt with in Chapter VI.

9. Production of anthocyanin in vegetative organs is characteristic of certain halophytes (*Salicornia*, *Suaeda*, *Atriplex*).

10. The red-leaved varieties of green-leaved types are due to the production of anthocyanin, which is either absent from the type, or present to only a slight extent (red-leaved varieties of *Fagus*, *Berberis*, *Brassica* and *Prunus*).

11. The petals and perianth are essentially the organs producing anthocyanin. To these must be added the perianth-like calyx (*Anemone*, *Delphinium*, *Aconitum*, *Begonia* and others). Also the bracts of the inflorescence may develop anthocyanin to a large extent, and may either assist (*Salvia*), or even take the place of, the corolla as organs of attraction (*Bougainvillaea*, *Euphorbia*, *Poinsettia*, spathes of Araceae). In general, we may say that the stems and pedicels of most inflorescences show an abundant formation of anthocyanin which no doubt increases the conspicuousness of these parts; yet, at the same time, the underlying cause is probably that the physiological conditions in all tissues of reproductive organs enhance the formation of pigment.

Anthocyanin is not confined to the perianth, but is frequently present in the carpels and style: of its occurrence in the stigma we have many examples among the Amentales (*Myrica*, *Alnus*, *Betula*, *Corylus*, *Carpinus*, *Salix*, *Populus*), as well as in *Rumex* and *Ricinus*.

In the stamens, too, it is found, especially in many anemophilous plants (*Populus*, *Fraxinus*) and in the Graminaceae (*Molinia*, *Phleum*, *Alopecurus*, *Cynosurus*, *Dactylis*). Pollen grains (Möbius, 86) may also be coloured with anthocyanin (*Campanula*).

12. Many fruits form abundant anthocyanin. The pigment may be temporary in some stages of development of capsules (many of the Leguminosae, *Hypericum*, *Sedum*) and schizocarps (Umbelliferae), the tissues of which eventually become dry. The more obvious cases are the drupes and berries, and of these, two types stand out, i.e. those in which the pigment is limited to the epicarp (*Prunus*, *Vitis*), and those in which it is present in both epicarp and mesocarp (*Rubus*, *Ribes*, *Ligustrum*,

Vaccinium, Atropa, Solanum, Sambucus, Viburnum, Lonicera). It may be formed also in the tissues of 'false fruits' (*Fragaria, Pyrus*). From other fruits (*Cucurbita, Solanum, Lycopersicum, Citrus*[1]) it is entirely absent.

13. Anthocyanin is sometimes developed in the cells of the testa of seeds (*Abrus, Phaseolus, Pisum*). Further instances are the seeds of *Adenanthera, Erythrina, Lepeirosia* (Potonié, 56), *Trifolium pratense* (Preyer, 66) and many Podalyrieae (Lindinger, 76). More frequently pigmentation in the seed-coat is due to impregnation of the cell-wall by pigments which cannot be classified as anthocyanins, and of which very little is known.

14. Roots and underground stems often form anthocyanin in considerable quantity. Three cases of this type of pigmentation may be differentiated:

(*a*) The pigment develops under ground apart from the influence of any outside factors. Well-known examples for roots are *Beta vulgaris* and *Raphanus sativus*, and for underground stems, *Solanum tuberosum*; in the latter species the pigment may be confined to the epidermal layers or may be present in the inner tissues (Salaman, 647). To the above may be added other cases such as:

Root-tips of species of Saxifragaceae and Crassulaceae (Irmisch, 27).

Roots of *Parietaria diffusa* and *Gesneria* sp. (Zopf, 48).

Roots of species of Pontederiaceae, Haemodoraceae and Cyperaceae in the ground (Ascherson, 43).

Roots of *Wachendorfia* in the ground (Hildebrand, 44).

(*b*) The pigment develops in roots normally exposed.

Aerial roots of *Ficus indica* (Möbius, 68).

Floating roots of Pontederiaceae (Hildebrand, 44; Ascherson, 43).

Aerial roots of Aroideae (van Tieghem, 33; Lierau, 51).

Aerial roots of Orchidaceae (Leitgeb)[2].

(*c*) The pigment develops in roots abnormally exposed.

Exposed roots of *Alnus* (Möbius, 68).

Roots of *Salix* grown in water in glasses (Schell, 374).

Adventitious roots of *Echeveria metallica* (Pirotta, 42).

Roots of *Zea Mays* growing in water or in moss and exposed to light (Dufour, 392; Devaux, 395).

Exposed roots of *Saccharum officinarum* (Benecke, 399).

Adventitious roots above ground of *Poa nemoralis* (Beyerinck, 390).

[1] Except in the red-fleshed variety, the so-called 'Blood Orange.'

[2] Leitgeb, H., 'Die Luftwurzeln der Orchideen,' *Denkschr. Ak. Wiss.*, Wien, 1865, XXIV, p. 179 (p. 204).

To Kerner (498) we owe the observation that rhizomes of *Dentaria bulbifera*, *Lathraea Squamaria* and of species of *Cardamine* and *Viola* become violet when exposed in water to sunlight.

Gertz (19) lays stress upon the fact that numbers of roots and underground stems form anthocyanin when, for some reason, they have been exposed, as for instance when they grow out from banks of streams, or in shallow water, or when exposed during autumnal ploughing. He appends the following list of such cases including *Tradescantia zebrina*, *Eriophorum angustifolium*, *Phragmites communis*, *Rumex Acetosella*, *Geranium molle*, *Potentilla anserina*, *Gunnera scabra*, *Stachys palustris*, *Plantago major*, *Anthemis arvensis* and *Artemisia vulgaris*.

15. A development of anthocyanin appears to be characteristic of many parasites and saprophytes, though possibly the absence of chlorophyll has the effect of making the red pigment more conspicuous (species of *Cuscuta*, *Orobanche*, *Lathraea*). Mirande (419) gives an interesting account of the development of anthocyanin in the genus *Cuscuta*.

CHAPTER III

THE HISTOLOGICAL DISTRIBUTION OF ANTHOCYANINS

WHEN tissues containing anthocyanin are examined microscopically, in the majority of cases the pigment is found to be in solution in the cell-sap, which, of course, occupies the vacuole or vacuoles of the cell. We know, as Gertz (19) points out, from the researches of de Vries (155) and Pfeffer[1] that in living cells the protoplasm is impermeable to anthocyanin in the cell-sap. But when the protoplasm is dead and semi-permeability ceases, anthocyanin exosmoses out of the cell, and at the same time it is often taken up, like a stain, by the protoplasm and nucleus[2].

From the condition of solution in the cell-sap, two deviations are possible: first, in tissues which gradually die or become woody, the anthocyanin may be absorbed into the substance of the cell-wall, thereby causing the latter structure to become coloured; secondly, the concentration of anthocyanin in living cells may become so great that the pigment separates out either in crystalline or amorphous form. In this solid state, anthocyanin may be pure, or may merely be absorbed by substances in the cell, such as proteins, tannins.

The following account is taken from Gertz (19) who has particularly investigated this portion of the subject. Several cases of pigmented

[1] Pfeffer, W., *Osmotische Untersuchungen*, Leipzig, 1877 (pp. 134, 135).

[2] Staining of the nucleus under these conditions has been noted by Hartig, de Vries and Molisch.

Hartig, Th., 'Chlorogen,' *Bot. Ztg.*, Berlin, 1854, XII, pp. 553–556 (p. 555).

Hartig, Th., 'Ueber das Verfahren bei Behandlung des Zellenkerns mit Farbstoffen,' *Bot. Ztg.*, Berlin, 1854, XII, pp. 877–881 (p. 879).

Hartig, Th., 'Ueber die Bewegung des Saftes in den Holzpflanzen,' *Bot. Ztg.*, Leipzig, 1862, XX, pp. 89–94 (p. 93).

Vries, H. de, 'Plasmolytische Studien über die Wand der Vacuolen,' *Jahrb. wiss. Bot.*, Berlin, 1885, XVI, pp. 465–593.

Molisch, H., *Untersuchungen über das Erfrieren der Pflanzen*, Jena, 1897 (pp. 28, 29).

On account of this property, anthocyanin has been used as a stain for histological purposes. The red anthocyanin of Cabbage has been employed in this way by Tait, Gierke, Flesch (174), and Claudius (199); Lavdowsky (175) also recommends anthocyanin from *Myrtillus* berries.

Tait, L., 'On the Freezing Process for Section-Cutting,' *J. Anat. Physiol.*, London, 1875, ser. 2, VIII, pp. 249–258.

Gierke, H., 'Färberei zu mikroskopischen Zwecken,' *Zs. wiss. Mikrosk.*, Braunschweig, 1884, I, pp. 62–100, 372–408, 497–557 (p. 99).

cell-walls are recorded by Gertz among Monocotyledons, but no very definite statements are made as to whether the colour is actually due to anthocyanin. The cases quoted are in *Phormium tenax* (Engelmann), *Eriophorum angustifolium* (Wulff), *Sorghum* (Kraus), *Tillandsia diantoidea* (Tassi), *Pontederia crassipes* (Hildebrand), *Angraecum superbum* and *Macroplectrum sesquipedale* (Hering); Gertz also includes examples which he has observed himself. Among Dicotyledons coloured cell-walls are more rare. Red bast (*Oxalis Ortgiesii*) and red epidermis (*Epilobium montanum*) are quoted; in *Jacquinia smaragdina* (Vesque[1]) the cuticle is crimson. According to Gertz, the coloured cell-walls described in *Homalomena rubrum* (Nägeli & Schwendener[2]) and in *Geranium Robertianum*, *Begonia maculata*, *Quercus palustris*, and *Rumex Acetosella* (Wigand, 12) are artificially produced; such artifacts can be brought about by putting anthocyanin-containing cells in a medium absorbing water, but in which anthocyanin is insoluble, such as ether. The phenomenon can be demonstrated in leaves of *Begonia* spp. In addition to the above cases, red pigment is found in walls of testa cells in *Abrus precatorius* (Nägeli & Schwendener[2]) though Möbius (68) found cell-sap too. Sclerotic cells of testa of *Erythrina* and *Goodia* have blue, violet and red cell-walls (Tschirch[3]); also in seeds of *Trifolium pratense* (Preyer, 66).

As regards the presence of solid and crystalline anthocyanin in cells the following list of cases compiled by Gertz (19) is very useful:

Tillandsia amoena, Hildebrand (30).
Aechmea sp., Dennert (14).
Allium Schoenoprasum, Courchet (101).
Lilium Martagon, Overton (420).
Dracaena Jonghi, Hassack (493).
Convallaria majalis, Weiss (95).
Iris pumila, Dennert (14).
Strelitzia Reginae, Mohl (88), Hildebrand (30).
Orchis mascula, Nägeli (89).
Epidendrum cochleatum, Schlockow[4].
Cattleya quadricolor, Malte[5].

[1] Vesque, J., 'Mémoire sur l'anatomie comparée de l'écorce,' *Ann. sci. nat. (Bot.)*, Paris, 1875, sér. 6, II, p. 82.

[2] Nägeli, C. L., und Schwendener, S., *Das Mikroskop, Theorie und Anwendung desselben*, Leipzig, 1877.

[3] Tschirch, A., *Angewandte Pflanzenanatomie*, Wien und Leipzig, 1889, Bd. I, pp. 62, 66.

[4] Schlockow, A., *Zur Anatomie der braunen Blüten*, Inaugural-Dissertation zu Heidelberg, Berlin, 1903.

[5] Malte, M. O., 'Untersuchungen über eigenartige Inhaltskörper bei den Orchideen,' *Bihang till K. Svenska Vet.-Akademiens Handlingar*, Bd. XXVII, No. 15, p. 32.

Laelia Perrinii, Malte[1].
Oncidium sphacelatum, Schlockow[2].
Dianthus Caryophyllus, Molisch (108).
Aquilegia atrata, Molisch (108).
Delphinium elatum, Weiss (95), Molisch (108).
D. formosum, Zimmermann[3].
D. tricolor, Fritsch (98).
D. sp., Molisch (108).
Glaucium fulvum, Schimper (97).
Papaver sp., Hildebrand (30).
Brassica oleracea, Molisch (108).
Hydrangea hortensis, Ichimura (207).
Rubus caesius, *R. corylifolius*, *R. glandulosus*, *R. laciniatus*, *R. lasiocarpus*, } Trécul (93).
Geum sp., Weiss (95).
Rosa sp., Molisch (108).
Pyrus communis, Sorauer (391).
Neptunia oleracea, Rosanoff (96).
Lathyrus heterophyllus, *L. silvestris*, *Cytisus Laburnum*, *C. Alschingeri*, *C. scoparius*, *Medicago sativa*, *Hedysarum coronarium*, *Ononis natrix*, *Baptisia australis*, } Molisch (108).
Amorpha fruticosa, Hildebrand (30).
Erodium Manescari, Molisch (108).
Pelargonium zonale, Molisch (108).
P. Odier hortorum, Molisch (108).
Vitis sp., Molisch (108).
Ampelopsis quinquefolia, Mer (371).
A. humulifolia, *A. muralis*, } Lengerken[4].
Viola tricolor, Nägeli (89).
Passiflora acerifolia, Böhm (92), Weiss (95).
P. alata, Weiss (95).
P. laciniata, Unger (90).

[1] Malte, M. O., *op. cit.*
[2] Schlockow, A., *op. cit.*
[3] Zimmermann, A., 'Die Morphologie und Physiologie der Pflanzenzelle' (Schenck's *Handbuch der Botanik*, Breslau, 1887, Bd. III (2), pp. 104, 105).
[4] Lengerken, A. von, 'Die Bildung der Haftballen an den Ranken einiger Arten der Gattung Ampelopsis,' *Bot. Ztg.*, Leipzig, 1885, XLIII, pp. 376, 378.

Passiflora limbata, Böhm (92).
P. suberosa, Unger (90), Böhm (92), Weiss (95).
Begonia discolor, Unger (90).
B. maculata, Molisch (108).
Centradenia floribunda, Buscalioni & Pollacci (17).
Primula sinensis, Bokorny (103).
Anagallis arvensis, Courchet (101).
A. arvensis var. *ciliata*, }
A. arvensis var *coerulea*, } Molisch (108).
Acantholimon sp., }
Swertia perennis, }
Amsonia salicifolia, } Müller (61).
Hoya carnosa, }
Gilia tricolor, Hildebrand (30).
Nemophila sp., Molisch (108).
Salvia splendens, Mohl (88).
Atropa Belladonna, Trécul (93).
Hyoscyamus niger, Müller (61).
Solanum americanum, Nägeli (94).
S. guineense, Trécul (93).
S. Melongena, Weiss (95).
S. nigrum, Hartig (87), Trécul (93), Weiss (95).
Antirrhinum majus, Molisch (108).
Gesneria caracasana, Dennert (14).
Columnea Schiedeana, Weiss (95).
Thunbergia alata, Fritsch (98).
Justicia speciosa, Pim (99).
Coffea arabica, Tschirch (104), Kroemer (106).
Viburnum Tinus, Fritsch (98).

To the above list Gertz adds a number of plants in which he himself has observed solid anthocyanin as well as anthocyanin in solution, and in these, with a few exceptions, the solid form of anthocyanin could not be obtained by chemical reagents; but in other plants he was able to obtain 'anthocyanin bodies' in the cells by treatment with chemical reagents such as alcohol, alkalies, sulphuric and hydrochloric acids.

There has been much doubt in some cases as to whether the coloured deposits are of pigment only, or of other substances such as proteins, tannins, etc., impregnated with pigment. Very beautiful crystals are figured by Molisch (108) from cells of red Cabbage leaves and leaves of *Begonia maculata*. Molisch notes that a low temperature induces the formation of the crystals which may disappear at higher temperatures. In *Brassica* and *Begonia*, appearances are certainly in favour of the crystals being pigment. In other plants where the solid deposit is in the form of coloured spheroids or globules, the latter often being viscous in consistency, it seems more probable that the matrix of the bodies is of a tannin or protein nature, and has become impregnated with pigment.

Gertz maintains several possibilities for the natural occurrence of anthocyanin in cells. (1) In crystals as mentioned above either of pure anthocyanin or of colourless substances infiltrated with anthocyanin. (2) In very young cells containing several vacuoles, anthocyanin in the vacuoles may appear as red or blue globules having a superficial likeness to anthocyanin bodies. (3) Anthocyanin may occur in amorphous or globular form, or bound up with amorphous or globular bodies which are of a different nature from the pigment.

To demonstrate how solid anthocyanin bodies may arise, Gertz describes two examples of what may happen artificially in the cell. If tissues of *Strobilanthes Dürianus* be placed in alcohol, coloured drops appear which afterwards disappear again. The explanation offered is that the alcohol first causes the cells to plasmolyse, and substances in the cell may be precipitated and take up anthocyanin. Later, when semipermeability ceases, the alcohol redissolves out the pigment. Secondly, in *Scopolia orientalis*, on additon of sulphuric acid, there is a precipitation of coloured bodies which does not disappear. In this case Gertz supposes that the precipitated substances are changed by the reagent, or are insoluble in the new medium, and so remain. As already stated, the above examples are artificial, but Gertz suggests that in a similar way in the normal cell, substances may be precipitated and absorb anthocyanin. This precipitation may be brought about by increased concentration, or by lowering of temperature. The latter was found by Molisch to be the cause of crystallisation of anthocyanin in *Brassica*.

As to the chemical nature of the substances which absorb anthocyanin in the cell, according to Gertz, little can be said. Often they appear to be tannin-like compounds, impregnated with anthocyanin, in the form of solid bodies (in young leaves of *Juglans*, *Quercus*, *Rubus*, *Ribes* and others), or globules as described in the Bromeliaceae (Wallin, 105), *Cissus* (D'Arbaumont[1]), *Euphorbia* (Gaucher[2]) and *Dicentra* (Zopf, 48). Possibly in some other cases the ground substance is protein; coloured aleurone grains have been recorded by Hartig (91), which are red in

[1] D'Arbaumont, 'La tige des Ampélidées,' *Ann. sci. nat.* (*Bot.*), Paris, 1881, sér. 6, XI, pp. 186–253 (pp. 241, 242).

[2] Gaucher, L., 'Recherches anatomiques sur les Euphorbiacées,' *Ann. sci. nat.* (*Bot.*), Paris, 1902, sér. 8, XV, pp. 161–309 (p. 179).

Laurus nobilis and blue in *Cheiranthus annuus*, and blue grains have been noted by Spiess (107) in *Zea*. There is also the possibility, even when the anthocyanin deposits are crystalline, of the stroma being of some other substance. Nägeli (94) has suggested that coloured protein crystalloids occur in *Solanum americanum*.

We may next consider the distribution of anthocyanins in the tissues of leaves, stems, flowers, fruits and seeds.

As regards leaves, reference may be made to the work of Morren (26), Pick (490), Hassack (493), Engelmann (494), Stahl (62, 505), Berthold (64), Griffon (418), and Buscalioni & Pollacci (17). No very general statement can be made about the histological distribution in leaves beyond that the pigment is often present in the epidermis, quite frequently in the subepidermal tissues also, and in many cases in the more internal tissues as well. Neither can it be predicted with certainty for any plant where the localisation will be, and it may vary considerably in the same plant according to whether we select for investigation leaves in the young, autumnal or winter condition, or leaves during the normal vegetative period; or again leaves under the influence of attacks of insects or Fungi. Accounts of the histological distribution in leaves have been made by various authors (Hassack, Engelmann, Pick, Stahl, Massart), and a very detailed classification is recorded by Buscalioni & Pollacci.

A less detailed classification due to Gertz is quoted as follows:

A. Permanent anthocyanin.

1. In the epidermis. Ex. *Orchis*, *Canna*, *Maranta*, *atropurpurea*-forms of *Fagus*, *Corylus* and *Acer*, and also of *Beta vulgaris*, *Atriplex hortensis* and *Croton* sp.

2. In peripheral layer of the ground tissue. Ex. *Dracaena*, *Eucomis punctata*, *Erythronium Dens-Canis*, *Arum maculatum*, *Arisarum vulgare*, *Berberis vulgaris atropurpurea*.

3. In both epidermis and ground tissue. Ex. *Aerua sanguinolenta*, *Aeschynanthus atropurpureus*, *Iresine Herbstii*, *Coleus* sp.

4. In median ground tissue. Ex. *Higginsia refulgens*, *Sinningia purpurea*, *Gesneria Donkelaari*, *Pellionia Daveauana*.

B. Periodic anthocyanin.

(*a*) In young leaves and spring leaves.

1. In the epidermis. Ex. *Rubus*, *Rosa*, *Rhus*, *Silene*, *Malva*, *Veronica*, *Odontites*, *Ajuga*, *Salvia*.

2. In the ground tissue. Ex. *Salix*, *Populus*, *Fagus*, *Acer*, *Cydonia*, *Vicia*, *Stellaria*, *Cerastium*, *Campanula*, *Taraxacum*.

3. In hairs. Ex. *Quercus rubra, Q. macrocarpa, Castanea vesca, Chenopodium album, Mallotus japonicus, Vitis alexandrina, V. Labrusca.*

(*b*) In older and autumnal leaves.

1. In the epidermis. Ex. *Philadelphus, Deutzia, Euonymus Europaeus,* and some Oenotheraceae.

2. In ground parenchyma. Ex. *Populus, Salix, Quercus, Rhus, Acer, Vitis, Ampelopsis, Prunus, Cornus, Viburnum.*

(*c*) In winter leaves.

1. In the epidermis. Ex. *Silene, Oenothera, Gaura, Veronica, Lamium.*

2. In the ground tissue. Ex. *Secale, Hedera, Calluna, Empetrum.*

One or two further observations of interest are mentioned by Gertz. This author notes that in some leaves there is an indication of division of labour as regards anthocyanin. When the pigment appears principally in the epidermis, it is often found in addition in isolated mesophyll cells which are poor in chlorophyll (*Spiraea, Rosa* and *Rubus*).

Although, as we have seen, from the observations of the above-mentioned authors, anthocyanin formation is by no means limited to any particular tissue of the leaf, yet Parkin (77), in a recent investigation on the localisation of anthocyanin, points out the prevalency of the idea among botanists that anthocyanin is usually formed in the epidermis. Parkin's researches are interesting because they give an indication as to the particular localisation of the pigment when formed under varying circumstances and under the influence of different factors. He examined 400 cases and classified his results as follows:

1. Transitory anthocyanin of young leaves. This is a feature of the young foliage of many plants in the tropics and temperate regions. The number of species examined was 235, and the distribution was classified as:

Epidermis	Mesophyll	Epidermis and mesophyll
20 %	64 %	16 %

2. Autumnal anthocyanin. The number of species examined was 81 and the distribution:

Epidermis	Mesophyll	Epidermis and mesophyll
11 %	78 %	11 %

3. The permanent anthocyanin of mature leaves. In this case the pigment appears as the leaf matures, and persists throughout the life of the leaf as a normal character. This class includes (*a*) leaves with uniformly red lower surfaces; (*b*) leaves with definite pigmented areas in the form of spots, blotches or zones; and (*c*) leaves of horticultural

varieties with coloured foliage. The number of species examined was 54, and the distribution:

Epidermis	Mesophyll	Epidermis and mesophyll
70 %	17 %	13 %

4. The accidental anthocyanin of mature leaves. This is not normally present in mature leaves but arises only under exceptional conditions such as: (*a*) excessive insolation, followed by cool nights, as seen in Alpine plants, and in evergreens during winter; (*b*) the result of injury, when a reddish zone often appears round a wound in a leaf; (*c*) through the accidental exposure of lower surface to the full rays of the sun. The number of cases examined was 30, and in the majority the anthocyanin was confined to the mesophyll. Parkin sums up by saying that anthocyanin of young and autumnal leaves is usually confined to the mesophyll; of mature leaves, when a normal feature, to the epidermis; and, when exceptional, to the mesophyll. The mesophyll, he considers, to be the usual, and, perhaps, more primitive seat for anthocyanin.

As regards the distribution in stems, we have the work of Berthold (64) and Buscalioni & Pollacci (17). Two types of distribution can be recognised:

1. Anthocyanin in the epidermis. Ex. *Gentiana*, some Labiatae (Berthold, 64), some *Solidago* and *Aster* spp.

2. Anthocyanin in subepidermal assimilating cells. Ex. Alsinaceae, Papilionaceae and Convolvulaceae.

The histological distribution in petals has been investigated by Hildebrand (30), Koschewnikow (40), Müller (61) and Buscalioni & Pollacci (17), and in anthers by Chatin (34). In the corolla, anthocyanin is chiefly located in the epidermis, either upper or under, or both, though its appearance in the subepidermal and inner tissues is by no means uncommon. The table on the following page, selected from Buscalioni & Pollacci's (17) records, gives an idea of the relative frequency with which pigments are formed in the various regions of the corolla and perianth.

Chatin (28) notes that anthocyanin occurs in deep-seated tissues in thicker petals, such as *Ulloa* and *Asclepias*, whereas in thinner petals it is usually confined to the epidermis.

The most interesting feature in connection with the histology of coloured petals is the combination of colour effects produced by the simultaneous presence of two, or even more, pigments in the cells. Strictly speaking, the plastid pigments, i.e. the yellow, orange-yellow and orange colouring matters which are bound up with special protoplasmic structures—the plastids, have no place in this account; and the same may be said for the soluble yellow pigments (mostly flavones). But both these classes occur so often with anthocyanin, and so frequently modify its colour, that some mention of them at this point will not be out of place. References can be made, in addition, to other authors; Hildebrand (30) wrote an early account of flower pigmentation, including combinations of plastid and sap colours; there is also an interesting paper by Bidgood (18) on flower colours, and many detailed observations by Dennert (14) and by the author (226).

	Name of plant	Hairs on lower epidermis	Lower epidermis	Lower hypo-dermal layer	Parenchyma	Vascular bundle	Parenchyma	Upper hypo-dermal layer	Upper epidermis	Hairs on upper epidermis
1	*Cypripedium* sp.	+								
2	*Lamium purpureum*	+	+	...	...	...	...	...	+	+
3	*Epacris*	...	+							
4	*Erica*	...	+							
5	*Cineraria*	...	+							
6	*Tropaeolum*	...	...	...	...	...	...	...	+	
7	*Magnolia*	...	+	...	...	...	...	...	+	
8	*Dianthus*	...	+	...	...	...	...	...	+	
9	*Dielytra spectabilis*	...	+	...	...	...	...	...	+	
10	*Hibiscus*	...	+	...	...	...	...	...	+	
11	*Scilla amoena*	...	+	...	...	...	...	...	+	
12	*Tulipa*	...	+	...	...	...	...	...	+	
13	*Camellia*	...	+	...	...	...	...	...	+	
14	*Cydonia japonica*	...	+	...	...	...	...	...	+	
15	*Clerodendron*	...	+	...	...	...	...	...	+	
16	*Corydalis bulbosa*	...	+	...	...	...	...	...	+	
17	*Azalea indica*	...	+	...	+	...	+	...	+	
18	*Muscari comosum*	...	+	...	+	...	+	...	+	
19	*Dendrobium tirsiflorum*	...	+	...	+	+	+	...	+	
20	*Helleborus niger*	...	...	+						
21	*Hyacinthus orientalis*	...	...	+						
22	*Dalechampia*	...	...	+	...	...	...	+		
23	*Echeveria grandiflora*	...	...	+	...	...	...	+		
24	*Begonia floribunda*	...	...	+	+	+	+	+		
25	*Vanda suavis*	...	...	...	+	...	+			
26	*Rhododendron hybridum*	...	...	...	...	...	+			
	Total	2	17	5	5	2	6	3	15	1

One of the most frequent combinations of pigments is purple, purplish-red, or red anthocyanin and yellow plastids. The resultant colour may be brown (*Cheiranthus Cheiri*, *Tagetes signata*), crimson (*Zinnia*), scarlet (*Geum coccineum*), or orange-red (*Tropaeolum majus*); there are of course a great many other cases of combination of these two pigments, and a correspondingly large number of shades of brown, crimson, or scarlet, as the case may be. A less frequent combination is dark brown, brownish-black, or black resulting from purple anthocyanin and chloroplastids. This effect is produced in some *Cypripedium* flowers, and we have already noted it in the case of leaves (see p. 22). The brown or black effect is due to the fact that the two pigments are complementary as regards the rays they absorb; those which are not absorbed by chlorophyll are absorbed by anthocyanin, and so the result is negative as regards colour. But black or brown is not always due to this combination; the black spots on *Adonis* and *Papaver* flowers owe their appearance to deep blue cell-sap, and in the dark markings on some varieties of Bean (*Phaseolus*) seeds the cells contain purple anthocyanin. There are, in addition, true brown and black pigments which appear in some flowers as in the spots on the alae of *Vicia Faba*[1].

Another class of combinations is the outcome of the mixture of anthocyanin and a soluble yellow pigment in the same cell. Such a combination occurs in the crimson flowers of *Mirabilis Jalapa*, though, on the whole, it is usually found to be characteristic of varieties which have arisen under cultivation, as for instance in the crimson varieties of *Antirrhinum majus*, *Dahlia variabilis* and *Portulaca grandiflora*. In these species the flower of the original wild type had some shade of magenta anthocyanin only. Three variations are then characteristic—variation to ivory-white, to yellow and to crimson, the latter being a mixture of magenta anthocyanin and soluble yellow pigment, such a combination as would not normally occur in nature.

It is interesting to note that colour of type and variation in the anthocyanin-soluble-yellow series are reversed in the anthocyanin-plastid series. In the latter, the original type is either crimson (*Zinnia*), yellow striped with brown (*Cheiranthus*), or orange-red (*Tropaeolum*); a variety characteristic of this series is one in which the yellow pigment almost disappears from the plastids, or is replaced by a much paler yellow substance. When anthocyanin is present with these pale yellow pigments, or the yellow pigment is altogether absent, a purple or magenta variety results, as in the purple *Cheiranthus* and magenta *Zinnia*. When the anthocyanin is more red, as in the orange-red *Tropaeolum majus*, a carmine variety arises when these pale plastids only are present. Thus, in the anthocyanin-soluble-yellow series we have a magenta (or purple) type with crimson, ivory-white and yellow varieties, whereas in the anthocyanin-plastid

[1] Möbius, M., 'Das Anthophaeïn, der braune Blüthenfarbstoff,' *Ber. D. bot. Ges.*, Berlin, 1900, XVIII, pp. 341–347. Schlockow, A., *Zur Anatomie der braunen Blüten*, Inaugural-Dissertation zu Heidelberg, Berlin, 1903.

series we have a crimson type, with magenta (or purple), ivory and yellow varieties.

Bidgood (18) mentions one or two unusual colour combinations: the lurid colour of some Delphiniums he attributes to the presence of cells containing red anthocyanin side by side with cells containing blue. *Crocus aureus*, also, on the outer side of the perianth leaves, shows green stripes which are due to a combination of blue sap colour on the outer side of the perianth leaves and yellow soluble pigment on the inner side. Dennert (14) gives many examples of the different ways in which plastid and sap pigments occur arranged in the tissues, and reference should be made to his paper if details are required. Generally speaking, the anthocyanin pigments occur in the epidermis of the corolla, and the chromoplastid either in the inner tissues, or in the epidermal, or both, and the great variety of such combinations accounts to a large extent for the numerous shades and tints. Dennert also observed that when both plastid and soluble pigments occur in the same cell, especially in the papillae of the epidermis, the plastids tend to occupy the base of the cell, while the soluble pigment is uniformly diffused.

Investigations on the distribution of anthocyanins in fruits have been undertaken by Borbás (38); and in seeds by Sempolowski (36), Preyer (66), Brandza, Lindinger (76) and Coupin (82).

Finally it may be as well to quote the results which Gertz (19) has given of a comparative examination of the localisation of anthocyanin in members of many of the Natural Orders, and from which he finds evidence of a certain amount of agreement between systematic relationship and distribution of pigment. His results are shortly summarised as follows:

Helobiae. In *Alisma, Butomus* and *Hydrocharis* subepidermal, in *Vallisneria* epidermal.

Gramineae. Epidermal in *Panicum, Oplismenus, Pennisetum, Calamagrostis, Setaria, Holcus, Weingaertneria, Catabrosa, Melica, Briza*; subepidermal in *Pharus, Phleum, Alopecurus, Baldingera, Bromus, Secale, Triticum, Avena, Aira*.

Aroideae. Subepidermal all through and similarly for Lemnaceae.

Bromeliaceae. Chiefly in hypodermis but in some forms also in epidermis.

Commelinaceae. Epidermal.

Juncaceae, Liliaceae and Amaryllidaceae. Subepidermal. In outermost scales of bulb of *Allium* often epidermal.

Scitamineae. Epidermal.

Orchidaceae. Epidermal in *Cypripedium*, *Orchis*, *Epipactis*, *Limodorum*, *Oncidium*; subepidermal in *Haemaria*, *Pogonia*, *Goodyera*, *Microstylis*, *Restrepia*, *Cattleya*, *Laelia*, *Dendrobium*, *Phalaenopsis*.

Piperaceae. Anthocyanin in subepidermal water tissue or in spongy parenchyma.

Salicaceae. Subepidermal all through.

Betulaceae. Periodic anthocyanin in ground tissue. Permanent in epidermis.

Fagaceae. In general subepidermal. In certain species of *Quercus* in hairs. In leaves of Copper Beech in epidermis.

Moraceae. Subepidermal in *Ficus*.

Aristolochiaceae. Epidermal.

Polygonaceae. In young leaves chiefly epidermal. In older, subepidermal, but types somewhat mixed so that localisation rather indefinite, and still more indefinite in Amarantaceae.

Nyctaginaceae, Aizoaceae and Portulacaceae. Epidermal.

Caryophyllaceae. In Alsinoideae always subepidermal; in Silenoideae usually epidermal, at least as regards origin. Exceptions are species of *Dianthus*, and, according to Wulff, *Silene acaulis*. Also subepidermal in stem of *Saponaria*.

Ranunculaceae. Subepidermal nearly all through. Epidermal in *Paeonia coriacea*, *Anemone Hepatica*, *A. japonica* and *A. Pulsatilla*.

Berberidaceae and Papaveraceae. Subepidermal.

Cruciferae. Subepidermal except in *Camelina silvestris*, *Braya*, *Draba verna*, *Alyssum*, *Farsetia*, *Malcomia*.

Crassulaceae. Anthocyanin epidermal; in epidermal idioblasts and in parenchymatous sheaths of vascular bundles.

Saxifragaceae. Localisation more indefinite in *Saxifraga*, but appears to start from the epidermis, at least in young leaves. In spring leaves of *Ribes*, epidermal; in autumnal, subepidermal.

Rosaceae. In spring leaves in general, anthocyanin in epidermis; in autumnal leaves, anthocyanin in the ground tissue. Subepidermal in spring leaves of *Cotoneaster*, *Cydonia*, *Pyrus*, *Photinia*, *Amelanchier*.

Leguminosae. In general subepidermal but found in the epidermis in *Acacia*, *Mimosa*, *Cercis*, *Gymnocladus*, *Gleditschia*, *Lupinus*, *Medicago*, *Trifolium*, *Indigofera*, *Glycyrrhiza*, *Lourea*, *Onobrychis*.

Geraniaceae. In young leaves epidermal, in older, subepidermal.

Tropaeolaceae, Linaceae, Polygalaceae. Subepidermal.

Euphorbiaceae. Generally subepidermal, epidermal in *Croton*, *Ricinus* and some *Euphorbia* spp.

Anacardiaceae. In *Pistacia* and *Rhus*, anthocyanin epidermal in spring leaves, subepidermal in autumnal leaves.

Aceraceae. Subepidermal in *Acer*, but epidermal in the red-leaved variety.

Vitaceae. Spring leaves epidermal, autumnal subepidermal.

Tiliaceae. Subepidermal.

Malvaceae. Epidermal.

Theaceae, Hypericaceae and Violaceae. Ground tissue.

Begoniaceae. Epidermal in leaves, and sometimes in certain spongy parenchyma cells.

Lythraceae and Myrtaceae. Subepidermal.

Oenotheraceae. In *Oenothera*, *Epilobium*, *Godetia*, *Gaura*, *Circaea*, usually epidermal. Subepidermal in autumnal leaves of *Jussieua*, *Chamaenerium*, *Fuchsia*.

Umbelliferae. Subepidermal, but epidermal in *Eryngium amethystinum*, and in stem of *Chaerophyllum temulum*, and *Conium maculatum*.

Cornaceae, Ericaceae and Epacridaceae subepidermal. In Primulaceae in ground tissue only except in *Androsace* and *Cyclamen* where it is epidermal.

Oleaceae. Usually subepidermal but in spring leaves of *Forsythia*, *Syringa*, *Chionanthus*, *Jasminum* in epidermis.

Gentianaceae. Epidermal except in *Menyanthes*.

Apocynaceae, Asclepiadaceae and Convolvulaceae. Subepidermal, but in *Cuscuta* often epidermal.

Boraginaceae. Altogether in ground tissue except in *Symphytum*.

Labiatae. Marked epidermal localisation. Subepidermal in certain *Coleus* varieties, and in autumnal leaves of *Prunella vulgaris*, *Ballota nigra*, *Betonica officinalis*, *Lycopus europaeus*.

Solanaceae. Subepidermal.

Scrophulariaceae. Subepidermal in *Verbascum*, *Linaria*, *Nemesia*, *Limosella*, or epidermal in *Pentstemon*, *Veronica*, *Digitalis*, *Euphrasia*, *Odontites*, *Rhinanthus*, *Pedicularis*, *Melampyrum*.

Bignoniaceae and Orobanchaceae. Epidermal.

Gesneriaceae. Chiefly epidermal but often also hypodermal and in the spongy parenchyma.

Lentibulariaceae. Epidermal in *Pinguicula*.

Acanthaceae. Varying localisation, usually epidermal.

Plantaginaceae. Epidermal.

Rubiaceae. Mesophyll in most cases.

Caprifoliaceae. Subepidermal except in *Sambucus racemosa* and *S. Ebulus*.

Valerianaceae, Dipsaceae and Campanulaceae. In peripheral ground parenchyma.

Compositae. Both epidermal and subepidermal localisation. In older leaves nearly always the latter. In younger, subepidermal in *Solidago, Achillea, Matricaria, Centaurea, Leontodon, Taraxacum, Tragopogon* and *Scorzonera*. Anthocyanin in hairs in *Hebeclinium janthinum, Eupatorium atrorubens* and *Gynura aurantiaca*.

CHAPTER IV

THE PROPERTIES AND REACTIONS OF ANTHOCYANINS

FROM time to time the question has been under discussion as to whether all the varieties of red, purple and blue plant pigments are merely one and the same compound, the different shades being due to the presence of other substances in the cell-sap, i.e. acids, alkalies, etc., or whether the term anthocyanin includes many different members of a great group. The earliest expression of opinion on this point is possibly that made by James Smithson (126) in the *Transactions* of the Royal Society in 1818; here he remarks, on the slightest experimental basis: "The colouring matter of the violet exists in the petals of red clover, the red tips of those of the common daisy of the fields, of the blue hyacinth, the holly hock, lavender, in the inner leaves of the artichoke, and in numerous other flowers. It likewise, made red by an acid, colours the skin of several plumbs, and, I think, of the scarlet geranium, and of the pomegranate tree. The red cabbage, and the rind of the long radish are also coloured by this principle. It is remarkable that these, on being merely bruised, become blue; and give a blue infusion with water. It is probable that the reddening acid in these cases is the carbonic; and which, on the rupture of the vessels which enclose it, escapes into the atmosphere." In the same way Marquart (5), and Fremy & Cloëz (140) originally recognised only one blue pigment, from which the red pigments were believed to be derived by action of acids. Of a similar opinion was Wigand (150), who writes: "Die rothe und blaue Farbe der Blüthen sind, wie sich theils aus den Uebergangserscheinungen, theils aus dem Auftreten beider Farben als homogene Färbung der Zellenflüssigkeit, theils aus dem übereinstimmenden Verhalten beider gegen chemische Reagentien ergiebt, unwesentlich verschiedene Zustände eines und desselben Stoffes, des *Anthocyans.*" Hansen (11), also, believed most red flower-colours to be due to one substance.

Later, N. J. C. Müller (184), on the basis of spectroscopic examination, announced that the red and blue pigments were of various kinds, but it is doubtful whether his materials were pure. Wiesner (149) also appeared to be uncertain as to whether all anthocyanins are alike, whereas Weigert (194) definitely distinguishes two groups of anthocyanin of which more will be said later. From this time onward, as investigations increased, there seemed to be little doubt that a number of substances are responsible for the different colours. Overton (420) held this point of view, and we may quote the words of Molisch (108) to the same effect: "dass der Begriff Anthokyan, wie er bisher in der Literatur gefasst wurde, kein einheitliches chemisches Individuum darstellt, sondern eine Gruppe von mehreren verschiedenen, wahrscheinlich verwandten Verbindungen." The more recent suggestion of Grafe (224) comes nearer the truth, for he says we must regard anthocyanin as a term to be used for a whole series of pigments, which may have a similar fundamental nucleus, but which differ in the complexes attached to the nucleus. Colour and other chemical reactions would then depend on a particular grouping common to all or most of the pigments. This view has been supported by the recent researches of Willstätter (260, 273, 274, 286–294, 300–307); he has isolated anthocyanin from the flowers or fruits of twenty (or more) plants, and has shown that they have the same fundamental structure. Some of the number, though derived from plants quite unrelated, appear to be identical; others differ in the number of their hydroxyl groups; others again have their hydroxyls replaced by the methyl radical. Thus we may now correctly consider the word anthocyanin to stand, as a collective term, for a class of substances comparable to the sugars, tannins, fats, proteins, etc.

The appearance of anthocyanin as crystals in the living tissues has been discussed in Chapter III. Outside the cell anthocyanin has also been obtained in crystalline form. Molisch (108) prepared crystals very readily from the petals of the scarlet Geranium, *Pelargonium zonale*, by placing a petal in distilled water under a cover-slip on a slide. The pigment diffuses out, and on slow evaporation deposits groups of beautiful needle-shaped crystals. By a similar method, using acetic acid instead of water, Molisch (108) obtained good crystals from petals of garden roses and of *Anemone fulgens*. Several other investigators crystallised anthocyanin from extracts. From the red pigment of Vine leaves, Gautier (190) isolated two substances termed by him α- and β-ampelochroïc acids. From a solution in hot water, the α-acid was obtained on cooling as a red crystalline powder. The β-acid also was deposited from water in red crystals on slow evaporation. Griffiths (206) prepared crystals from pigment of 'Geranium' (*Pelargonium*) flowers. From flowers of *Althaea rosea*, Grafe (212) isolated a deep red pigment which separated out from alcohol in minute crystalline plates. Portheim & Scholl (219) also succeeded in crystallising the anthocyanin from the testa of seeds of *Phaseolus multiflorus*. Later, *Pelargonium* flowers were again employed by Grafe (237) as material for the purification and analysis of anthocyanin, and like Molisch, Grafe found that this pigment very readily crystallised.

Willstätter's pigments were all obtained as crystalline oxonium salts with acids; only in the case of delphinin and cyanidin has the colour base itself been prepared in crystalline form. The potassium salt of the glucoside, cyanin, moreover, was obtained in fine crystals. It is probable that the crystalline products of the earlier workers also were salts.

Before describing the solubilities of anthocyanin, it should be mentioned that the pigment usually exists in the plant in the form of a glucoside. In this form it has been isolated from fruits of the Bilberry (Heise, 193), from flowers of *Althaea*, *Pelargonium* (Grafe, 212, 237), and from all the plants investigated by Willstätter, who found the non-glucosidal pigment only in certain varieties of the Grape. Later, Rosenheim (325) noted the presence of non-glucosidal pigment in young Vine leaves, and Jonesco (328, 329, 340) in fruits of *Ruscus aculeatus*, *Solanum Dulcamara* and a number of red flowers and leaves.

As a glucoside, anthocyanin is readily soluble in water, and since it is in this form that the pigments chiefly occur in the cell, they can be extracted with water. After hydrolysis, in the non-glucosidal state, the pigment is far less soluble in water, and in some cases almost or quite insoluble, i.e. *Antirrhinum* (Wheldale & Bassett, 270) and Bilberry (Heise, 193). To the consideration of these glucosides we shall return again later in the chapter.

In ether, anthocyanin is insoluble, as also in benzene, carbon bisulphide, chloroform and similar solvents in which plastid pigments are soluble. In alcohol, the greater proportion of anthocyanins are soluble; there are definite exceptions, such as those of the Amarantaceae, Chenopodiaceae and Phytolaccaceae which are entirely insoluble in this solvent. To these may be added the blue pigment of *Centaurea* (Willstätter, 260), the glucosidal pigment of *Althaea* (Grafe, 212) and probably others. In many flowers it is difficult to extract the petals completely with alcohol. This is possibly due in some cases to the presence of more than one pigment, certain of which are insoluble in alcohol; or there may also be retention of the pigment to some extent by the coagulated cell-contents.

A curious phenomenon is connected with the alcohol solutions of most anthocyanins; such solutions, though at first coloured red or purple as the case may be, somewhat rapidly lose their colour and eventually become quite colourless; the same effect is produced by immersing petals in strong alcohol. The colour returns on evaporation of the alcohol or, in many cases if the solution is sufficiently strong, on adding water. A few drops of acid, also, restore the colour completely; similarly a few drops of alkali produce the green (or yellow) reaction characteristic of anthocyanin. This phenomenon was first remarked upon by Nehemiah Grew (1): "Again though no *Blew Flowers*, that I know of, will give a *Blew Tincture* to *Spirit* of *Wine*: yet having been for some days infused in the said *Spirit*, and the *Spirit* still remaining in a manner *Limpid*, and void of the least *Ray* of *Blew*; if you drop into it a little *Spirit* of *Sulphur*, it is somewhat surprizing to see, that it immediately strikes it into a full *Red*, as if it had been *Blew* before: and so, if you drop *Spirit* of *Sal Armoniac* or other *Alkaly* upon it, it presently strikes it *Green*....It is likewise to be noted, That both *Yellow* and *Red Flowers* give a stronger and fuller *Tincture* to *Water*, than to *Spirit* of *Wine*; as in the *Tinctures* of *Cowslip*, *Poppys*, *Clove-July-Flowers* and *Roses*, made both in *Water* and *Spirit* of *Wine*, and compared together, is easily seen." Loss of colour in alcohol was also mentioned by Morot (136), Filhol (139) and Fremy & Cloëz (140). In 1884, Hansen (11) commented on it, and suggested that anthocyanin, in absolute alcohol, forms a colourless anhydride. Keeble & Armstrong (254) have offered another, though unsatisfactory, explanation of this decolorisation (see p. 119). Willstätter (260) has shown the loss of colour to be due to the formation of a colourless isomer (see p. 63).

The appearance, colour, melting point, crystalline form, etc., of the pigments extracted by Willstätter are given in the tables on pp. 78–81.

Qualitative Reactions.

With respect to qualitative reactions, the necessity, already mentioned in connection with the properties of anthocyanin, for a guarantee of the purity of the pigment used is of paramount importance. Crude extracts invariably contain other substances which may modify, or completely alter, the reactions of the pigment itself. In the next chapter, accounts are given of special methods for purification and analyses of anthocyanin, and on studying these, the futility of applying tests to any but pure material will be recognised at once. It is only after careful extraction, and purification by means of analyses, that the reactions of any pigment can be determined with certainty, and qualitative tests on impure extracts are to a large extent worthless. Nevertheless, numbers of observations have been made on more or less impure material, and the following account deals with the general reactions (special reactions are given on pp. 78–81):

With alkalies. When an aqueous or alcoholic extract of anthocyanin is treated with alkali, the pigment turns green[1] and often finally yellow. Sometimes a blue colour precedes the green; with very dilute alkali, or with solutions of salts having a weakly alkaline reaction, a blue colour only may appear. Similarly red, purple and blue flowers placed in ammonia vapour as a rule turn green.

With acids. Anthocyanins almost invariably turn bright red with acids, though the shade may vary in different cases.

With lead acetate. Anthocyanin extracts are generally precipitated by lead acetate, and the colour of the precipitates is usually some shade of green or blue; occasionally it is red.

The reactions of anthocyanins with acids and alkalies have formed a subject for discussion from time to time. The whole matter is so bound up with the views of those who have worked on the pigments that something of the nature of a historical summary must be included.

The question first to be considered is whether the green coloration given when tissues and crude extracts containing anthocyanin are treated with alkalies is a reaction of anthocyanin alone, or is the combined result of reactions with anthocyanin plus reactions with other substances present in the cell or solution, and this can only be determined satisfactorily by testing pure pigments. In most cases, when white flowers are treated with alkalies, or exposed to ammonia vapour, a bright yellow colour is developed, indicating a reaction of alkalies with a class of substances known as flavones which are almost universally distributed in plants. The flavone pigments in bulk are yellow, but exist in the cell-sap in such small quantities as to be inconspicuous except when treated with alkali, when an intense yellow colour is developed; with ferric chloride solution they give a green or brown coloration. They, moreover, occur in the plant largely as glucosides, in which form they are readily soluble in alcohol and water, and hence they are present in all crude aqueous and alcoholic extracts of anthocyanin. Extracts of white flowers, or in fact of any non-anthocyanic parts, give as a rule yellow or orange-yellow precipitates with lead acetate, which are insoluble flavone salts of lead.

That there is some substance in white flowers which turns yellow with ammonia was noticed by Boyle (121): "we thought fit to make Trial upon the Flowers of *Jasmin*, they being both White as to Colour, and esteem'd to be of a more Oyly nature than other Flowers. Whereupon having taken the White parts only of the Flowers, and rubb'd them somewhat hard with my Finger upon a piece of clean Paper,...a strong Alcalizate Solution, did immediately turn the almost Colourless Paper

[1] In some cases such solutions are slightly dichroïc, green and red.

moisten'd by the Juice of the *Jasmin*,...a Deep, though somewhat Greenish Yellow,...when we try'd the Experiment with the Leaves of those purely White Flowers that appear about the end of Winter, and are commonly call'd *Snow drops*, the event, was not much unlike that, which, we have been newly mentioning." Later, in the *Comptes Rendus* of 1854 and 1860, Filhol (139, 146) published papers of considerable interest in connection with this point. Filhol found that when white flowers of *Viburnum Opulus*, *Philadelphus coronarius* and other plants were exposed to ammonia, they turned yellow; as yellow, he says, as the flowers of Laburnum. The same results he observed in leaves in the parts free from chlorophyll. When the flowers, after treatment with alkali, were placed in acidified water, they became white again. The substance which gives the yellow colour was found to be soluble in water and alcohol and slightly so in ether, and Filhol terms it *xanthogène*. When coloured, red or purple, flowers were treated with ammonia, they turned green as a rule, but in some cases blue (*Papaver*, *Pelargonium*, *Salvia splendens*). In one particular experiment when he added aluminium hydroxide to an extract of Vervain flowers, the aluminium hydroxide became yellow but the supernatant liquid retained the purple colour. Thus he comes to the conclusion that the green coloration with alkalies in *Viburnum* and other flowers is due to a mixture of a blue colour given by anthocyanin plus a yellow colour given by *xanthogène*. But in the case of the flowers of *Pelargonium*, *Papaver* and *Salvia* the *xanthogène* is absent and so the anthocyanin becomes blue or violet with alkalies. In the later paper (146) he remarks on the resemblance of *xanthogène* to luteolin (which we now know to be a flavone), but he was unable to establish the identity. Similar views were advanced by Wiesner (149) in 1862. From a series of reactions given by various flower pigments he concludes that colourless sap contains, as a rule, a tannin giving a green reaction with iron salts, and a yellow colour with alkalies. Like Filhol he believes that anthocyanin itself gives a blue, never a green, reaction with alkalies, and the green colour is due to admixture with yellow given by the tannin. In plants free from tannins giving the yellow reaction, anthocyanin turns blue with alkali. Wiesner's tannins are probably for the most part flavones, since true tannins are rare in flowers. These views on the reactions of anthocyanin gave rise to a certain amount of controversy, for Wigand (150) and Nägeli & Schwendener (152) held the opinion that the green coloration is given by anthocyanin itself, and may appear when the iron-greening tannins are absent. Some of the arguments involved in the discussion are given by Wiesner in a later paper (156). The alkali reaction of anthocyanin is also mentioned by Overton (420), who considers the blue colour to be due to the formation of an acid salt, the green colour to a neutral salt of the pigment, anthocyanin itself being a dibasic acid. This view is accepted by Grafe (212) as being in accordance with the reactions of *Althaea* pigment which he prepared in a pure state. Anthocyanin from *Antirrhinum* (Wheldale & Bassett, 270), purified from accompanying substances, still gives the green alkali reaction. Willstätter (260), as far as can be determined from his publications, considers the reactions of anthocyanin to be as follows. The pure blue anthocyanin from the Cornflower is not altered by alkali, i.e. pure anthocyanin gives a blue colour with alkalies, but if a solution of the blue pigment has stood for a time, the colour reaction with alkalies is green. This is due to the fact that from the pigment a colourless isomer (see pp. 54, 63, 66) has been formed, and this gives a yellow colour with alkalies; hence the blue plus yellow results in a green reaction. Crude extracts from the flowers, he says, also give a green colour owing to the presence of yellow pigments—obviously flavones. (Die reine blaue Farbstofflösung zeigt auf Zusatz von wenig Soda zunächst keine Farbänderung, eine gestandene Lösung wird hingegen grünblau oder blaugrün weil die Lösung nun auch die farblose Modifikation enthält. Dies ist das Verhalten eines wässrigen Blütenauszugs, worin sich überdies noch gelbe Farbstoffe befinden, deren Alkalisalze intensiv gelb sind.) The blue colour given by alkalies with the pure anthocyanin pigment will, Willstätter found, become green or yellow by decomposition on standing, or with excess of alkali. Further consideration will be given below (pp. 54, 55) to these reactions. With anthocyanins from other plants Willstätter notes various reactions with alkalies (see p. 57). Hence we have at present the following suggestions. Pure anthocyanin from *Centaurea* gives a blue colour with alkalies; the green colour given in solutions and crude extracts is due to mixture with the yellow colour produced by the colourless isomer or accompanying flavones, or both. Pure anthocyanin from *Antirrhinum* gives a green colour with alkalies, and this cannot be due to admixture with flavones, as the latter are removed by purification; it is difficult to suppose it due to admixture with a colourless isomer, since this is not formed in a strongly acid solution such as that from which the pigment separates out in preparation. In many cases the green reaction either rapidly or slowly changes to yellow, and the original colour does not return on neutralisation, so that evidently some anthocyanins are completely destroyed by alkalies.

The same difficulty arises with regard to the precipitates with lead acetate, for the accompanying flavones produce yellow or orange-yellow precipitates with lead acetate, and hence the actual lead salt of anthocyanin is not identifiable except from the pure pigment with which the results are the same as with alkalies. The colour of the lead precipitate varies very considerably in crude solutions. In extracts from white flowers containing little anthocyanin, it is greenish-yellow; from flowers containing much anthocyanin, bright green or blue-green, whereas from red varieties (Wheldale, 226), it is practically red with a greenish tinge. The latter must be distinguished from the red precipitates given by the special pigments of the Amarantaceae and Phytolaccaceae which are mentioned below.

On the basis of qualitative reactions, Weigert (194), in 1895, attempted a classification of anthocyanins into two groups:

The 'Weinroth' group, of which the pigments give blue-grey or blue-green precipitates with basic lead acetate, give the Erdmann reaction (see below), are precipitated and give a bright red colour with concentrated hydrochloric acid, and turn green on addition of alkalies. Ex.—Pigments from *Vitis*, *Ampelopsis quinquefolia*, *Rhus typhina*, *Cornus sanguinea*, etc.

The 'Rübenroth' group, of which the pigments give a red precipitate with basic lead acetate, do not give the Erdmann reaction, turn dark violet with concentrated hydrochloric acid and violet with ammonia, but with other bases yellow. Ex.—Pigments from *Beta*, *Iresine Lindeni*, *Achyranthes Verschaffeltii*, *Amaranthus*, *Atriplex hortensis*, *Phytolacca decandra*. According to Gertz (19), the following should be added to Weigert's 'Rübenroth' group: several Chenopodiaceae (*Blitum virgatum*, *Atriplex litoralis*, *Corispermum canescens*), Amarantaceae (except *Mogiphanes brasiliensis*), Nyctaginaceae (*Oxybaphus nyctagineus*), Phytolaccaceae (*Phytolacca decandra*, leaves), Aizoaceae (*Tetragonia crystallina*, *Mesembryanthemum nodiflorum*), Portulacaceae (*Portulaca grandiflora*) and Basellaceae (*Basella rubra*). The remaining families of the group Centrospermae, i.e. Polygonaceae and Caryophyllaceae, seem to be distinguished by anthocyanin of the 'Weinroth' group as also the greater number of the Chenopodiaceae. Kryz (323) adds the pigments of Fuchsia and Cactus to the beet-red group.

The reaction of Erdmann (163) was originally employed in order to detect true wine pigment and may be described as follows: The pigment solution (wine) is diluted with four times its volume of water, eight drops of concentrated hydrochloric acid are added, and the mixture shaken up with 16 c.c. of amyl alcohol. The amyl alcohol separates out with a fine violet-red colour, the underlying acid solution being yellow or cherry-red. If the amyl alcohol is separated off, and an equal volume of water added, together with two drops of concentrated ammonia, the amyl alcohol decolorises and the underlying solution becomes bright green. If the original acid solution below the amyl alcohol is placed in a porcelain dish and carefully neutralised with dilute ammonia, at neutralisation point an indigo-blue colour is produced which afterwards becomes green.

The reactions of anthocyanins with acids and alkalies, the Erdmann reaction, etc., have received a new interpretation through the recent researches of Willstätter (260). These views are based upon work devoted in particular to the anthocyanin of *Centaurea* but also to anthocyanins in general. As far as can be gathered from the various publications, the following represent, in the main, the views of Willstätter (see also Chapter v).

1. Red, purple and blue pigments occur in the plant almost entirely as glucosides (anthocyanins), which can be hydrolysed artificially with the formation of sugar and non-glucosidal pigments (anthocyanidins). From a dilute acid solution anthocyanidins are completely removed by amyl alcohol, whereas anthocyanins are either not taken up at all by this solvent, or, if they are to some extent, they can be removed by again shaking with dilute acid. This may be demonstrated experimentally by extracting fresh material containing anthocyanin with dilute sulphuric acid. After filtration and addition of amyl alcohol, no pigment is taken up by the alcohol. But if the solutions are heated for one-half to three-quarters of an hour on a water-bath, the anthocyanins are hydrolysed, and on addition of amyl alcohol, the anthocyanidins are quantitatively removed.

2. Anthocyanin is itself an acid and in the free state is purple.

3. There is a blue modification which is the potassium salt of the purple.

4. There is a red modification, which is the oxonium salt of anthocyanin; the pigment may be combined with either organic or inorganic acids.

5. In *Centaurea*, as well as in some other plants, all three forms of anthocyanin readily change to a colourless isomer; with the red form this only occurs in absence of excess of acid. The change to a colourless isomer can be prevented by adding neutral salts to the anthocyanin solution; the anthocyanin forms additive compounds with these substances, thereby preventing the isomeric change.

As regards the colour reactions with alkalies, Willstätter, as we have already seen, gives the following explanation. With alkalies a blue salt is formed, which may become green owing to mixture with flavones or the colourless isomer (see pp. 52, 63), if these are present in the extract. The blue salt is also unstable, and with excess of alkali passes to a greenish or even yellow decomposition product. But if a neutral mineral salt is present, the blue salt is rendered stable. This may be brought about by either (1) acidifying the anthocyanin and then neutralising or (2) by adding the neutral salt ($NaCl$, $NaNO_3$). Thus, for example, if pigment of Bilberries or Grapes (which presumably contain little flavone) is treated with alkali, it gives a greenish colour. But if treated first with some salt, it gives a blue colour, and the same result may be brought about by acidification and subsequent neutralisation.

It is now possible to explain the Erdmann reaction, which, according to Willstätter, has hitherto been misunderstood. The following account is given more or less in Willstätter's words. The Erdmann reaction is based on the fact that new, or fairly new, wine gives a green colour on neutralisation with ammonia, but after treatment with hydrochloric acid and neutralisation, a dark greenish-blue colour. This behaviour was incorrectly explained by Erdmann as being due to the splitting of the wine pigment into two pigments by the hydrochloric acid. Erdmann separated the above two pigments by shaking with amyl alcohol; the violet-red amyl alcohol layer gives with dilute ammonia first a bright green, then a brownish-green colour, while the acidified water solution becomes indigo-blue on neutralisation. As a matter of fact no breaking up is brought about by the acid, but during fermentation of the grape juice, a portion of the anthocyanin has been hydrolysed to anthocyanidin. The effect of ammonia on the wine pigment is to give a blue coloration rapidly passing to a green decomposition product. But, as explained by Willstätter, if the anthocyanin is first acidified and then made alkaline, the blue colour is more stable. On shaking up the acidified wine with amyl alcohol, the latter takes up the small portion of anthocyanidin as an oxonium salt. If the wine is heated with hydrochloric acid, or if the hydrochloric layer of the Erdmann reaction is heated, the whole is hydrolysed, and the anthocyanidin can be extracted quantitatively with amyl alcohol. But the hydrolysis does not happen in the cold. Willstätter further points out a source of error in the reaction. Grape juice from fresh berries gives, after acidification with hydrochloric acid, a little pigment in the amyl alcohol. If the amyl alcohol is then washed with dilute sulphuric acid, it is nearly decolorised. The small amount of pigment in the amyl alcohol is not hydrolysed, but is due to the fact that hydrochloric-amyl alcohol takes up a little of the anthocyanin oxonium salt, whereas sulphuric-amyl alcohol takes practically none. Hence it is advisable to use sulphuric acid for the test. The statement by Weigert, that the 'Weinroth' group gives the Erdmann reaction, is regarded by Willstätter as erroneous, for the latter maintains that, apart from the small amount of glucoside salt which can be washed out again from the amyl alcohol, the pigment of flowers, berries and leaves remains completely in the water-acid layer.

The sensitiveness of anthocyanin to acids and alkalies has suggested its use as an indicator. In fact these were the first reactions to attract the attention of chemists (see p. 10). Its use in this way has been revived from time to time by Pellagri (160), Sacher (230) and others, but without any permanent success.

Reactions with iron salts. Here again the actual colour reaction is difficult to estimate unless the pure pigment is tested. The flavones give green or brownish-green colorations with iron salts, the tannins green or blue. Though anthocyanins on the whole appear to give either a blue, or no colour, if tannin or flavone be present it is obvious that no reliance can be placed on the results.

Reactions with sodium bisulphite. An interesting reaction of anthocyanins is that given with sulphur dioxide and bisulphites. It was well known at a very early date that flowers containing anthocyanin, or extracts of the pigment, are bleached by sulphur dioxide gas, and that the colour is again restored by stronger acids. Boyle (121) writes: "That Roses held over the Fume of Sulphur, may quickly by it be depriv'd of their Colour, and have as much of their Leaves, as the Fume works upon, burn'd pale, is an Experiment, that divers others have tried, as well as I. But (*Pyrophilus*) it may seem somewhat strange...That, whereas the Fume of Sulphur will,...Whiten the Leaves of Roses; That Liquor, which is commonly call'd Oyl of Sulphur...does powerfully heighten the Tincture of Red Roses." Further observations on this bleaching action were made by Kuhlmann (132), Hünefeld (134) and Schönbein (137). Solutions of anthocyanin decolorised by sulphur dioxide have been employed by Kastle to test the relative 'strengths' of acids. Kastle (211) is of the opinion that the decolorisation is not caused by reduction, and the same view is held by Grafe (212, 237), who prepared bisulphite derivatives from the anthocyanins of both *Althaea* and *Pelargonium* by addition of sodium bisulphite. Both products were colourless, but the red colour returned on addition of a trace of a stronger acid. Grafe concludes that the anthocyanins contain aldehyde colour-producing groups, which form additive compounds with bisulphites, whereby the linkings in the molecule and the resultant colour are changed.

Action of nascent hydrogen. A reaction which would appear to be one of reduction is that produced by treating acid solutions of anthocyanin with zinc dust. The colour rapidly disappears and the solution remains colourless if air be excluded. On exposure to air, if the reducing action is not very violent, the colour returns, the surface of the liquid becoming coloured before the deeper layers. Kastle (211) does not consider the reaction to be of the nature of reduction, since the colour did not return on treatment with oxidising enzymes. Grafe (237) holds the view that the loss of colour is due to changes brought about in the aldehyde groups (which he postulates) by the action of nascent hydrogen. The fact that the return of colour in air is not equally great with all acids (Wheldale & Bassett, 271) may indicate that the reaction is not a simple reduction process.

Compounds with acids. In all cases both anthocyanins and anthocyanidins appear to form definite compounds with acids (Willstätter's oxonium salts), since such compounds occur in crystalline form.

Spectrum of anthocyanin. A considerable amount of attention has been devoted to the spectroscopic examination of flower and leaf pigments. Sorby (153, 158), Müller (184), Engelmann (494), Lepel (166) and Formánek (201) may be mentioned as workers on these lines; but the results are of little value for identification, or otherwise, on account of the impurity of the products employed, that is the doubt as to the number of pigments present, etc.

Willstätter (260) has distinguished various groups of anthocyanins by their different behaviour to reagents, though the observations do not pretend to include any kind of systematic classification. The following represent some of the classes:

1. Red in acid solution, blue with soda and a blue precipitate with lead acetate; pigment readily isomerises to a colourless modification (*Centaurea, Rosa, Lathyrus*).
2. Red in acid solution, blue with soda and a blue precipitate with lead acetate; pigment decolorises less readily or not at all (Grapes, Bilberries, flowers of *Delphinium*).
3. Yellowish-red in acid solution, blue with soda, red precipitate with lead acetate (Radish).
4. Yellowish-red in acid solution (*Pelargonium*) and blue-red (*Papaver*); both violet with soda and decolorised by isomerisation.

5. Red in acid solution; with soda red in dense layers, blue-green in thin layers (Pinks) or red-violet to red-brown (Aster).

6. Violet in acid solution, red with soda, red precipitate with lead acetate (Beet-root, *Atriplex*).

The author (226, 227) has also made observations on crude extracts of anthocyanins from a very large number of flowers, using the colour reactions given by various chemical reagents, i.e. sulphuric, hydrochloric and oxalic acids, ammonia, caustic potash, lime water, ferric chloride, ferrous sulphate, potassium ferrocyanide, uranium, lead, copper and sodium acetates, stannous chloride and others. Although a certain amount of differentiation was possible on this basis of qualitative reactions, it was soon found that there were too many aberrant and peculiar forms of pigment to arrive at any satisfactory classification.

Before we close the chapter, there is yet another extract which may well be quoted from the writings of Boyle (121), since it dealt two hundred and fifty years ago with some of the phenomena which have formed the basis of Willstätter's constitutional formulae for the cyanidin of the Cornflower, i.e. the reactions of anthocyanin with acids and alkalies, and its instability in water solution.

Boyle writes: "There is a Weed, more known to Plowmen than belov'd by them, whose Flowers from their Colour are commonly call'd *Blew-bottles*, and *Corn-weed* from their Growing among Corn. These Flowers some Ladies do, upon the account of their Lovely Colour, think worth the being Candied, which when they are, they will long retain so fair a Colour, as makes them a very fine Sallad in the Winter. But I have try'd, that when they are freshly gather'd, they will afford a Juice, which when newly express'd, (for in some cases 'twill soon enough degenerate) affords a very deep and pleasant Blew. Now, (to draw this to our present Scope) by dropping on this fresh Juice, a little Spirit of Salt, (that being the Acid Spirit I had then at hand) it immediately turn'd (as I predicted) into a Red. And if instead of the Sowr Spirit I mingled with it a little strong Solution of an Alcalizate Salt, it did presently disclose a lovely Green;...And I remember, that finding this Blew Liquor, when freshly made, to be capable of serving in a Pen for an Ink of that Colour, I attempted by moistning one part of a piece of White Paper with the Spirit of Salt I have been mentioning, and another with some Alcalizate or Volatile Liquor, to draw a Line on the leisurely dry'd Paper, that should, e'vn before the Ink was dry, appear partly Blew, partly Red, and partly Green."

CHAPTER V

THE ISOLATION AND CONSTITUTION OF ANTHOCYANINS

SEVERAL general and rather vague views have been held as to the constitution of anthocyanin, without any particular experimental evidence. Thus Wigand (150) believed these pigments to arise by oxidation from a colourless tannin-like chromogen; Overton (420) considered them to be tannin-like substances combined with sugar, and Katić (441), too, found that anthocyanin gave the reactions of a tannin. Palladin (218), again recognised in anthocyanin a respiratory pigment, and yet other suggestions have been advanced by Filhol, Mirande and Combes.

But in the following cases, definite isolation of the pigments has been attempted, and analyses have been made; the methods and results, however, are so varied that a separate account is essential in each case. We may enumerate the cases thus:

1849. Morot (136). The flower-pigment of the Cornflower (*Centaurea Cyanus*).
1858. Glénard (143, 144). The colouring matter of wine.
1877. Church (161). The pigment from leaves of *Coleus*.
1877. Senier (162). The flower-pigment of the Rose (*Rosa gallica*).
1878. Gautier (164). The colouring matter of wine.
1889. Heise (182). The colouring matter of grapes.
1892. Gautier (190). The pigment from red Vine leaves.
1892. Glan (191). The flower-pigment of the Hollyhock (*Althaea rosea*).
1894. Heise (193). The pigment from fruits of the Bilberry.
1903. Griffiths (206). The flower-pigment of *Pelargonium*.
1906 and 1909. Grafe (212, 224). The flower-pigment of *Althaea rosea*.
1911. Grafe (237). The flower-pigment of *Pelargonium*.
1913 and 1914. Wheldale (259, 270). The flower-pigment of the Snapdragon (*Antirrhinum majus*).
1913. Willstätter (260). The flower-pigment of *Centaurea Cyanus*.
1914–1916. Willstätter and others (273, 274, 286–294, 300–307). The flower-pigment of deep-red Chrysanthemum (*Chrysanthemum indicum*), scarlet- and deep-red Dahlia (*Dahlia variabilis*), black Hollyhock (*Althaea rosea*), violet variety of Larkspur (*Delphinium Consolida*), wild Mallow (*Malva sylvestris*), blue-black Pansy (*Viola tricolor*), violet-red Peony (*Paeonia*), purple-blue Petunia (*Petunia hybrida hort.*), double purple-scarlet variety of Poppy (*Papaver Rhoeas*), Rose (*Rosa gallica*), Scarlet Geranium (*Pelargonium zonale*), Scarlet Salvia (*Salvia coccinea* and *S. splendens*), and purple-red Summer Aster (*Callistephus chinensis*). The pigment from fruit of the Bilberry (*Vaccinium Myrtillus*), Cranberry (*V. Vitis-Idaea*), Grape (*Vitis vinifera*), Sloe (*Prunus spinosa*) and Sweet Cherry (*P. avium*).

1918. Schudel (318). The pigment from root of the Beet (*Beta vulgaris*) and Radish (*Raphanus sativus*).

1922. Currey (338, 339). The flower-pigment of scarlet *Pelargonium* and red *Rosa gallica*.

1923. Anderson (350). The pigment from fruit of the Grape.

Except for the case of the flowers of *Antirrhinum majus*, of which the pigment has not been subsequently isolated, no detailed account will be given of the work up to 1913, since practically all the plants used by earlier workers have been again investigated more satisfactorily by Willstätter.

The methods adopted by earlier workers usually followed the plan of precipitating the pigments as a lead salt, decomposing this with removal of lead, and finally precipitating the pigment from alcoholic solution by ether. In the light of later analyses, it is obvious that these earlier preparations were not pure. Among them, however, the work of Grafe stands out as especially excellent; he prepared crystalline products from Hollyhock and *Pelargonium* flowers, but, here again, when compared with Willstätter's results, it is obvious that his products were not sufficiently pure.

In the following pages the isolation of the pigment of *Centaurea* flowers by Willstätter is given in detail, as it is, in general, representative of the whole series of investigations.

The flower-pigment of the Snapdragon (Wheldale, 259, 270). The anthocyanin pigment, as in most flowers, is only present in the epidermis, while the inner tissues contain a flavone. Pigment was prepared separately from the following varieties (see p. 162): magenta (various shades together), ivory tinged with magenta, crimson, rose doré (various shades) and bronze (various shades). In the magenta and rose doré series, apigenin is present in addition to anthocyanin; in the crimson and bronze, both apigenin and a second pigment, luteolin (see p. 115). All the flowers have, in addition, a patch of deep yellow pigment on the palate. The latter pigment can be eliminated, if desired, by tearing away the lower half of the flower, and using the upper half only for extraction; this device was adopted in the preparation of some samples of pigment. Thus, in any method of extraction, we have to deal, not only with an anthocyanin pigment, but also with one or more accompanying flavones. Two anthocyanins were found to be responsible for the colour varieties, viz. a true red anthocyanin in the rose doré and bronze series, and a magenta (blue-red) anthocyanin in the magenta and crimson series. The following method was employed for obtaining the pigment in quantity.

The flowers were boiled with water in saucepans, and the water extract filtered through large filters into lixiviating jars. The anthocyanins and flavones were then precipitated as lead salts by adding solid lead acetate to the hot solution until no more precipitate is formed. (The colour of the precipitates varies according to the flowers used; it is blue-green for full-coloured magenta, yellow-green for tinged ivory, dirty red for rose doré, and so forth. The colour of the lead salt of the anthocyanin is obviously modified by the amount and colour of the lead salts of the accompanying flavones.) The lead precipitate was filtered off, a vacuum pump being used for filtration; the solid cake of lead salt was then decomposed with 5–10 % sulphuric acid. The lead sulphate was filtered off, and a bright red solution of the pigments was obtained. This solution contains all the pigments in the flower used, both anthocyanin and flavones, in the form of glucosides in dilute acid solution. The solution was then boiled in a large Jena flask, fitted with a reflux condenser, for several hours. On cooling, the anthocyanin and flavones, now less soluble and no longer in the form of glucosides, separate out as a dark purplish- or brownish-red powder, according to the flower-colour used. The crude pigment was filtered off through as small a funnel as possible by means of a vacuum pump, washed, and dried over calcium chloride. The well-dried pigment was then finely ground, and placed in a Soxhlet thimble. The thimble was suspended just above the surface of ether contained in a wide-necked Erlenmeyer flask fitted with a condenser, and the ether kept boiling upon an electric heater. This process was continued until the ether ceased to extract any yellow colour. In this way the anthocyanins are obtained practically free from flavones, since the latter are soluble in ether. The anthocyanin residue in the thimble was then taken up in absolute alcohol, and filtered, and was, in this way, freed from a quantity of brown substance, which was insoluble in alcohol, and which was probably formed during the hydrolysis of the glucoside with sulphuric acid. The absolute alcohol solution, evaporated to its minimum bulk, is then poured into a large volume of ether, by which means the anthocyanin is precipitated, but any flavone present as impurity is retained in solution. The dried precipitate of anthocyanin was again extracted with ether to remove traces of flavone. The method of precipitation gives better results, as regards the purity of anthocyanin, than crystallisation, for, on crystallising a mixture of anthocyanin and flavone, both substances crystallise out together, and one is unaware of the presence of flavone in the product obtained.

The two forms of anthocyanin, red and magenta, were extracted and purified in this way from the flowers of different varieties of *Antirrhinum* mentioned above.

Pure red anthocyanin is an indian-red powder. It is readily soluble in absolute alcohol, almost insoluble in water, and slightly soluble in dilute acids and ethyl acetate; insoluble in ether, chloroform and benzene. In concentrated sulphuric acid it forms a reddish solution with a slight green fluorescence. It is soluble in alkalies to a greenish-yellow solution. With ferric chloride it gives a brownish-green coloration. With lead acetate, a brownish-yellow precipitate.

Pure magenta anthocyanin is a magenta-red powder with similar properties and solubilities to the red. In concentrated sulphuric acid it gives a red solution with a slight greenish fluorescence. It is soluble to a green solution in alkalies. With ferric chloride solution it gives a brownish-green coloration. With lead acetate it forms a greenish-black precipitate of a lead salt. In alkaline solutions it is strongly fluorescent, green by transmitted, red by reflected light.

The results of combustion of the pure anthocyanin were:

	C	H	O
Red anthocyanin: from rose doré ...	51·93 %	5·02 %	43·05 %
,, ,, ,, bronze ...	51·37 %	5·05 %	43·58 %
,, ,, ,, ,, ...	52·12 %	4·97 %	42·91 %
Magenta ,, ,, magenta ...	50·26 %	4·89 %	44·85 %
,, ,, ,, ivory tinged with magenta	50·68 %	5·54 %	43·78 %
,, ,, from crimson ...	50·56 %	4·90 %	44·54 %

Attempts were made to determine the molecular weight by depression of freezing point, using phenol as a solvent, but the results, though consistent for a series of experiments, were obviously far too low. Acetic acid, and various other solvents, did not dissolve enough of the pigment to give measurable depressions. Attempts to determine the molecular weight by elevation of the boiling point in absolute alcohol gave mean values of 572 for the red, and 717 for the magenta. The elevation of the boiling point was so slight that the error in the value obtained may be very considerable.

The combustion results give, as simplest formulae, $C_{15}H_{18}O_{10}$ for the magenta, and $C_8H_9O_5$ for the red. The boiling point determination of the molecular weight would appear to indicate that the molecule is 2 ($C_{15}H_{18}O_{10}$), i.e. $C_{30}H_{36}O_{20}$, which has a molecular weight of 716 for the magenta, and 3 ($C_8H_9O_5$), i.e. $C_{24}H_{27}O_{15}$, which has a molecular weight of 555 for the red.

An attempt was made to estimate the number of hydroxyl groups present in the anthocyanin molecule by means of Zerewitinoff's modification of Hibbert & Sudborough's method. This consists in dissolving the substance in thoroughly dried pyridine, treating it in a suitable apparatus with a considerable excess of methyl magnesium iodide, and collecting the gas evolved. Each hydroxyl group causes the evolution of a molecule of methane. It should be noticed that 'hydroxyl groups,' as determined by this method, include those forming part of the carboxyl groups, and also such ketone groups as can give rise to hydroxyl by tautomeric change. The values obtained indicate that the red anthocyanin, taking the formula as $C_{24}H_{27}O_{15}$, contains twelve hydroxyl groups as defined above, while the magenta, taking the formula as $C_{30}H_{36}O_{20}$, contains fifteen hydroxyl groups.

The flower-pigment of the Cornflower (Willstätter, 260, 273). In *Centaurea* flowers, according to Willstätter, there are three modifications of one anthocyanin pigment: a purple pigment (cyanin), which is itself a free acid; a blue pigment, which is the potassium salt of the purple, and which constitutes the greater part of the colouring matter of the flower; a red pigment, which is the oxonium salt of the purple with some organic acid (other oxonium salts can be obtained artificially with inorganic acids, such as hydrochloric acid). Cyanin, moreover, isomerises to a colourless form which is an acid too, and forms colourless alkali salts[1]; there is also a colourless isomer of the blue pigment. When the anthocyanin of the flower is extracted with water, the deep blue solution rapidly loses its colour; this is due to the above isomerisation, and the colourless solution, on addition of acid, will become as red as a solution of the original blue pigment would on acidification. The red modification also loses colour in absence of acid, i.e. if sufficiently diluted with water or alcohol. On concentration, a colourless solution will return to its original colour, blue, red or violet, as the case may be.

For preparation of the blue pigment on a large scale, dried flowers ground to a fine powder were employed. The blue pigment can be extracted rapidly with water or very dilute alcohol, but change to the colourless isomer tends to take place. This change can be prevented, however, by addition of much sodium nitrate or chloride to the pigment solution.

[1] "Das Cyanin isomerisiert sich zu einer farblosen Modifikation, welche gleichfalls sauer ist und farblose Alkalisalze bildet." The alkali salts of the isomer are, however, definitely stated to be yellow later in the paper.

The cyanin salt can then be precipitated from the water solution with alcohol in which it is insoluble. It is further purified by fractional precipitation with alcohol from water solution.

Cyanin is a glucoside, but on hydrolysis it gives the free pigment, cyanidin, and sugar. The pigment, cyanidin, like the glucoside, cyanin, forms crystalline oxonium salts with hydrochloric acid.

The method of isolation of the blue pigment, in greater detail, was as follows. The powder of ground flowers was mixed with sand, extracted with water or 20 % alcohol, and filtered, and finely powdered sodium nitrate was added. The deep blue solution was then mixed with 96 % alcohol (2·5 vols. alcohol : 1 vol. extract), when the pigment was precipitated in blue flakes, which were separated by a centrifuge. The pigment was further purified by taking up in water, and precipitating with alcohol. For the exact sequence of operations, the original paper should be consulted. The filtrate from the precipitates contained much of the colourless isomer. The pigment was also extracted by another method in which the use of sodium chloride or nitrate was eliminated. This is necessary if one wishes to obtain the naturally-occurring cyanin salt, for if the method first described is employed the sodium replaces the potassium originally present. In the second method, the flower powder was mixed with sand, and extracted with dilute alcohol (80 vols. water : 20 vols. 96 % alcohol). It was then filtered and precipitated with alcohol (3 vols. filtrate:5 vols. 96 % alcohol), and the precipitate separated by a centrifuge. All the operations should be carried out as rapidly as possible in order to avoid isomeric change. The crude pigment was again purified by reprecipitation. It was found that the product obtained by the first method contained sodium nitrate as impurity. This was removed by extracting with 75 % alcohol, in which the impurity is soluble, though not the pigment. The product was, however, still further purified by precipitation, and in the end contained both sodium and potassium, the former as a result of the use of sodium nitrate. The product obtained by the second method (i.e. when the use of sodium salts for protection against isomerisation was avoided), after purification, contained no sodium.

By dialysis of the cyanin salt in a 20 % sodium chloride solution, dark blue crystals were obtained. These Willstätter regards as an addition product of the cyanin salt with sodium chloride.

With regard to properties, the blue cyanin salt is insoluble in alcohol, but soluble in water; a concentrated water solution shows loss of colour only after a day or two, but a dilute solution decolorises in an hour or so. The isomerisation, as already mentioned, is best prevented by sodium chloride or nitrate, but potassium nitrate or chloride has little or no effect. The pure blue product shows practically no change in colour on addition of a little sodium hydroxide solution, but a solution of pigment which has stood becomes blue-green, or green-blue, on account of the presence of the colourless modification.

The next operation was the preparation of a crystalline salt of cyanin with hydrochloric acid. The cyanin alkali salt, obtained by the methods described, was dissolved in 20 % hydrochloric acid. From this solution some accompanying carbohydrates (pentosans), which are present as impurity, are precipitated by addition of absolute alcohol. After filtration the pigment chloride was precipitated by ether. The crude product was then taken up in absolute alcohol, which frees it from colloidal substance and other impurities. The purified alcoholic solution was then acidified with strong hydrochloric acid, and concentrated in a vacuum desiccator. Cyanin chloride separates out as an amorphous product which is soluble in both alcohol and water to a bright red solution; the solution becomes rapidly paler, but in absolute alcohol the loss of colour is much less rapid than in the presence of water. To obtain the crystalline form, the amorphous product was dissolved in absolute alcohol, filtered, and mixed with a third of its volume of a 7 % solution of hydrochloric acid in water, and set to crystallise. The crystals are deep blue rhomboidal plates with a golden lustre: in a powdered form, the colour is brown-red. The chloride crystallises out with three molecules of water of crystallisation, and the formula obtained by elementary analysis is $C_{27}H_{31}O_{16}Cl \cdot 3H_2O$. The water-free product on analysis gave $C_{27}H_{31}O_{16}Cl$. The crystalline cyanin chloride is almost insoluble in water, soluble with difficulty in cold alcohol, acetone and chloroform; insoluble in benzene; slightly soluble in dilute hydrochloric and sulphuric acids. It is stable in acid solution: in water solution it decolorises, with the formation of the isomer, especially if dilute, but the colour returns on acidification. The colourless isomer is also formed by warming with absolute alcohol; such a solution then gives with lead acetate a green precipitate, on account of the mixture of the blue salt of the pigment with the yellow alkali salt of its isomer. Pure cyanin chloride gives with calcium carbonate a violet solution; with sodium hydroxide, a blue solution; with lead acetate, a blue precipitate, which gives a green lead salt on standing; and with sodium carbonate, a violet colour which eventually becomes yellow. The chloride is reduced with zinc and hydrochloric acid, and also decolorised with sodium bisulphite, the colour in the latter case returning on addition of an acid. As a glucoside it reduces Fehling's solution.

The glucoside cyanin is rapidly hydrolysed with 20 % hydrochloric acid, cyanidin being formed which separates out as the chloride from the hot solution. The combined sugar was identified as glucose. The crystals, in the form of needles, of cyanidin chloride are brown-red under the microscope, give a violet streak, and a brown-red powder. They have no water of crystallisation, and an elementary analysis gave the formula $C_{15}H_{11}O_6Cl$. The chloride is soluble in alcohol with a fine violet-red colour; it is also soluble with difficulty in dilute hydrochloric acid. It crystallises from a mixture of alcohol and dilute hydrochloric acid. It is also soluble in amyl alcohol, and when such a solution is shaken with caustic soda solution, the pigment goes into the watery layer with a blue colour. With sodium carbonate it gives a violet or blue colour (green when the colourless isomer is also present); with lead acetate it gives a blue precipitate.

The violet modification of anthocyanin was obtained apparently by precipitating a solution of cyanin chloride with lead acetate, and decomposing the lead salt with excess of sulphuretted hydrogen. On filtration and evaporation the violet form of the pigment is obtained.

The colourless modification of the pigment is soluble in ether; on evaporation of the ether it remains in the form of colourless crystals, which give an intense red colour on heating with hydrochloric acid. With alkalies it gives a yellow colour, but is apparently not a flavone (see p. 110).

As a result of his analyses, Willstätter came to the conclusion that the violet form of the pigment has the constitution of an inner oxonium salt of the following type:

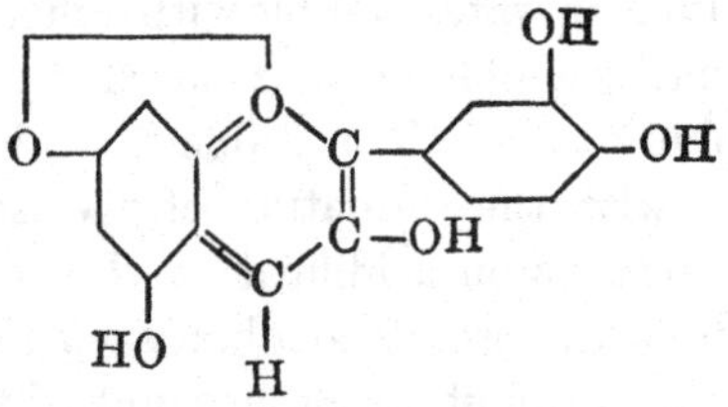

Red cyanidin chloride would be represented as:

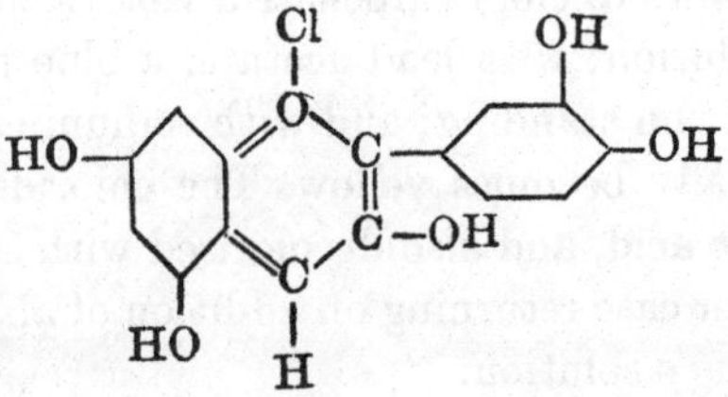

whereas the blue pigment, which constitutes the greater part of the colouring matter of the flower, is a salt of cyanidin with an alkali metal, the position of the metal being unknown.

On isolating pigments from other material, Willstätter was next able to show that the pigment of the flowers of the Larkspur is a glucoside of a compound of a similar type which he termed delphinidin. The neutral form is represented as:

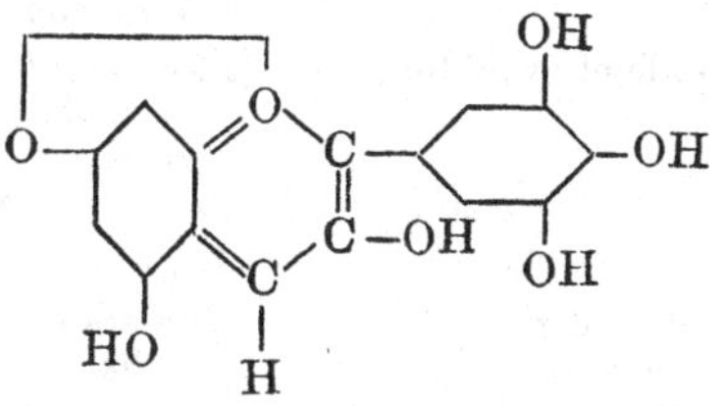

Similarly, from the scarlet Geranium he obtained a glucoside of pelargonidin, of which the neutral form is:

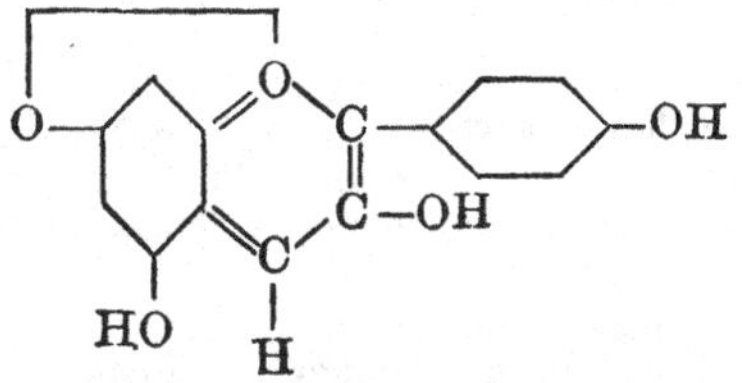

Like cyanidin, pelargonidin and delphinidin were obtained as crystalline chlorides, and they can be prepared from any of the naturally-occurring glucosides by hydrolysis. They both also isomerise to colourless pseudobases which may be obtained in crystalline form. Later, Willstätter (275) achieved the synthesis of pelargonidin.

All pigments subsequently investigated have been found to be derivatives of these three fundamental compounds, pelargonidin, cyanidin and delphinidin. The modifications are due to the presence of different sugars in the glucosides (usually glucose, sometimes rhamnose or galactose) and to the substitution of various hydroxyl groups by the methyl group.

The following represent a list of the plants investigated by Willstätter, and the pigments found

Derivatives of Pelargonidin.

Callistephin ...	Monoglucoside of pelargonidin	Flowers of Summer Aster (*Callistephus chinensis*)
Pelargonin ...	Diglucoside of pelargonidin	Flowers of Scarlet Geranium (*Pelargonium zonale*), pink variety of Cornflower (*Centaurea Cyanus*) and certain varieties of Dahlia (*D. variabilis*)

Salvianin ...	Diglucoside of pelargonidin + malonic acid	Flowers of Scarlet Salvia (*S. coccinea* and *S. splendens*)

Derivatives of Cyanidin.

Asterin ...	Monoglucoside of cyanidin	Flowers of Aster (*Callistephus chinensis*)
Chrysanthemin	Monoglucoside of cyanidin	Flowers of Chrysanthemum (*C. indicum*)
Idaein ...	Monogalactoside of cyanidin	Fruit of Cranberry (*Vaccinium Vitis-Idaea*)
Cyanin ...	Diglucoside of cyanidin	Flowers of Cornflower (*Centaurea Cyanus*), *Rosa gallica* and certain varieties of Dahlia (*D. variabilis*)
Mekocyanin ...	Diglucoside of cyanidin	Flowers of Poppy (*Papaver Rhoeas*)
Keracyanin ...	Rhamnoglucoside of cyanidin	Fruit of Cherry (*Prunus avium*)
Peonin ...	Diglucoside of peonidin (cyanidin monomethyl ether)	Flowers of Peony (*Paeonia officinalis*)

Derivatives of Delphinidin.

Violanin ...	Rhamnoglucoside of delphinidin	Flowers of Pansy (*Viola tricolor*)
Delphinin ...	Diglucoside of delphinidin + *p*-hydroxybenzoic acid	Flowers of Larkspur (*Delphinium Consolida*)
Ampelopsin ...	Monoglucoside of ampelopsidin (delphinidin monomethyl ether)	Fruit of Virginian Creeper (*Ampelopsis quinquefolia*)
Myrtillin ...	Monogalactoside of myrtillidin (delphinidin monomethyl ether)	Fruit of Bilberry (*Vaccinium Myrtillus*)
Althaeïn ...	Monoglucoside of myrtillidin	Flowers of deep purple variety of Hollyhock (*Althaea rosea*)
Petunin ...	Diglucoside of petunidin (delphinidin monomethyl ether)	Flowers of Petunia (*P. hybrida*)
Malvin ...	Diglucoside of malvidin (delphinidin dimethyl ether)	Flowers of Mallow (*Malva sylvestris*)
Oenin ...	Monoglucoside of oenidin (delphinidin dimethyl ether)	Fruit of Grape (*Vitis vinifera*)

There are some additional facts of interest in connection with the occurrence of pigments (Willstätter, 288):

(*a*) Presence of the pigment, in different form, in varieties of the same species. *Centaurea Cyanus*. The dark purple-red variety contains neutral cyanin (13–14 %, dry weight of flower), and the blue type the potassium salt of cyanin (0·7 %). *Delphinium.* The violet variety contains delphinin; the blue type the potassium salt of delphinin.

(*b*) The presence of different pigments in different varieties of the same species, and the presence of more than one pigment in the same flower. *Centaurea*. The purple and blue flowers contain cyanin, the pink variety, pelargonin (4 %). *Pelargonium.* The pigment of the scarlet *P. zonale*, var. 'Meteor' is pelargonin (6–14 %); so also is that of the bluish-pink *P. peltatum* (1 %). But a violet-red variety of *P. zonale* was found to contain cyanin, chiefly, accompanied by a little pelargonin. *Dahlia.* The deep brown-red double (Cactus) varieties, 'J. H. Jackson,' 'Harold,' 'Matchless,' 'Night,' and 'Othello' form cyanin (one dark-red variety 20 %), but pelargonin is formed in the scarlet-red varieties, 'Rakete' (5·6 %) and 'Alt-Heidelberg' (4 %). Pelargonin, together with a little cyanin probably, was also found in a dark-violet variety. *Aster.* The purple-red Summer Aster (*Callistephus chinensis,* Nees, syn. *Aster chinensis,* L.) contains callistephin and asterin.

From qualitative reactions, Willstätter states that pelargonidin derivatives occur in scarlet-red *Gladiolus*; also a trace in other varieties. *Zinnia elegans* contains both pelargonidin and cyanidin derivatives. Cyanidin derivatives occur in *Gladiolus, Gaillardia bicolor, Helenium autumnale, Tulipa Gesneriana, Tropaeolum majus,* berries of Red Currant, Raspberry and Mountain Ash.

Willstätter is of the opinion that variation in flower-colour largely depends on the presence of acids, alkalies, salts, etc., in the cell-sap. For instance, he says, the scarlet *Pelargonium* contains an acid salt of pelargonin, but in the flowers of the bluish-pink *P. peltatum* and in the pink *Centaurea,* the pelargonin is probably in either an acid-free or basic form.

The constitution of the various methyl derivatives is represented as follows:

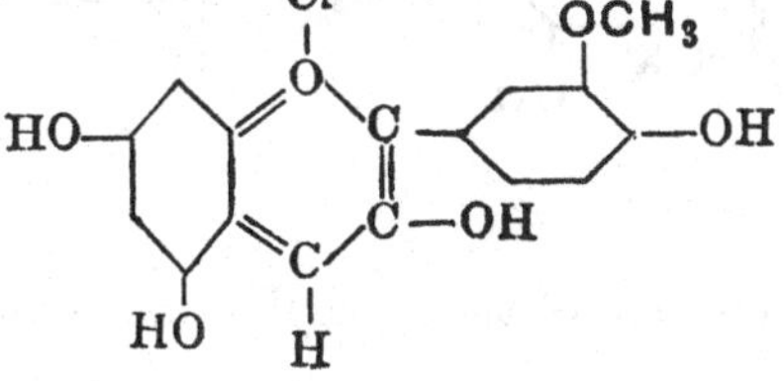

Cl, OCH_3, HO, OH, C—OH, HO, H

Peonidin (probably)

Cl, OH, HO, OCH_3, C—OH, OH, HO, H

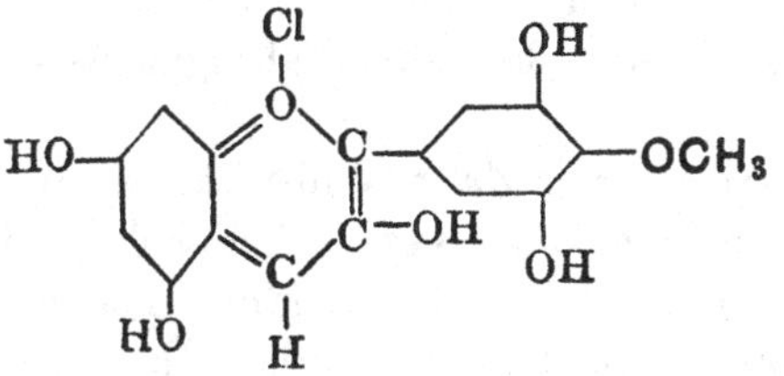

Ampelopsidin

Cl, OX, HO, OH, C—OX′, OH, HO, H — Myrtillidin

Cl, OX, HO, OH, C—OX′, OH, HO, H — Petunidin

Cl, OH, XO, OCH_3, C—OH, OH, X′O, H — Malvidin

Cl, OX, HO, OCH_3, C—OX′, OH, HO, H — Oenidin

In the last four cases, the position of the methyl group is uncertain; it may *either* be at X *or* X′.

In brief outline, some of the details of the extraction of the pigments are given as follows; greater detail, if required, may be obtained from original papers. The properties of the individual pigments are given on pp. 78-81.

The flower-pigment of the Summer Aster (Willstätter & Burdick, 302). The pigments isolated were termed *callistephin* (monoglucoside of pelargonidin) and *asterin* (monoglucoside of cyanidin). The glacial acetic extract of the petals was precipitated with ether, the syrupy product taken up in very dilute hydrochloric acid, filtered, and the pigments in the filtrate precipitated with lead acetate. The precipitate was treated with glacial acetic acid, which dissolves the lead salts of the pigments only, filtered, and the lead salts again precipitated with ether. The product was then decomposed with propyl-alcohol containing 25 % of methyl-alcohol-hydrochloric acid, filtered and the mixture of chlorides of the two pigments precipitated with ether. This was again repeated. Asterin and callistephin chlorides were separated by fractional crystallisation from alcohol-hydrochloric acid. On hydrolysis, callistephin chloride yields one molecule of pelargonidin chloride and one molecule of glucose.

The flower-pigment of the Scarlet Geranium (Willstätter & Bolton, 286). The pigment isolated was *pelargonin* (diglucoside of pelargonidin). The fresh flowers were digested for several days with 96 % alcohol, and to the extract 20 % alcohol-hydrochloric acid was added together with ether. The precipitate of pigment was dissolved in boiling 2 % methyl-alcohol-hydrochloric acid, and to the solution 10 % hydrochloric acid was added, whereby the pelargonin chloride was precipitated. The crystallised product gave, on hydrolysis with hydrochloric acid, as previously mentioned, pelargonidin and two molecules of glucose.

The flower-pigment of the Scarlet Sage (Willstätter & Bolton, 300). The pigment isolated was termed *salvianin*. Its exact composition is uncertain, but it is probably a diglucoside of pelargonidin in combination with malonic acid. Fresh petals were extracted with glacial acetic acid, and the crude pigment, as acetate, precipitated by ether from the extract. It was purified by converting the crude acetate into the picrate, and then treating the latter with methyl-alcohol-hydrochloric acid to obtain the chloride. The latter, however, was not obtained crystalline but only as a solidified oil.

In presence of hydrochloric acid, salvianin is gradually converted into *salvin* and *salvinin*. Salvin chloride crystallises out and has probably the composition $C_{27}H_{27}O_{13}Cl$ (i.e. pelargonin chloride $- 2H_2O$). From the mother liquors, by saturating with amyl alcohol, salvinin chloride is also obtained crystalline. It is a normal diglucoside of pelargonidin, and, on hydrolysis, it yields one molecule of pelargonidin and two molecules of glucose.

The flower-pigment of the Chrysanthemum (Willstätter & Bolton, 301). The pigment isolated was termed *chrysanthemin* and was found to be a monoglucoside of cyanidin. The petals were extracted with glacial acetic acid. The pigment, as acetate, was precipitated from the filtrate from the petals by ether. The pigment was purified by adding saturated picric acid solution to an aqueous solution of the crude acetate. This was converted into the chloride by treatment with methyl-alcohol-hydrochloric acid, and the chloride precipitated by ether. It was crystallised from alcohol-water-hydrochloric acid. On hydrolysis it gave one molecule of cyanidin chloride and one molecule of glucose.

The pigment from fruit of Cranberry (Willstätter & Mallison, 287). The pigment isolated was termed *idaeïn*, and was found to be a monogalactoside of cyanidin. The skins of the berries were digested with glacial acetic acid for several days, and the extract precipitated with ether. The crude precipitate was again taken up in glacial acetic, and fractionally precipitated with ether; the second fraction was dissolved in water and picric acid added. A precipitate of the picrate of the pigment was formed which was extracted with methyl alcohol, reprecipitated by ether, and finally crystallised from water. The picrate was converted into the chloride by treatment with methyl-alcohol-hydrochloric acid. The difference in the attached sugar molecule, i.e. galactose instead of glucose, causes idaeïn to differ in properties (solubilities and behaviour with alkalies) from cyanin.

The flower-pigment of the Rose (Willstätter & Nolan, 292). This pigment was found to be identical with the *cyanin* of the Cornflower. The dried and powdered petals were digested with 2 % methyl-alcohol-hydrochloric acid. From the extract the colouring matter was precipitated with ether, and the precipitate contained impure cyanin chloride. After purification, the crystalline product was found on analysis to yield cyanidin and two molecules of dextrose.

The flower-pigment of the Poppy (Willstätter & Weil, 305). The pigment isolated was termed *mekocyanin*, and was found to be a diglucoside of cyanidin. The fresh petals were extracted with glacial acetic acid. To the filtered extract, 10 % methyl-alcohol-hydrochloric acid was added, and the pigment precipitated with ether. This precipitation was again repeated. The crude pigment was precipitated as a syrup from which the mekocyanin was extracted with 3 % alcohol-hydrochloric acid. It was next fractionally precipitated by glacial acetic acid followed by ether. The middle fraction was converted into the crystalline ferrocyanide, and reconverted into chloride. The latter was eventually obtained as a crystalline product from aqueous hydrochloric acid by precipitation with acetone. On complete hydrolysis it yielded one molecule of cyanidin chloride and two molecules of glucose. On partial hydrolysis, it yielded a monoglucoside, identified as chrysanthemin chloride.

The pigment from fruit of Cherry (Willstätter & Zollinger, 306). The pigment isolated was termed *keracyanin* and was found to be a rhamnoglucoside of cyanidin. The skins of the fruit were extracted with glacial acetic acid, and the crude acetate of the pigment precipitated by ether. It was purified by gradually precipitating with ether, and then converting into the lead salt. The latter was decomposed, and the pigment eventually obtained in the crystalline state by again fractionally precipitating from methyl-alcohol-hydrochloric acid by ether. On hydrolysis, keracyanin yielded cyanidin, glucose and rhamnose.

The flower-pigment of the Peony (Willstätter & Nolan, 293). The pigment isolated was termed *peonin*, and was found to be a diglucoside of *peonidin* (monomethyl ether of cyanidin). The dried and powdered flowers were digested with 2 % methyl-alcohol-hydrochloric acid. The crude pigment was purified by various processes, and finally crystallised as the chloride from *N*/2 hydrochloric acid. Peonin, on hydrolysis by boiling with 20 % hydrochloric acid, yields peonidin and two molecules of glucose. Peonidin is a monomethyl ether of cyanidin, and is converted into the latter on treatment with hydroidic acid.

The flower-pigment of the Pansy (Willstätter & Weil, 304). The pigment isolated and termed *violanin* is probably a rhamnoglucoside of delphinidin. The petals were extracted with 2 % methyl-alcohol-hydrochloric acid, filtered, and the pigment precipitated as chloride by ether. It was crystallised from alcohol-hydrochloric acid. On hydrolysis it yielded delphinidin chloride, rhamnose and glucose.

The flower-pigment of the Larkspur (Willstätter & Mieg, 290). The pigment isolated was termed *delphinin* and found to be a diglucoside of delphinidin. The dried and powdered flowers were extracted with aqueous alcoholic-hydrochloric acid. To the extract ether was added, thereby producing impure delphinin chloride, which was then purified by further treatment with methyl and ethyl alcohols and ether. The amorphous delphinin chloride was finally again precipitated by concentrated hydrochloric acid, and it does not readily crystallise. It is hydrolysed by water with the precipitation of the violet base. With hot, moderately concentrated hydrochloric acid it is hydrolysed into two molecules of dextrose, two molecules of *p*-oxybenzoic acid and delphinidin chloride.

The pigment from fruit of Bilberry (Willstätter & Zollinger, 294, 307). The pigment isolated was termed *myrtillin* (monogalactoside of myrtillidin). The skins of the berries were used after being dried and ground. The pigment was extracted rapidly by warming with ethyl alcohol which contained a small percentage of hydrochloric acid, and the solution was precipitated with ether. The precipitate, mixed with a large quantity of a colourless product, was separated, by taking up in water, from many of its impurities. By addition, with cooling, of a double weight of concentrated hydrochloric acid, the chloride was precipitated almost pure, and quite pure by repetition of the operation. For crystallisation, a third of its volume of 9 % hydrochloric acid was added to the solution of the pigment in wood spirit; on slow evaporation the pigment separated out in beautiful flat prisms. Myrtillin gives, on hydrolysis, one molecule of galactose and one molecule of myrtillidin. Myrtillidin was also isolated from flowers of the Hollyhock.

The flower-pigment of the Hollyhock (Willstätter & Martin, 289). The pigment isolated was termed *althaein* (monoglucoside of myrtillidin). The dried flowers of a very deep red variety were extracted with methyl-alcohol-hydrochloric acid, and ether was added to the extract which precipitates the pigment, althaein, as chloride. The crude product was purified by converting into the crystalline picrate, reconverting into chloride which was eventually crystallised from methyl-alcohol-hydrochloric acid. On hydrolysis, althaein chloride gives one molecule of myrtillidin and one molecule of glucose.

The flower-pigment of the Petunia (Willstätter & Burdick, 303). The pigment isolated was termed *petunin*, and was found to be a diglucoside of a monomethyl ether of delphinidin which was termed *petunidin*. Fresh petals were extracted with glacial acetic acid, and the filtered extract precipitated with ether. The crude precipitate was treated with very dilute hydrochloric acid, and from this the chloride crystallised out on standing. It was recrystallised from dilute acid. On hydrolysis it yields one molecule of petunidin and two molecules of glucose.

The flower-pigment of Mallow (Willstätter & Mieg, 291). The pigment isolated was termed *malvin* and was found to be a diglucoside of *malvidin* (delphinidin dimethyl ether). The dried and powdered flowers were digested with 2 % methyl-alcohol-hydrochloric acid and about $\frac{1}{33}$ of the volume of concentrated hydrochloric acid, and the extract was precipitated with ether. The precipitate, after some purification, was warmed with aqueous picric acid (1·5 %). The pigment, malvin, crystallised out as picrate, and was converted into chloride by treatment with methyl-alcohol-hydrochloric acid. Malvin yields, on hydrolysis with boiling 20 % hydrochloric acid, malvidin and two molecules of dextrose.

The pigment from Grapes (Willstätter & Zollinger, 294, 307). The pigment isolated was termed *oenin* and was found to be a monoglucoside of *oenidin* (delphinidin dimethyl ether). The skins of dark blue grapes were extracted in the cold with glacial acetic acid, and the dark red filtrate precipitated with ether. A sticky precipitate was obtained which, after washing with ether, was put into an excess of water picric acid solution, and warmed for a short time. On cooling, the picrate crystallised out from the solution in long prisms of a fine red colour. By changing the solution to methyl-alcohol-hydrochloric acid, it yielded the solution of the pigment chloride, which was precipitated with ether-petrol-ether, and crystallised from water-alcohol-hydrochloric acid in the form of hard beetle-green prisms. On hydrolysis, oenin decomposes into oenidin and one molecule of glucose.

The pigment from the fruit of the Sloe (Willstätter & Zollinger, 306). The pigment isolated was termed *prunicyanin*. Its exact composition is uncertain, but it is probably cyanidin in combination with rhamnose and a hexose. The skins of the fruit were extracted with glacial acetic acid, and the pigment precipitated by ether. Various methods of purification were tried, but the chloride of the pigment was not obtained crystalline. On hydrolysis, it yielded cyanidin chloride, rhamnose and a hexose (unidentified).

The pigment from root of the Beet (Schudel, 318). A pigment in the form of a glucoside, termed *betanin*, was isolated as a crystalline picrate. Unlike all other anthocyanin pigments it was found to contain 8·6 % of nitrogen. The sugar-free pigment, *betanidin*, was only obtained as the ethyl derivative. It was shown, qualitatively, to contain nitrogen.

The pigment from root of the Radish (Schudel, 318). The chief pigment of a scarlet variety is a glucoside of pelargonidin, though it may contain small quantities of a cyanidin glucoside. The chief pigment, on the other hand, of a purple or violet variety is a glucoside of cyanidin, though it may contain small quantities of the pelargonidin glucoside.

The flower-pigment of the Scarlet Geranium and the Rose (Currey, 338, 339). In the main, this work only confirms previous results.

The pigment from Grapes (Anderson, 350). Pigment from several varieties found to be identical with each other and to closely resemble oenidin.

A short summary of the formulae[1] given by various authors for anthocyanins may be stated as follows:

Flower-pigment of *Althaea rosea*	$x(C_4H_5O_2)$	(Glan).
,, ,, ,, ...	$C_{14}H_{16}O_6$	(Grafe).
,, ,, ,, (glucoside and oxidation product)	$C_{20}H_{30}O_{13}$	,,
,, ,, ,, ...	$C_{16}H_{12}O_7$	(Willstätter).

[1] The formulae are so arranged rather as a matter of interest than for comparison, since the earlier workers laboured under disadvantages which would obviously detract from the value of their numbers.

Pigment	Formula	Authority
Flower-pigment of *Antirrhinum* (magenta)	$x(C_{15}H_{18}O_{10})$	(Wheldale).
,, ,, ,, (red)	$x(C_{8}H_{9}O_{5})$	,,
,, ,, *Centaurea Cyanus*	$x(C_{2}H_{3}O_{2})$	(Morot).
,, ,, ,, ,, ...	$C_{15}H_{10}O_{6}$	(Willstätter).
,, ,, *Delphinium*	$C_{15}H_{10}O_{7}$	,,
,, ,, *Malva*	$C_{17}H_{14}O_{7}$	,,
,, ,, *Pelargonium*	$C_{15}H_{10}O_{6}$	(Griffiths).
,, ,, ,, ...	$C_{18}H_{26}O_{13}$	(Grafe).
,, ,, ,, (glucoside and oxidation product)	$C_{24}H_{44}O_{20}$	,,
,, ,, ,, ...	$C_{15}H_{10}O_{5}$	(Willstätter).
,, ,, *Rosa gallica*	$C_{7}H_{11}O_{10}$	(Senier).
,, ,, ,, ...	$C_{15}H_{10}O_{6}$	(Willstätter).
Fruit-pigment of *Vaccinium Myrtillus*	$C_{14}H_{14}O_{7}$	(Heise).
,, ,, ,, ,, ...	$C_{16}H_{12}O_{7}$	(Willstätter).
,, ,, ,, *Vitis-Idaea*...	$C_{15}H_{10}O_{6}$	,,
,, ,, *Vitis vinifera* (in wine)	$C_{20}H_{10}O_{10}$	(Glénard).
,, ,, ,, ,, ,, ...	$C_{20}H_{20}O_{10}$	(Gautier).
,, ,, ,, ,, ,, ...	$C_{21}H_{20}O_{10}$	,,
,, ,, ,, ,, ...	$C_{17}H_{14}O_{7}$	(Willstätter).
Red leaf-pigment of *Coleus*	$C_{10}H_{10}O_{5}$	(Church).
,, ,, *Vitis vinifera*	$C_{19}H_{16}O_{10}$	(Gautier).
	$C_{26}H_{24}O_{15}$	,,
	$C_{17}H_{18}O_{10}$	,,

TABLES OF PROPERTIES OF CHLORIDES OF PIGMENTS DERIVED FROM PELARGONIDIN, CYANIDIN AND DELPHINIDIN

TABLE OF PROPERTIES OF CHLORIDES OF PIGMENTS

	Pelargonidin chloride	**Callistephin chloride**	**Cyanidin chloride**
Composition	$C_{15}H_{11}O_5Cl . H_2O$; water given off with difficulty	$C_{21}H_{21}O_{10}Cl . 2\text{–}2\frac{1}{2}H_2O$	$C_{15}H_{11}O_6Cl . H_2O$; water given off with difficulty
Components	—	$C_{15}H_{11}O_5Cl + 1$ glucose	—
Crystallisation	Platelets or prisms (red-brown)	Fine needles (orange-red)	Long needles (brown-red) or platelets; M.P. 220°
Solubility in water and alcohols	Readily sol. in hot water; readily sol. in methyl and ethyl alcohols	Readily sol. in water and ethyl alcohol	Insol. in water; v. readily sol. in alcohols
Reaction with soda ...	Violet, then blue	Red-violet	Violet, then blue
Reaction with $FeCl_3$ ...	No reaction	No reaction	Blue (in alcohol); violet (in dil. alcohol)
Isomerisation	Isomerises	Isomerises	Isomerises
	Pelargonin chloride	**Salvinin chloride**	**Mekocyanin chloride**
Composition	$C_{27}H_{31}O_{15}Cl . 4H_2O$	$C_{27}H_{31}O_{15}Cl . 5H_2O$	$C_{27}H_{31}O_{16}Cl . 3H_2O$; $1H_2O$ given off *in vac.* at 105°
Components	$C_{15}H_{11}O_5Cl + 2$ glucose	$C_{15}H_{11}O_5Cl + 2$ glucose	$C_{15}H_{11}O_6Cl + 2$ glucose
Crystallisation	Fine needles (scarlet-red); M.P. 180°	Needles (purple-red)	Needles
Solubility in water and alcohols	Somewhat sol. in water; sol. in methyl and ethyl alcohols, less in latter	Fairly sol. in water and ethyl alcohol; more sol. in methyl alcohol	Very readily sol. in water; sol. with difficulty in ethyl alcohol; rather more sol. in methyl alcohol
Reaction with soda ...	Only violet	—	Blue
Reaction with $FeCl_3$ ...	No reaction	No reaction	Blue (in alcohol); violet (in water)
Isomerisation	Readily isomerises	—	Isomerises

DERIVED FROM PELARGONIDIN AND CYANIDIN.

Asterin chloride	**Chrysanthemin chloride**	**Cyanin chloride**	**Idaeïn chloride**
$C_{21}H_{21}O_{11}Cl \, . \, 1\frac{1}{2}H_2O$	$C_{21}H_{21}O_{11}Cl \, . \, 1\frac{1}{2}H_2O$	$C_{27}H_{31}O_{16}Cl \, . \, 2\frac{1}{2}H_2O$; $\frac{3}{4}$ mol. H_2O given off *in vac.* at 105°	$C_{21}H_{21}O_{11}Cl \, . \, 2\frac{1}{2}H_2O$; water given off completely in desiccator
$C_{15}H_{11}O_6Cl + 1$ glucose	$C_{15}H_{11}O_6Cl + 1$ glucose	$C_{15}H_{11}O_6Cl + 2$ glucose	$C_{15}H_{11}O_6Cl + 1$ galactose
Prisms (red-brown)	Rhomboidal plates (red-violet). No M.P.	Rhomboidal plates (grey-violet); M.P. 203–204°	Monoclinic prisms (brown red with green metallic lustre); M.P. 210°
Sol. in water	Readily sol. in water and methyl alcohol; difficultly sol. in ethyl alcohol	Almost insol. in cold water; readily in hot; v. difficultly sol. in alcohol	Readily sol. in water and ethyl alcohol
Blue	Blue	Violet, then blue	Blue to green to yellow
Blue to violet	Blue (in alcohol); violet (in dil. alcohol)	Blue (in alcohol); violet (in water)	Blue (in alcohol); violet (in dil. alcohol)
Isomerises	Isomerises	Readily isomerises	Isomerises

Keracyanin chloride	**Peonidin chloride**	**Peonin chloride**
$C_{27}H_{31}O_{15}Cl \, . \, 3$ or $4H_2O$; water given off *in vac.* at 105°	$C_{16}H_{13}O_6Cl \, . \, H_2O$; water given off in desiccator	$C_{28}H_{33}O_{16}Cl \, . \, 5H_2O$; water given off completely in desiccator
$C_{15}H_{11}O_6Cl + 1$ glucose $+ 1$ rhamnose	—	$C_{16}H_{13}O_6Cl + 2$ glucose
Needles or prisms (bronze-brown or brown-yellow)	Long needles (reddish-brown)	Needles (red-violet); M.P. 165°
Readily sol. in hot water; v. readily sol. in methyl alcohol; fairly sol. in ethyl alcohol	Rather readily sol. in water; v. readily sol. in alcohol	V. readily sol. in water; considerably sol. in alcohol
Blue	Violet, then blue	Violet, then blue
Blue (in alcohol); blue-violet (in water)	Violet-red (in alcohol)	Violet tinge (in alcohol); violet-red (on dilution)
Isomerises	Isomerises	Rapidly isomerises

TABLE OF PROPERTIES OF CHLORIDES OF

	Delphinidin chloride	**Violanin chloride**	**Delphinin chloride**
Composition	$C_{15}H_{11}O_7Cl \cdot 2H_2O$; half water given off with difficulty	$C_{27}H_{31}O_{16}Cl \cdot 6H_2O$	$C_{41}H_{39}O_{21}Cl \cdot 12H_2O$; $2H_2O$ held fast
Components	—	$C_{15}H_{11}O_7Cl$ + 1 glucose + 1 rhamnose	$C_{15}H_{11}O_7Cl$ + 2 glucose + 2 *p*-oxybenzoic acid
Crystallisation	Platelets or prisms (red-brown); do not melt	Hexagonal or tetrahedral tablets (blue - violet with greenish metallic lustre)	Prismatic tables (bluish-red); M.P. 200–203°
Solubility in water and alcohols	Readily sol. in alcohol and water	Sol. with difficulty in water. Fairly sol. in ethyl alcohol	Decomposed by water with pptn. of violet base; difficultly sol. in abs. alcohol; sol. in methyl alcohol
Reaction with soda ...	Violet, then blue	Blue	Violet, then blue
Reaction with $FeCl_3$ ...	Blue (in alcohol); blue-violet (in dil. alcohol)	Blue	Blue (in water and alcohol)
Isomerisation	Isomerises	Isomerises	Does not isomerise
	Althaein chloride	**Petunidin chloride**	**Petunin chloride**
Composition	$C_{22}H_{23}O_{12}Cl \cdot 4H_2O$	$C_{16}H_{13}O_7Cl \cdot 2H_2O$; H_2O given off *in vac.* at 105°	$C_{28}H_{33}O_{17}Cl \cdot 2H_2O$
Components	$C_{16}H_{13}O_7Cl$ + 1 glucose	—	$C_{16}H_{13}O_7Cl$ + 2 glucose
Crystallisation	Prisms (brown-red with bronze lustre)	Prisms or leaflets (grey-brown or yellow-brown	Right-angled tables (violet with copper metallic lustre); M.P. 178°
Solubility in water and alcohols	Readily sol. in cold water; readily sol. in methyl alcohol, less so in ethyl alcohol	Readily sol. in ethyl alcohol	Sol. in water; v. readily sol. in methyl alcohol; less readily in ethyl alcohol
Reaction with soda ...	Violet, then blue	—	Violet, then blue
Reaction with $FeCl_3$ ...	Violet-blue (in alcohol); violet (on dilution)	Intense blue (in alcohol)	Blue (in water and alcohol)
Isomerisation	Isomerises	—	Isomerises

PIGMENTS DERIVED FROM DELPHINIDIN

Myrtillidin chloride	**Myrtillin chloride**
$C_{16}H_{13}O_7Cl \cdot 2H_2O$	$C_{22}H_{23}O_{12}Cl \cdot 4H_2O$; water given off *in vac.* at 108°
—	$C_{16}H_{13}O_7Cl + 1$ galactose
Needles or prisms (yellow-brown)	Rhomboidal plates (bronze-brown)
Considerably sol. in water; v. readily sol. in ethyl and methyl alcohols	Readily sol. in water and methyl alcohol
Violet, then blue	Blue
Blue (in alcohol); violet (on dilution)	Blue (in alcohol)
Isomerises	Readily isomerises

Malvidin chloride	**Malvin chloride**	**Oenidin chloride**	**Oenin chloride**
$C_{17}H_{15}O_7Cl \cdot 2H_2O$; 1 mol. water given off with difficulty	$C_{29}H_{35}O_{17}Cl \cdot 8H_2O$ water-free in desiccator	$C_{17}H_{15}O_7Cl \cdot 1\frac{1}{2}H_2O$; $\frac{1}{2}$ mol. water given off first *in vac.* at 100°	$C_{23}H_{25}O_{12}Cl \cdot 4H_2O$ and $6H_2O$; becomes almost free from water in desiccator
—	$C_{17}H_{15}O_7Cl + 2$ glucose	—	$C_{17}H_{15}O_7Cl + 1$ glucose
Long prisms and needles (olive-brown); does not melt at 300°	Long prisms and needles (purple-red); M.P. 165°	Prisms or needles (brown)	Prisms (brown-red)
Considerably sol. in water; readily in ethyl alcohol; difficultly in methyl alcohol	Difficultly sol. in water; readily sol. in methyl alcohol; less in ethyl alcohol	Readily sol. in water; v. readily sol. in methyl and ethyl alcohols	Readily sol. in water and methyl alcohol; moderately sol. in ethyl alcohol
Violet, then blue	Blue	Violet	Blue-violet
No reaction	No reaction	No reaction	No reaction (in water)
Isomerises	Readily isomerises	—	Readily isomerises

CHAPTER VI

PHYSIOLOGICAL CONDITIONS AND FACTORS INFLUENCING THE FORMATION OF ANTHOCYANINS

Connection with photosynthesis.

An examination of the relative distribution of anthocyanin and chlorophyll at once suggests that these pigments are more or less complementary as regards their appearance in the plant tissues. In leaves, the chief seat of chlorophyll, anthocyanin is found to a less extent than in any other organ, and under *normal* circumstances the development is most frequently confined to the epidermis (see p. 39), or to a few sub-epidermal layers, often only where they overlie the midrib or main veins. Moreover, when red pigment is present in the epidermis, the guard cells of the stomata, which contain chlorophyll, are generally free from pigment. In petioles and stems also, anthocyanin is on the whole limited to the epidermis, or to a few sub-epidermal layers. Bracts of all kinds, on both the inflorescence and other parts of the plant, frequently contain anthocyanin and correspondingly little chlorophyll; although flowers in the bud stage and unripe fruits have fairly abundant chlorophyll, as the flowers and fruits mature the chlorophyll disappears and anthocyanin develops.

Since chloroplasts are primarily concerned with photosynthesis, one would naturally conclude that the latter process and the formation of anthocyanin are to some extent mutually exclusive. The existence of some such alternation is further emphasised by the appearance of anthocyanin which accompanies lessened photosynthetic activity, as in plants towards the end of their vegetative season, in autumnal reddening, in leaves in an unhealthy condition and in evergreens during winter. Since all metabolic activity ultimately depends on photosynthesis, it is not a convincing argument that a decline in general metabolism (apart from photosynthesis) may directly bring about the formation of pigment. There is without doubt good evidence for believing that anthocyanin is not readily produced where carbon assimilation is most active, and that decreased photosynthesis from any outside cause is favourable to its formation.

Hence in any consideration of the direct bearing of outside factors, such as temperature and light, on anthocyanin formation, recognition should be first given to possible indirect effects produced by these factors through the medium of photosynthesis.

Connection with accumulation of synthetic products.

Even at the height of summer when the vegetative organs are in a condition of maximum activity, quite a number of individual plants may be found having isolated leaves or shoots which are either entirely red or have developed patches or blotches of anthocyanin. In the majority of cases, one will find on careful examination that there has been some injury to the leaf, petiole or stem, as the case may be, and it is to the distal side of the injured spot that the reddening occurs. Such injuries may be classified as: (1) mechanical, caused by chance cutting or breaking; (2) attacks of insects, including gall insects and caterpillars; (3) infection by Fungi.

(1) Let us deal first with mechanical injury. Frequently leaves may be found in which the lamina is partially severed transversely, and the severed portion has reddened. Or the petiole or stem is partially broken, and the leaf or leaves above the point of injury have turned red. In *Rumex*, *Oenothera*, *Pelargonium*, *Plantago* and many other plants, it is easy to bring about such reddening artificially by pinching the lamina or petiole, and in other cases by decortication. Or sometimes isolated leaves, as for instance those of *Rheum*, left lying on the ground in a damp place will eventually redden. In other genera and species it is difficult, or impossible probably, to induce reddening by such means. Reference to these phenomena has often been made by various authors: Gautier (190), Kraus (398), Berthold (64), Linsbauer (428), Küster (437), Daniel (424) and finally Combes (461, 472). Further Combes has made a series of experiments on decortication of stems of many species with a view to investigating the phenomena more fully. By these means he has distinguished three types of results:

(*a*) Those in which anthocyanin appeared more or less rapidly in the branches, petioles and above all in the leaves. Ex. *Spiraea* spp., *Mahonia aquifolium*, *Prunus Pissardi*.

(*b*) Those in which anthocyanin appeared more or less rapidly in the stems and petioles but not in the leaves. Ex. *Ceanothus azureus*, *Catalpa bignonioides*.

(*c*) Those in which pigment never appeared as a result of decortication. Ex. *Rhodotypos kerrioides*, *Robinia Pseudacacia*, *Pinus excelsa*.

The time elapsing before the appearance of pigment also varied considerably. In addition it was noted, as would be expected, that no pigment appeared on decortication in albino varieties, although the coloured type might produce abundant pigment under similar treatment.

(2) Injuries brought about by insects present no points of special interest but may be regarded rather as cases of (1). Frequently the midrib or petiole is partly eaten away, and reddening occurs on the distal side; or holes are eaten in the lamina, and red blotches are formed in their vicinity. Mirande (449) has observed that excursions of leaf-boring larvae in leaves of *Galeopsis Tetrahit* result in the production of anthocyanin. Under this heading also may be included the development of anthocyanin in or near galls. For details and examples the works of Hieronymus (401), Küstenmacher (413), Küster (437) and Guttenburg (440) may be consulted.

(3) It is frequently found that the pathological conditions called forth by the attacks of Fungi are accompanied by abnormal development of anthocyanin. In leaves of *Tussilago*, for instance, infected by *Puccinia* a circular band of anthocyanin often appears surrounding the aecidium spots. Other references to this matter may be looked for in the works of Sorauer (391), Tubeuf (416), Lüdi (429) and Rostrup (432).

There is little doubt that, in the above cases of injury and decortication, the formation of pigment is directly connected with an interference with the translocation current. Injury to the living tissues of the conducting system of the veins, midrib or petiole of the leaf, or of corresponding tissue in the stem, leads to an accumulation of synthetic products in the leaves[1]. Of such products several authors—Mirande (452), Combes (222)—have maintained that it is the carbohydrates and glucosides which most influence the formation of anthocyanin, and Combes has shown by analyses that leaves reddened by decortication contain a higher percentage of sugars and glucosides. It seems likely also that parasitic growths may interfere with the progress of the translocation current through the small veins of the leaf, thereby causing congested areas to arise in which the sugar contents are above normal. But it is conceivable that the pathological condition resultant on fungal attacks may be the direct cause, in some way, of pigment formation. The more intimate connection between anthocyanin and sugars will be discussed in a later chapter.

The experiments of decortication, etc., lead also to the conclusion that the chromogen of anthocyanin is synthesised in the leaves. For in cases

[1] Also lack of water. See effect of drought on p. 93.

where leaves and shoots have reddened owing to the blocking of the translocation current, less development of pigment has often been noticed in flowers and fruits. Gautier (190) made various experiments on vines in order to illustrate his view that the chromogen of the grape pigment is synthesised in the leaves, and is oxidised after passing into the fruit. Vine branches were deprived of their leaves, and this was shown to prevent a development of pigment in the fruit. In another experiment, the petioles of the leaves were ligatured with the result that the fruits remained green and the leaves themselves reddened. Ravaz (467), on the other hand, grafted a vine with purple grapes on to a white-fruited variety, and found that although the white variety produced no pigment in the leaves, the fruit of the graft was coloured every year. Hence Ravaz concludes that the pigment is synthesised in the fruit itself, though the latter may be nourished by the leaves. A direct connection between leaves and flower-colour may be demonstrated by removing a developing inflorescence from a plant, such as *Digitalis purpurea*, when the leaves will generally turn red. Gertz (19) observed the same result in a plant of *Geum rivale* from which the flowers had been cut off.

There is also reason to believe that special richness in nutriment, or synthetic products, is connected with anthocyanin formation. An experiment is quoted by Berthold as illustrating this point. From two- or three-year old individuals of *Acer pseudoplatanus* all buds were removed but the terminal one, which, as a result, received an excessive amount of nutriment and on development was strongly reddened. Some such explanation, as Gertz suggests, may account for the excessive reddening of leaves of adventitious shoots arising at the base of felled trees (*Populus, Tilia, Acer*).

Bonnier (381, 394, 415) and Heckel (387) have noted that a greater intensity of flower-colour is produced in many species when grown at high altitudes, as compared with the colour of the flowers in lowland regions. It is conceivable that this increase in intensity of colour is brought about by increase of synthetic products, due to greater insolation by day accompanied by low night temperature, and to stunted growth, rather than to any direct action of light on pigment formation, but at present there is no conclusive evidence.

In the same way lack of synthetic products due to poor conditions of the plant reduces pigment formation. The experiments of Sachs (358), Askenasy (369), Vöchting (412) and Klebs (447), in which leaves of plants were kept in darkness while the flowers were exposed to light, are not so conclusive as those previously mentioned. The method is so drastic, and may influence the whole nutrition of the plant to such an extent, that the non-development of flower-colour cannot be regarded as having any great significance.

A phenomenon of considerable interest in connection with nutrition and formation of red pigment is that pointed out by Mirande (419) as occurring in the genus *Cuscuta*, in which anthocyanin is widely produced. From observations made upon the development of many different species of *Cuscuta* on various hosts, Mirande concludes that the amount of pigment varies not only in different species but also in each species according to the host on which it grows. For instance, the same species growing on *Sambucus nigra* (poor) and *Forsythia viridissima* (rich in sugar) becomes green on the former but very red on the latter. Hence Mirande correctly deduces the fact that not only good development of the parasite but also the formation of red colour is correlated with good nutrition. In nature, species of *Cuscuta* passing from one host to another are seen to show different amounts of pigmentation. Chemical tests made by Mirande on extracts from the host plants showed that the greatest production of colour was found when the host plants were capable of producing most sugar.

The conclusions which may be drawn from all the instances quoted above are that an unnatural accumulation of synthetic products may cause colour to be developed in organs not normally coloured. At the same time a good supply of nutrition will intensify colour in parts of plants which are normally coloured. There is reason to believe that, of the accumulated substances, the most potent in bringing about colour production are sugars and glucosides, and this will be found to be borne out by observations connected with the effect of other factors considered in this chapter. Also an increase of the chromogens from which anthocyanins are produced probably takes place under the conditions mentioned above.

The relationship between anthocyanin development and sugar-feeding is reserved for a later paragraph (see p. 93).

Effect of temperature.

Most general observations bear out the conclusion that increase of anthocyanin is correlated with lowering of temperature. The most obvious demonstrations are the autumnal coloration of leaves and, to a lesser extent, the reddening of evergreen leaves in winter (*Ligustrum*, *Hedera*, *Mahonia*). The question of the relationship between low temperature and anthocyanin formation has been specially considered by Overton

(420). This author had observed that *Hydrocharis* plants grown in cane sugar solutions became very strongly reddened, and the idea occurred to him that an excess of sugar in the cell-contents might similarly be the cause of autumnal and winter coloration of leaves. In view of this suggestion, it is difficult to estimate any direct effect of low temperature on anthocyanin formation, because of the indirect effect produced by the same conditions on (1) photosynthesis, (2) formation of starch from sugar, (3) growth in general and probably (4) translocation. A decrease of activity of (1) leads to a decrease of sugar contents in the cell; but a decrease of (2) and (4) has the opposite effect. The process of removal of synthetic products from the leaves is, according to Sachs, greatly retarded by low temperature. Hence similar conditions of clogging to those brought about by injury, which were mentioned in the previous section, might arise and there would be a resultant production of pigment. The synthesis of starch from sugar is also a process which is retarded by low temperature. Thus Müller-Thurgau[1] has shown that at temperatures below 5° C. quite a considerable portion of the starch contents of the potato is changed to sugar, and with a rise of temperature the greater portion of starch is again regenerated. According to Lidforss[2], evergreen leaves in winter are also completely starch-free but contain very considerable quantities of glucose, which is again to a large extent changed back to starch if the leaves are artificially warmed. Overton himself examined the sugar content of autumnal leaves and found considerable quantities present, appreciably more, at any rate, than in the same species at midsummer.

From the above statements it will be seen that low temperature may greatly affect the sugar contents of the tissues, and hence may in this way cause the reddening, apart from any more direct effect.

Overton made some observations on the effect of temperature on reddening of *Hydrocharis* leaves, and found that the higher the temperature the less anthocyanin is formed and *vice versa*; but obviously in this case it is impossible to eliminate the effect of temperature on the photosynthetic activity of the leaves, and on growth and respiration in general with the resultant employment of synthetic materials. Klebs also (447) gives an account of the effect of temperature on the colour of the flowers of *Campanula Trachelium*. From cultures at various times of the year

[1] Müller-Thurgau, H., 'Ueber Zuckeranhäufung in Pflanzentheilen in Folge niederer Temperatur,' *Landw. Jahrb.*, Berlin, 1882, XI, pp. 751–828.

[2] Lidforss, B., 'Zur Physiologie und Biologie der wintergrünen Flora,' *Bot. Centralbl.*, Cassel, 1896, LXVIII, pp. 33–44.

in green-houses, etc., kept at different temperatures, he obtained a variation in flower-colour from white (in heat) through pale blue to deep blue (in cold). He also observed that *Primula sinensis* produces tinged flowers in a hot-house and full-coloured flowers in the cold. Klebs is of the opinion that the colour changes induced by changes of temperature are not directly due to the effect of temperature on pigment formation but indirectly to the effect of temperature on metabolism. At high temperatures, growth is so rapid that the substances used in pigment formation are not present in sufficient quantity.

Effect of light.

As regards the effect of light on anthocyanin formation, there have been numerous observations, some more or less conflicting. As with temperature, the main question at issue is again, whether light directly affects anthocyanin formation, or whether its influence is only indirect, in so far as it affects photosynthesis and the accumulation of the products of this process, among which may be the chromogens from which pigment is formed[1].

The effect of the absence of light on the development of pigment in flowers may first be considered. As early as 1799, Senebier (2) noted that the Crocus and Tulip develop coloured flowers in the dark. The same observations were recorded by Marquart (5) in 1835 for *Crocus sativus*. Later Sachs made definite experiments on various plants either by growing them entirely in the dark, or by enclosing in a dark chamber certain shoots or branches only, the other parts of the plant being in the light. Of plants grown entirely in the dark, Sachs (356) was able to distinguish two classes: (*a*) flowers which develop colour normally in the dark without being previously exposed to light (*Tulipa*, *Iris*, *Hyacinthus*, *Crocus*); (*b*) flowers which only develop colour in the dark if the buds have been fully exposed to light until just before opening (*Brassica*, *Tropaeolum*, *Papaver*, *Cucurbita*). Of *Tulipa Gesneriana* Sachs specially remarks: "Die schön gefärbten und normal entfalteten Blüthen auf den etiolirten Pflanzen machten einen höchst sonderbaren Eindruck." And of *Iris pumila* he says: "der zart hellbläuliche Grundton der Perigonzipfel, die dunkelviolette Aderung, welche gegen den Grund der Zipfel hin in

[1] I am indebted to Dr F. F. Blackman for drawing my attention to a fact which is interesting in this connection, i.e., that the development of chlorophyll may be affected by nutrition. The statement is founded on the observation that certain Algae develop chlorophyll in the dark when provided artificially with protein; when supplied with nitrates under the same conditions, however, the pigment does not appear.

das bläulich Purpurescirende übergeht, das Orangegelb der Bärte, das schön warme Blau der Narben und die himmelblaue Färbung des Pollens, alle diese Färbungen waren bei der Blüthe der etiolirten Pflanze eher glänzender und gesättigter als bei den am Lichte entfalteten." In the case of *Tropaeolum*, *Papaver*, etc., if the plants were darkened just before the buds unfolded, normally coloured flowers were produced, but later buds developed flowers with decreasing amounts of pigment. In the cases where shoots only were darkened, the flowers borne upon them were normal as regards size, but the coloration was less intense. Further experiments were made by Askenasy (369) who confirmed Sachs' results for *Tulipa* and *Crocus*, though he found *Hyacinthus* flowers rather less coloured in the dark. Other plants (*Pulmonaria*, *Antirrhinum*, *Silene*, *Prunella*) showed less production of anthocyanin in the dark if the buds had not been previously exposed to light; in *Prunella*, almost white flowers were formed. Sorby (158), Beulaygue (426), Gertz (19) and others have confirmed these results for the flowers of various species, showing that considerably less, very little, or no colouring matter at all, is formed in the dark.

As regards coloration of fruits, observations do not entirely agree. Senebier notes that apples do not redden unless exposed directly to light. However, as pointed out by Gertz, the reddening of apples is more or less accessory, but in fruits of which anthocyanin production is a distinguishing feature (*Crataegus*, *Rosa*, *Sambucus*), Askenasy (369) found by partial darkening when green, the pigment developed equally well in both illuminated and darkened parts. Both Laurent (402) and Müller-Thurgau (385) agree that the coloration of grapes can take place in the dark.

In purely vegetative organs observations are more conflicting. Of anthocyanin formed in the dark, the following examples may be quoted:

Leaves of *Beta* (Morren).

Seedlings of *Phalaris* and *Secale* (Hallier, 359).

Intense rose-red coleoptile in *Secale* (Gertz, 19).

Red coloration in shoots of *Opuntia robusta* and *O. leucotricha*.

Spots on leaf of *Orchis latifolia* (Gertz, 19).

Red-veined leaves of *Crepis paludosa* (Gertz, 19).

Faint reddening in stolons of *Solanum tuberosum* (Gertz, 19; Sachs, 356).

On the other hand innumerable cases may be quoted in which light appears necessary for the formation of the pigment:

Reddening of seedlings is entirely absent in the dark in *Polygonum tartaricum*, *Celosia*, *Beta* (Weretennikow, 360). This has been confirmed by Schell (372) for seedlings of *Polygonum*, *Rumex*, *Rheum* and *Amaranthus* and by Batalin (380) for *Fagopyrum*. The most casual observation will also afford instances of cases where anthocyanin is developed on the sides of stems, twigs and petioles which are exposed to the sun, the opposite side remaining green[1]. Such phenomena are specially mentioned in stems of *Cornus sanguinea*, *C. sibirica*, species of *Tilia*, *Rosa* and *Rubus* (Gertz, 19), of *Cuscuta* (Mirande, 419), of *Helianthus*, *Crataegus* and stolons of *Fragaria* (Dufour, 392).

The development of autumnal coloration often only takes place in the parts of leaves and stems exposed to light, as was noted long ago in *Viburnum Lantana* (Voigt, 351) and *Rhus Coriaria* (Macaire-Princep, 3). Gertz (19) also points out that, in *Viburnum Opulus*, *V. Lantana*, *Cornus sanguinea*, *C. sibirica* and *Prunus Padus*, the autumnal leaves may show quite clear natural photographs of the leaves covering them, because anthocyanin is absent from the covered surface. Similar lines and spots may be observed in winter-reddened leaves (*Silene*, *Viscaria*, *Armeria*, *Hieracium*, *Pilosella*).

Development of pigment in roots exposed to light has been observed in *Salix* (Hallier, 359; Schell, 374) and *Zea* (Dufour, 392; Devaux, 395).

It is difficult to draw conclusions from the rather conflicting statements given above. One fact stands out definitely, namely, that absence of light itself is no bar to the formation of anthocyanin in many cases, such as the root of *Beta* and the flowers of *Tulipa* and *Iris*. But in the majority of such cases there is obviously a plentiful supply of synthetic products in the storage tissues of the bulbs, corms, seeds and fruits concerned, from which the chromogen for the pigment can be synthesised, if there is not also already a storage of this substance. But where the supply of chromogen is directly dependent on the photosynthetic activity of the leaves, darkening of the flower-buds, unless they are already practically mature and supplied with chromogen, prevents or diminishes the formation of anthocyanin. The reddening of leaves and stems where exposed to light may in some way be due to local accumulation of synthetic products, though the direct effect of light is superficially more probable.

The effect of light in autumnal coloration is even less well explained on this accumulation hypothesis, except that in the last stages of the leaf's existence, photosynthesis must be best carried out in the parts

[1] It has been noted by Parkin (77) that the under surface of leaves is more sensitive to light and reddens more easily than the upper.

most exposed to light. There is also a general tendency to accumulation of synthetic products owing to low night temperature.

The formation of anthocyanin in normally uncoloured roots when exposed to light appears to be the most convincing evidence at hand for the production of anthocyanin due to the direct action of light. Until more evidence has been collected from a number of carefully devised experiments, no definite inference can be drawn.

In conclusion we may mention some work on more systematic lines which was published by Linsbauer (458) in 1908. He endeavoured to find out the more precise relationships between light and the formation of anthocyanin. For this purpose he used seedlings of *Fagopyrum esculentum* which had been grown in the dark and were quite etiolated. Such seedlings were then exposed to light (lamp) of different intensities and for varying lengths of time. From his results Linsbauer concluded that the photo-chemical process of anthocyanin production in light is a typical stimulus reaction, and is dependent upon both the intensity and duration of light. He investigated also the relationship between the times of reaction and presentation, and found it analogous in many respects to that in other stimulus processes, i.e. geotropism, for instance. Whether, however, the appearance of anthocyanin in these seedlings is due to the direct action of light, or to the products of photosynthesis induced by light, cannot be readily ascertained.

Mirande (485, 486) has recently made some interesting observations on the effect of light on the development of anthocyanin in the detached scales from the bulbs of *Lilium candidum*. At whatever the altitude the experiment was carried out, the pigment is never produced in direct light; it is produced only in diffuse light, the amount required varying with the altitude. Only the rays of the luminous part of the spectrum are effective, and of these the blue and indigo are most active, the red less so; the green are inactive.

Connection with presence of oxygen.

That oxidation plays a part in the formation of anthocyanin has frequently been suggested. The production of red pigment through the oxidation of a chromogen was the hypothesis brought forward by Wigand (150) as early as 1862, and the same idea has been revived successively by Palladin (218), Wheldale (226, 227) and Combes (466). That the process is controlled by a specific oxidase has been postulated by Buscalioni & Pollacci (17), Mirande (452) and Wheldale (226, 227). The actual dependence of the process on the presence of oxygen is illustrated by the experiments of Mer (371), who mentions the fact that leaves of *Cissus* do not redden under water, and in 1889 Emery (396) noted that no colour is produced in submerged flowers.

More definite experiments were performed by Katić (441), incidentally, among a series of investigations primarily concerned with the effect of culture solutions on the formation of anthocyanin. Leaves in certain culture solutions of sugars, and other substances, were found to produce anthocyanin, but if enclosed in vessels, from which all oxygen had been removed by alkaline pyrogallol, no trace of colour was observed. Katić also found that colour was less rapidly developed in air under reduced pressure than in normal atmosphere. Also in certain culture solutions an increased pressure of oxygen produced greater development of colour.

Still more elaborate investigations were made by Combes (466). The experiments consisted in analysing the gaseous exchange of red and green leaves under similar conditions. Leaves were employed in which reddening had taken place and was proceeding from different causes, as for instance: leaves of *Ampelopsis hederacea* reddened by exposure to light; of *Rumex crispus* and *Oenothera Lamarckiana* by attacks of parasites; of *Spiraea prunifolia* and *Mahonia aquifolium* by decortication; of *Rubus fruticosus* with autumnal coloration; and finally of young leaves of *Ailanthus glandulosa* in which reddening, on the contrary, was disappearing.

From observations upon gaseous exchange in the above leaves, which are exemplified in the accompanying table, Combes drew the following conclusions:

Oxygen fixed (+) or lost (−) during one hour, consisting of half-hour of day and half-hour of night.

	Red leaves	Green leaves
Ampelopsis	− ·0872	− ·329
Rumex	+ ·0963	− ·8067
Oenothera	+ ·2442	+ ·1622
Spiraea	+ ·1357	− ·3226
Mahonia	+ ·0829	− ·1565
Rubus	+ ·2011	− ·0435
Ailanthus	− 1·959	− 2·589

The appearance of anthocyanin is accompanied by an accumulation of oxygen in the tissues; the disappearance of the pigment is, on the contrary, accompanied by a considerable loss of oxygen. Observations leading to similar conclusions have been made by Rosé (476) and Nicolas (479).

Effect of drought.

Here again the direct effect is difficult to estimate owing to the simultaneous effect on photosynthesis, but whether direct or indirect, there is ample evidence that drought increases anthocyanin formation. Molisch (403) found that leaves of *Peireskia aculeata*, *Tradescantia*, *Panicum variegatum* and *Fuchsia* reddened strongly if only watered a little. The author has made the same observations for pot plants of *Pelargonium*. Eberhardt (434) also found an increase of anthocyanin in leaves of *Coleus Blumei* and *Achyranthes angustifolia* when grown in a very dry atmosphere. According to Warming (414, 433) plants such as *Tillaea aquatica*, *Peplis Portula* and *Elatine* are green when growing in water, though individuals on land may be strongly red.

The physiological drought of salt marshes may similarly explain the development of anthocyanin in halophytes (*Salicornia*, *Suaeda*).

An interesting case of the connection between reddening and drought has been observed by Miyoshi (462). This author noticed that the leaves of certain tropical trees, especially *Terminalia Cattapa*, in the East Indies, Ceylon and Java, take on a beautiful red colour before the leaf-fall. The reddening is described as affecting at first a few leaves only, but later the number increases. Of the early stages the author says: "Vor der Ferne betrachtet erschienen die gefärbten Blätter wie rote Blüten in voller Pracht." Since the phenomenon takes place at the dry period of the year, Miyoshi suggests the term 'Trockenröte,' and considers the causes of reddening to be drought coupled with strong insolation.

Effect of sugar-feeding.

The term sugar-feeding means that the plant is supplied artificially with extra amounts of sugar. In the case of floating or submerged water plants, the whole plant can be immersed for experiment in dilute sugar solution; in the case of land plants, the stems of leafy branches, or the petioles of isolated leaves, can be put in the solution; or the leaves can be cut into pieces and floated in the liquid.

The first investigations of the results of this process were carried out by Overton (420). While conducting some experiments on osmosis with *Hydrocharis Morsus-ranae*, Overton noted that the leaves of this plant tended to become red when the plants were grown in 5 % cane sugar solution. Later, the idea occurred to him that autumnal coloration might be due to excess of sugar in the leaf tissues, and he claims to have shown that autumnal leaves contain more sugar than green leaves. On the basis of this idea, he then commenced some systematic investigations on sugar-feeding with a view to gaining more knowledge of the whole phenomenon.

Experiments were first made with *Hydrocharis* plants grown in various solutions, and the results were as follows:

Hydrocharis Morsus-ranae. Plants in 2 % invert sugar showed excess of anthocyanin over the control in four days. This was true of pigment in all parts—leaves, petioles, stolons and roots, and the intensity of pigmentation increased with time. In 2 % cane sugar the results were similar.

Plants in both the above solutions flowered rather earlier than control plants, but the flowers were unaffected by the cultures.

In 2 % glucose there was the usual reddening; also in 2 % laevulose. In 2 % galactose there was no reddening, and in 5 % lactose the reddening was very slight. In 2 %, 4 % and 10 % glycerine no colour developed; this was also the case in solutions of potassium nitrate, sodium chloride, sodium sulphate and of other salts.

In 3 % invert sugar in a dark room there was no trace of reddening.

Microscopically it was found that colour was produced in the mesophyll cells, and never in either the upper or under epidermis.

The following species were also used:

Elodea canadensis. In 2–3 % invert sugar a reddish colour developed in the younger leaves.

Vallisneria spiralis. In sugar cultures there was an increase of red colour which was located in the epidermis as well as in the inner tissues.

Potamogeton perfoliatus, P. pectinatus. In 2–3 % invert sugar there was no result. In *P. pusillus* a reddish colour appeared.

In *Najus major* there was no effect, but in *N. minor* there was a slight result.

Lemna minor, L. trisulca in various kinds of sugar solutions gave negative results. The same was true for *Pistia Stratiotes.*

Utricularia Bremii. In 2 % invert sugar reddening in the bladders appeared in two to three days. Finally the leaves and bladders became quite red. In 2 % cane sugar the result was similar. In *Utricularia minor*, reddening appeared in ·5 % cane sugar; also in glucose, invert and cane sugars (·5 %– 5 %). *U. vulgaris* behaved similarly to *U. Bremii* and *U. minor.* In 2–5 % lactose there was no colour in two weeks, but after four weeks a slight colour (due to hydrolysis probably). There was no colour in galactose; slight colour only in glycerine and none in salt solutions.

In *Utricularia*, reddening in sugar cultures developed just as little in complete darkness as in *Hydrocharis*.

Ceratophyllum demersum. In 2–3 % invert sugar there was some reddening in the cells of sub-epidermal and deeper tissues, though the epidermis was uncoloured.

Experiments were next made with land plants, using either leafy twigs or isolated leaves:

Lilium Martagon. A leafy stem was placed in 2 % invert sugar. This and control stems in distilled water were placed in a south-east window. After about seven days, the plant in sugar solution showed reddening in the leaf-tips, which afterwards spread, day by day, over the leaf, while the control showed no reddening. The pigment was found to be located in the inner leaf tissues, the upper and under epidermal cells being free. Experiments were made with about 20 other specimens, and in each case the results were the same. In 2 % glucose solution reddening could be detected in four days, becoming more intense in the course of time. In 2 % fructose there was a similar result. In 2 % cane sugar the red coloration came later and was less intense. In lactose, galactose and glycerine solutions there was no reddening, and the same was the case with various salt solutions. It was found also that ethyl and amyl alcohols, ketones and ether, in solution, caused development of red pigment, but Overton considers the result to be one of injury. He is inclined to believe that the alcohols, ketones, etc., acted as narcotics and so prevented translocation, rather than that they served as material for building up the pigment.

Fritillaria imperialis was found to produce no colour in sugar solution.

Ilex aquifolium. A twig of *Ilex* was put in 3 % invert sugar. In two days some reddening appeared, whereas control plants showed no colour. The pigment was found in the palisade, and to some extent in the spongy parenchyma, but none was present in the epidermis. In glucose and fructose there was also considerable reddening.

Hedera Helix. In 2–3 % invert sugar red colour appeared in a few days, but it was not intense nor was it uniformly distributed.

Mahonia Aquifolium. A twig of this plant gave a negative result in 2 % invert sugar.

Ligustrum vulgare. In 2 % invert sugar red pigment appeared after eight days. The pigment was localised in the palisade cells.

Ampelopsis hederacea. In 2 % invert sugar reddening commenced, but the experiment could not be continued as the leaves tended to fall from the leaf-stalks. In autumn, good results of artificial reddening were obtained, green leaves only, of course, being used.

Saxifraga crassifolia. In 3 % invert sugar the leaves reddened in a few days.

Aquilegia vulgaris. Young leaves with petioles in 2 % invert sugar showed distinct reddening in four days.

Taraxacum officinale. When the base of the leaves was placed in a 2 % solution of invert sugar, a very fine red colour developed in two or more days over the whole leaf, and the pigment was located in the inner tissues and not in the epidermis. Leaves, however, in distilled water may redden in time. Reddening of leaves was found to be characteristic of many Compositae, and as this result was often obtained to some extent in distilled water in a good light, they did not afford very suitable material. Leaves of *Eupatorium Cannabinum* and *Prenanthes purpurea* which did not redden to any extent in distilled water, became very red in sugar solution.

Epilobium spp. Leafy stems of *E. parviflorum* in 2 % invert sugar gave a good red colour.

Negative results in 2–3 % invert sugar were obtained with *Anthriscus silvestris*; also with *Rubus* species.

From the above researches Overton draws the conclusion that in many species of Monocotyledons and Dicotyledons, both water and land plants, sugar-feeding will bring about anthocyanin formation. There is also this further correlation, that, with the exception of submerged water plants, there is a negative result with sugar-feeding when the plant in normal circumstances produces anthocyanin in the epidermal cells. But if in normal circumstances the pigment is formed in the inner tissues, then sugar-feeding (especially with fructose and glucose) produces red pigment in a high percentage of cases, and this pigment is localised in the mesophyll and not in the epidermis.

Overton also tried the effect of putting the inflorescence of white-flowered varieties into sugar solution. Thus, for instance, the inflorescence stalk of white *Pelargonium zonale* was placed in 3 % solution of invert sugar, but no trace of red colour was formed in the flowers though the stalk showed reddening. Negative results were also obtained with *Anemone japonica.* Overton concludes that some other factor, apart from the presence of sugar, is necessary in these cases.

Further researches on sugar-feeding were made some years later by Katić (441), and of these the experiments on *Hydrilla verticillata* (Hydrocharitaceae) are given in the greatest detail as follows:

Inorganic culture media containing various salts of potassium, sodium, calcium, magnesium, ammonium, iron, aluminium and lithium had practically no effect on the formation of pigment. In glucose (·05–3 %) solutions, red pigment was formed in the light and in the dark, and in isolated leaves more quickly than in pieces of leafy stem. In laevulose (1–5 %) and cane sugar also pigment developed both in the light and in the dark, but the cane sugar (of concentration ·5–25 %) was most favourable to reddening. Only a slight coloration appeared with maltose (1–3 %) both in the light and in the dark. In lactose (1–5 %), raffinose (1–10 %), inulin (1–3 %) and glycerine (4–5 %) pigment was formed only in the light. Some colour was developed in (1–4 %) ethyl alcohol and in mannite (1–2 %), but none in galactose (up to 5 %), arabinose, formol (·001 %), dextrin, salicin or asparagin.

The effect of mixed solutions of carbohydrates and various salts, such as those mentioned above, was tried. It was found that potassium nitrate and mono-potassium phosphate quickened the formation of pigment in sugar solution. Various other potassium salts (except potassium bichromate) had the same effect in a less and varying degree. Katić is of the opinion that the effect of potassium salts is chemical and not osmotic. Sodium salts, on the whole, were found to have little effect; magnesium sulphate and nitrate some positive effect; calcium salts (except $CaH_4(PO_4)_2$), aluminium sulphate, ferric chloride and some ammonium salts a preventative effect.

Various alkalies, sodium carbonate, potassium hydroxide, calcium hydroxide and magnesium oxide were tried with sugar solutions; the result was to quicken the formation of anthocyanin. Acids, as for instance, tartaric, salicylic, citric and oxalic, on the contrary, appeared to slow down the formation; to this tannic acid formed an exception, since the reddening appeared more quickly in its presence.

Although anthocyanin was produced in the cultures in the dark, Katić found that the development was always earlier in the light. As regards temperature, it was found that 25° C. was the optimum for the appearance of the pigment in the light and 28° C. in the dark.

The absence of carbon dioxide had no effect on the formation of anthocyanin in sugar solutions, except in the case of glycerine, where the development was weaker. Oxygen, on the contrary, was found to be necessary for the reddening.

Katić made further observations with a number of other plants, and although his experiments are described in great detail in his dissertation, no more than a short statement will be given here as they were largely on the same lines as those with *Hydrilla*.

Elodea canadensis. In ·26–1 % cane sugar there was a slight coloration in the light, but none in the dark; in 1 % grape sugar there was colour both in the light and in the dark. In the case of cane sugar the development was increased by addition of potassium nitrate and calcium sulphate.

Hydrocharis Morsus-ranae. In cane sugar and laevulose some pigment appeared in the dark; in grape sugar and maltose there was none in the dark.

Sagittaria natans. In 5 % cane sugar there was a considerable development of colour, but only in the light.

Allium Cepa. The colourless scales of a variety which normally contained some anthocyanin turned red in sugar solution; a white variety, however, produced no colour under any conditions.

Canna indica. Etiolated leaves in 5–15 % cane sugar formed pigment.

Veronica Chamaedrys. In sugar cultures colour only developed in the light and was located in the epidermal cells.

Rosa 'Maréchal Niel.' Green leaves, or pieces of leaves, in 15 % cane sugar formed colour only in the light. It was found in both epidermis and spongy tissue.

Saxifraga cordifolia. Pieces of a leaf in 5 % cane sugar only gave a good development of anthocyanin in the light.

Pittosporum (*undulatum*?). Green leaves and pieces of leaves in 5–15 % cane sugar formed pigment only in the light, and it was found to be located in the inner tissues and not in the epidermis.

Bellis perennis. In 15 % sugar solution green leaves developed a weak red coloration only in the light, and it was confined to the epidermal cells.

Thus we see that Katić's results differ from those of Overton in two points. First, anthocyanin may develop in sugar cultures in the dark, and secondly, it is not necessarily confined to the epidermis.

The question of the results of sugar-feeding has been taken up more recently (1912) by Gertz (473). The following account is taken entirely from his paper. He deals with a point in Overton's results we have just mentioned, i.e. the statement that by sugar-feeding anthocyanin can be induced to form in the mesophyll, but not in the epidermal cells. Thus it would appear possible that the chloroplastids might have some influence on the formation of pigment, and the failure of petals to produce anthocyanin in sugar culture might be considered a corroboration of this view. In *Vallisneria spiralis* and *Elodea*, Overton found epidermal anthocyanin on sugar culture, but this is no proof, since the epidermis in those plants contains chlorophyll. General evidence is, moreover, against the view of the connection with chloroplasts, as the epidermis often contains anthocyanin, and this pigment is also developed in chlorophyll-free saprophytes and parasites and in perianth leaves.

In order to investigate this point, Gertz made some sugar culture experiments with parts of albino leaves which were free from chloroplastids. For this purpose he used first the leaves of *Oplismenus imbecillis* (Graminaceae), and found that anthocyanin may be induced to form in the complete absence of chlorophyll. Therefore the question of the activity of the chloroplastids is definitely solved, but as only this one species had been used, further experiments were made with other plants. A modification of Overton's method was employed; portions, 20 × 20 mm. square, were cut out from the leaves and floated upside down on the solutions in glass dishes, and in this way free transpiration and respiration could go on through the stomata. The solutions were also changed from time to time, and the edges of the leaves freshly cut. Cane sugar only was used of 5–10 % concentration. The following results were obtained.

Oplismenus imbecillis. The leaves are variegated—red, green and white. In the white portions there are no chloroplastids, not even in guard cells of the stomata. When white portions were used, anthocyanin appeared in quantity in the epidermis after four days. In the dark, however, there was only slight, though definite, coloration. Hence Gertz concluded that, though not absolutely necessary, light is a powerful agent in assisting the formation. Cultures in distilled water, both in the light and in the dark, produced no anthocyanin. To obtain absolutely comparable results, parts of the same leaf, *A*, *B*, *C* and *D*, were treated as follows. *A* and *B* were placed in 5 % sugar solution, *C* and *D* in distilled water; *A* and *C* were exposed to the light, *B* and *D* kept in the dark. The results were as before. Both *A* and *B* formed anthocyanin, but it was insignificant in *B*; *C* and *D* were free from anthocyanin.

Tradescantia Loekensis. Again the leaves are variegated—red, white and green. No chlorophyll occurs in the white parts except in the guard cells. In cane sugar the result was negative and no anthocyanin was formed. Gertz has no suggestions to offer except that possibly some unknown factor is essential to the formation of pigment.

Beta vulgaris. Only negative results were obtained.

Rumex domesticus. A variegated green and white form was used, of which the white parts, including the guard cells of the stomata, are entirely free from chlorophyll. In less than one day in 5 % sugar solution, anthocyanin appeared, and was found to be located in the lower as well as in the upper epidermis. In the dark, less pigment was formed. Experiments were made with ordinary *Rumex*, and were in complete agreement, since anthocyanin was formed chiefly in the epidermis (except stomata). Similar observations on *R. Patentia* have also been made by Palladin (218).

Cornus florida. The leaves have green and white parts, the latter being free from chlorophyll. After a week in 10 % cane sugar they showed only traces of anthocyanin which was located in the spongy parenchyma.

Euonymus radicans. In 10 % sugar the albino parts showed a faint reddening in the lower epidermis with the exception of the stomata.

Lonicera brachypoda. The leaves have white reticulations on a green ground. After a week in 10 % cane sugar solution anthocyanin was found in the palisade parenchyma.

Other experiments were made with green leaves as follows:

Plantago major. In 10 % sugar solution the pieces of leaves reddened strongly, and anthocyanin was located entirely in the lower epidermis except the stomata. Attempts were made to induce anthocyanin formation in sugar cultures in isolated epidermal layers, but they were unsuccessful.

Sium latifolium. In 10 % sugar solution anthocyanin appeared readily in pieces of leaves, as well as in entire leaves and shoots, and was found to be localised in cells of the mesophyll. With *Cerefolium sylvestre* negative results were obtained.

Epilobium parviflorum. In 5 % sugar solution anthocyanin developed plentifully in the lower epidermis and occasionally in the upper.

Euphorbia Cyparissias. In bracts in 10 % sugar solution anthocyanin was formed in both the upper and lower epidermis. In *Phytolacca decandra* it appeared in the epidermis (except stomata).

Thus Overton's view, which is expressed as follows—"...mit Ausnahme der untergetauchten Wasserpflanzen scheinen solche Versuche fast durchweg bei denjenigen Pflanzen negativ auszufallen, deren natürliche Rothfärbung...der Gegenwart von rothem Zellsaft in den Epidermiszellen zu verdanken ist," is, as Gertz points out, not correct; for colour due to anthocyanin appears entirely in the epidermis in *Oplismenus imbecillis, Euonymus radicans, Plantago major, Euphorbia Cyparissias* and *Phytolacca decandra*; both in the epidermis and mesophyll in *Rumex domesticus, Tussilago Farfara* and *Epilobium parviflorum.* On the other hand in *Cornus florida, Lonicera brachypoda* and *Sium latifolium* it appears in the ground parenchyma. These results are also in accordance with those of Katić.

Thus it would appear to be definitely settled that chloroplastids are not essential to anthocyanin formation from the results with leaves of

Oplismenus imbecillis, *Rumex domesticus*, *Cornus florida*, *Euonymus radicans* and *Lonicera brachypoda*, though the results with *Tradescantia* and *Beta* are negative. Yet Gertz does not seem to be quite assured on the point, partly because, as he points out, the epidermal leucoplastids are closely related to chloroplastids, and partly because the epidermis is connected with chloroplast-containing cells.

As regards the effect of light on the results of sugar-feeding, Overton maintained that colour was not produced in the dark. This was not found by Gertz to be the case with *Oplismenus* and *Rumex*, and Katić, moreover, demonstrated that *Hydrilla*, *Hydrocharis*, *Allium* and *Phalaris* develop pigment in sugar cultures in the dark. Gertz is of the opinion that under natural conditions the appearance of anthocyanin may not be very largely affected by illumination, as a whole constellation of factors may take part in its formation in the kind of way we have tried to indicate in the previous sections.

Gertz finally considers the formation of anthocyanin in petals resulting from sugar-feeding. As we have seen, Overton failed to get any result with *Pelargonium* and *Anemone*. Gertz considers that such a result, if it were possible, would give an even more important proof that anthocyanin production is independent of chloroplastid activity. A striking observation in this connection has been made by Goiran (397) on *Cyclamen persicum giganteum* of which the flower is white except for a patch of red anthocyanin at the base of the petal. Goiran made transverse cuts through part of the petal, and though the petal tips remained fresh, after a time they developed red pigment. This result is akin to the formation of anthocyanin in leaves when the conducting system is injured by cutting, or by insects, Fungi, etc. There is also an observation by Gertz on white petals of *Saxifraga* in which anthocyanin appeared as a result of insect attacks. Gertz, however, failed to get any result from sugar cultures of *Bellis perennis*, *Anemone japonica*, *Magnolia acuminata*, *Tropaeolum majus*, *Deutzia gracilis*, *Begonia* sp., and *Pisum sativum*. But eventually a positive result was obtained with white petals of *Viburnum Opulus* (the cultivated form with neuter flowers). These petals in sugar solution formed anthocyanin in either one or both epidermal layers. Thus Overton's prophecy that petals might be found which would redden in sugar culture appears to be justified in this case.

Effect of amount of organic acids in leaves.

There is evidence in support of the view that the conditions which bring about the formation of anthocyanin are frequently those associated with high sugar concentration and increased respiration. It is possible that the acid content of the leaf may also increase under these conditions. If anthocyanin pigment occurs naturally in the form of its colourless isomer, increased amount of acid might augment the coloration. It has been stated by Nicolas (479) that red leaves contain more acids than green. Kohler (333, 334), however, from analyses, concludes that the formation of anthocyanin is not correlated with acid content. She suggests that the acid content may be a balance resulting from formation and destruction, and that more elaborate analyses might show that acid formation increased during pigment formation.

Mirande (345), on the contrary, has shown, by estimation, that scales from the bulb of *Lilium Martagon* form increased amounts of acid after being detached. There is an increase in uncoloured scales kept in darkness, but the increase is still greater in those kept in light and consequently reddened. Injury also induces increase of acid formation, but this is small by comparison.

It is not an easy task to set forth any general explanation of all the facts recorded in the previous sections, but a short review, as far as possible, of the situation may be useful at this stage. Thus, we may say that two points emerge from the preceding considerations: (1) that the chromogen for anthocyanin is probably formed in the leaf; (2) that an accumulation of carbohydrates (sugars) and glucosides leads to the formation of anthocyanin.

One explanation of the reddening phenomenon, advanced by various authors, Palladin (218), Mirande (452) and Combes (466), is that presence of excess of carbohydrates increases oxidation processes, and hence, if anthocyanin is an oxidation product, the formation of this pigment. Such an explanation would meet practically all cases of anthocyanin formation, but there is no special evidence for its justification.

Another suggestion is that the acid products resulting from increased repiration form coloured oxonium salts with the colourless leucobases (isomers) of anthocyanins already present in the tissues.

Let us now consider how the various suggestions fit the chief cases.

(1) In a *normal* plant with green leaves, coloured flowers and tinged stems and petioles, chromogen and sugars are synthesised in the leaves and translocated away when formed. If the plant is kept in the dark or shade, photosynthesis stops or is lessened, the supply of synthetic products falls below normal, and the flower-colour may be pale and the stems and petioles green. Conversely, great photosynthetic activity produces

a plentiful supply of sugars and chromogen which results in rich flower-colour and appearance of pigment in the vegetative organs. (In addition, light itself may directly increase pigment formation.) The intense colours in the flowers and the development of anthocyanin in the vegetative parts of High Alpine plants may be explained by strong insolation, stunted growth employing little material, and slow translocation due to low night temperature[1]. The power of some plants to produce normally coloured flowers and fruits in the dark is due to a plentiful supply of reserve material (flowers of *Tulipa*).

(2) If there is more than a plentiful supply of chromogen and carbohydrates owing to the flowering parts being removed, a certain amount of *abnormal* reddening may take place, ex., leaves of plants deprived of

[1] Exception will probably be taken to this statement on the ground that many High Alpine plants can be grown in lowland regions, and their flowers do not then show any perceptible loss of colour, whereas if the statement were true, we might expect to find considerable diminution of colour under these conditions. As a matter of actual fact, on the basis of observation, this criticism is not altogether valid, for, as we have previously stated, Gaston Bonnier (415), Kerner (498) and others have shown that in many cases the flowers of plants growing at high altitudes and in high latitudes have a more intense colour than those of individuals of the same species grown either in the plains or in lower latitudes. There is also a more general aspect of the question, which may be outlined as follows. Every plant is the expression of a chemical (or, fundamentally, physical) entity, and this expression can only fluctuate within limits on account of the definiteness of the chemical (or physical) constitution underlying it. *Broadly speaking* these chemical (or physical) entities are adapted to their habitats, that is, they are only able to exist under those conditions in which the chemical reactions (or physical processes) essential to their existence can take place. Sometimes a plant, for example species of *Opuntia*, will live and flourish to a certain extent in a habitat to which it is not adapted, such as the temperate zones. The conditions for its metabolic processes do not in this case lie beyond the limits of such a climate. But the expression of its entity, of necessity, remains the same. We may assume that, for *Opuntia*, variation in the direction of the characters of plants of the temperate zones does not occur, since such changes would involve chemical reactions (or physical processes) which are outside the sphere of its constitution. Hence it remains a typical species of *Opuntia*. Other examples no doubt could be selected where there were greater fluctuations on change of habitat, but these again would depend either on the wideness of the sphere of chemical (or physical) activity of the particular plant, or on the relation of the particular fluctuation to the environment. The same line of argument we have applied to *Opuntia* can be applied to the members of any plant formation, and among them, to High Alpine plants. These are *on the whole* adapted to their environment, and intense flower-colour is one of the expressions of the entity of this particular type of plant. Transferred to lowland regions, such a plant is still able to grow and flourish, and of necessity it retains its entity. Yet fluctuations in colour intensity of the flowers and in the amount of pigment in the vegetative organs are, within limits, the kind of variation we should suppose possible, though not necessarily inevitable; for such fluctuations depend merely on the direction, or amount, of certain chemical reactions which are, in any case, part of the essential metabolism of the plant. If we are to believe the evidence of Gaston Bonnier and others, it is just these variations which may occur.

the inflorescence, adventitious shoots from tree trunks; a similar explanation may hold for reddening in sugar cultures.

(3) A still further increase of carbohydrates and chromogen such as is caused by blocking the translocation current leads to still more abnormal reddening, ex., injured leaves, decorticated stems, etc.

(4) In autumnal and winter leaves there may be an accumulation of sugar and chromogen owing to the slowing down of the translocation current, and the lack of starch formation at low temperatures. The effect of low temperature may also retard growth in general and the using up of synthetic products.

CHAPTER VII

REACTIONS INVOLVED IN THE FORMATION OF ANTHOCYANINS

NEHEMIAH GREW (1) evolved, as a result of his observations, an elaborate scheme to explain the origin of the various pigments in plants. It is expressed in the scientific language of the period and is difficult to translate into modern conceptions, but it is clear that he believed the presence of air to play an important part in the process. Speaking of colours in roots, he says these organs show less variety in this respect than leaves and flowers—"The *Cause* hereof being, for that they are kept, by the *Earth*, from a free & open *Aer*; which concurreth with the *Juyces* of the several *Parts*, to the *Production* of their several *Colours*. And therefore the upper parts of *Roots*, when they happen to stand naked above the Ground, are often deyed with several *Colours*."

A publication by Schübler & Franck (131) in 1825 perhaps contains the first definite hypothesis as to the origin of red and blue pigments. These investigators treated extracts from coloured flowers and leaves with acids and alkalies, and the results led them to the view that green pigment of leaves occupies a mean position, from which an oxidised yellow-red series is formed, on the one hand, by the action of acids, and a deoxidised blue-violet series, on the other hand, by the action of alkalies. The hypothesis was further reinforced in 1828 by Macaire-Princep (3), who maintained that chlorophyll, when treated with acids, becomes oxidised first to yellow, then to red and orange pigments, and this oxidised chlorophyll can be turned green again by alkalies. The red plant pigments are, in his opinion, oxidised chlorophyll, and the blue, a mixture of red chlorophyll with a vegetable alkali. All colours may thus be looked upon as simple modifications of chlorophyll.

These erroneous ideas were accepted, without any real experimental basis, by plant physiologists of the day as Von Mohl (7, 8) very clearly points out in 1838. "Les physiologistes s'occupaient plutôt généralement de faire des spéculations sur les couleurs des plantes, de les rapporter aux couleurs du prisme, que de chercher à connaître la nature des matières colorantes elles-mêmes. Comme dans le prisme, le vert tient le milieu et se trouve bordé, d'un côté, par le jaune et le rouge, et de l'autre côté, par le bleu et le violet, on croyait que le vert des plantes était de même le point indifférentiel entre la série de couleur rouge-jaune et celle du bleu, et c'est par l'oxygénation et la désoxygénation de la couleur verte, qu'on cherchait à expliquer l'origine de ces couleurs, en se fondant sur des expériences chimiques incertaines, sur des idées fausses d'oxygénation et de désoxygénation, sur l'action des alcalis et des acides." A more correct point of view was reached too by Leopold Gmelin (151) in 1828[1]; he refused to accept the fact that chlorophyll is reddened by acids, and that the products formed from chlorophyll by acids, or naturally in autumnal leaves, again become green with alkalies. He notes, also, that red autumnal leaves contain both yellow chlorophyll, and a blue colouring matter which is reddened by acids.

Gmelin's criticisms, as well as others of the same kind, appear to have been unconvincing; for again, in 1835, Clamor Marquart (5) returns to the origin of anthocyanin from a metamorphosis of chlorophyll. Though objecting to the oxidation and deoxidation hypothesis, he retains the idea of the formation of other pigments from chlorophyll, this time, however, by addition and subtraction of water. By the action of strong sulphuric acid, i.e. by subtraction of water from chlorophyll, a blue colouring matter[2] (anthocyanin) is obtained which turns red with acids and green with alkalies. This blue substance, he says, forms the basis of all blue, violet and red flower-pigments.

Marquart's evidence was thought to be insufficient by Von Mohl (7, 8); for, with regard to the sulphuric acid reaction, the latter very naturally remarks: "Si, dans ce cas, la couleur bleue doit annoncer la formation artificielle de l'anthocyane par la chlorophylle, il est impossible de concevoir pourquoi, malgré la présence de l'acide sulfurique libre, la chlorophylle reste bleue et ne devient pas rouge." Also, with respect to Marquart's evidence that cells which originally contained chlorophyll, later contain anthocyanin, as for instance petals which are primarily green, and then become blue or red. Von Mohl points out, among other evidence, first, that anthocyanin is characteristic of the epidermis while chlorophyll is found in the inner tissues; secondly, that chlorophyll may be found, apparently in no lessened quantity, in cells which have become reddened with anthocyanin. In the course of time, as investigations proceeded, it became clear that anthocyanin is not derived from chlorophyll; Macaire-Princep and Marquart had been led astray in their views by the close connection between the disappearance of chlorophyll, and the

[1] Earlier edition than that in the Bibliography.

[2] This is of course, as we now know, a reaction given by plastid pigments.

appearance of anthocyanin, and *vice versa*. Yet, in the light of the additional knowledge which we now possess, it can still be stated with truth that there is a certain relationship, though an indirect one, between the two pigments, or, more strictly perhaps, between the spheres of their chemical activity.

The hypothesis of Wigand (150) in 1862 came very near to some of those held at the present time. Wigand maintained that anthocyanin arises by oxidation from a colourless substance which occurs in solution in the cell, and which changes to red under certain circumstances, and may, after a time, become colourless again. He considers it to be a tannin on the grounds that:

1. The red colouring of spring and autumn appears only in tannin-containing plants, and though it is not always found in these, yet it never develops in plants free from tannin.
2. Only in those tissues or cells (especially the epidermal cells and veins) in which the tannin was previously present does the colouring matter develop later.
3. The red sap, like the tannin, is turned green or blue with iron salts, and yellow with potash or ammonia.

The tannins of Wigand were very probably, in many cases, flavones, from which it has been suggested that the anthocyanins may arise.

From analyses of pigments of wine and grapes, Gautier (164, 190) in 1892 also formed the opinion that anthocyanin is an oxidised product of tannic acids, but no suggestions are made as to the particular reactions involved. At a later date, in 1899, Overton (420) concluded, on the basis of many observations and experiments, that red pigment is formed when there is an accumulation of sugars, but beyond stating that anthocyanin is probably a tannin-like substance existing, in combination with sugar, as a glucoside, Overton makes no definite statement as to the mode of formation. Mirande (452) in 1907 suggested that the appearance of anthocyanin when leaves are injured by insects is due to an accumulation in the tissues of tannins and glucose, accompanied by the presence of an oxidase. Laborde (214, 215, 216) in 1908 also came to the conclusion that there is a relationship between tannins and anthocyanins. To quote Combes: "l'auteur (i.e. Laborde) assimile le phénomène du rougissement à une action diastasique qui donnerait naissance à une matière colorante rouge aux dépens d'un noyau chromogène de nature phénolique que possèderaient tous les tannins." Laborde was able to obtain red pigments from tannins by means of certain chemical reagents and other treatment.

The authors mentioned are only a selection of those who have believed in the origin of anthocyanin from tannins, and the hypothesis held the field for many years; we find it accepted in most text-books even of fairly recent date (Pfeffer)[1], though the evidence for its acceptance is far from satisfactory.

The next hypothesis of importance is that of Palladin (218, 225)[2]; he considered anthocyanin to be a member of a class of pigments which he himself termed 'respiration pigments.' Although this hypothesis is connected primarily with the *function* of anthocyanin, nevertheless certain reactions are involved which justify its consideration in this chapter. Since Palladin's views are largely based on the action of oxidising enzymes, some preliminary account of these substances may not be out of place at this point.

Certain organic compounds, such as guaiacum tincture, α-naphthol, paraphenylene-diamine, benzidine, etc., are used as tests for oxidases since they become oxidised to coloured products when treated with oxidising enzymes under certain conditions. When the juice or water extract of some plants is added, for instance, to guaiacum tincture, a blue colour is immediately developed (more rarely with benzidine), and the plant is said to contain a direct oxidase. Of other plants the juice or extract gives no colour with any of the reagents until hydrogen peroxide is added, and the plant is said to contain an indirect oxidase. If a systematic examination be made of all natural orders as to their oxidase content, it will be found that the direct oxidase reaction is characteristic of about 60 % of the Angiosperms (Compositae, Labiatae, Umbelliferae, etc.), and the indirect oxidase reaction of the remainder (Cruciferae, Ericaceae, Crassulaceae, etc.)[3]. Some orders, however, contain genera representative of both types. It may be noted, in addition, that the plants giving the direct oxidase reaction turn brown on injury, or on exposure to chloroform vapour, or often when placed in absolute alcohol. The juices and extracts of such plants also turn brown or reddish-brown on exposure to air.

[1] *The Physiology of Plants*, translated by A. J. Ewart, Oxford, 1900.

[2] Also Palladin, W., 'Das Blut der Pflanzen,' *Ber. D. bot. Ges.*, Berlin, 1908, XXVI*a*, pp. 125–132. 'Die Verbreitung der Atmungschromogene bei den Pflanzen,' *ibid.* pp. 378–389. 'Ueber Prochromogene der pflanzlichen Atmungschromogene,' *ibid.* 1909, XXVII, pp. 101–106. 'Ueber die Bedeutung der Atmungspigmente in den Oxydationsprozessen der Pflanzen,' *ibid.* 1912, XXX, pp. 104–107. 'Die Atmungspigmente der Pflanzen,' *Zschr. physiol. Chem.*, Strassburg, 1908, LV, pp. 207–222.

[3] Onslow, M. W., 'Oxidising Enzymes. IV. The Distribution of Oxidising Enzymes among the Higher Plants,' *Biochem. J.*, Cambridge, 1921, XV, pp. 107–112.

The results of investigations made by the author[1], considered in conjunction with those obtained by other workers, led to the suggestion that the direct oxidase reaction is due to a system consisting of two enzymes, oxygenase and peroxidase, together with a substance containing the catechol grouping. The oxygenase catalyses the autoxidation of the catechol substance with formation of peroxide. The peroxidase acts on the peroxide, transferring 'active' oxygen to the acceptor. The formation of brown or reddish-brown pigments in extracts or tissues of those plants which contain direct oxidase may be regarded as the outcome of the oxidation of a mixture consisting of oxygenase, peroxidase and a number of aromatic substances of which one at least is capable of acting as a peroxide by autoxidation.

It is upon reactions of the kind just mentioned that Palladin's hypothesis of 'respiration pigments' is based. Palladin makes extracts from a number of plants throughout the vegetable kingdom, and after boiling to destroy any enzyme in the extracts themselves, adds peroxidase solution (obtained from Horse-radish root) and hydrogen peroxide. In all cases, red, reddish-brown, brown or purple pigments are produced in the extracts; these pigments, moreover, are formed most readily and in greatest quantity in extracts of plants we know to contain, previous to heating, a direct oxidase. From his results Palladin deduces the fact that chromogens of an aromatic nature are universally distributed, and may be oxidised in the presence of oxidising enzymes and a peroxide (though these, let it be noted, he always adds to the extract). The whole series —chromogen, enzyme, peroxide and pigment—form a system for transferring oxygen to respirable materials, and hence the term 'respiration pigments.' The chromogens, moreover, are considered by Palladin to be present in the living plant as prochromogens of the nature of glucosides, the hydrolysis and synthesis of which are controlled by glucoside-splitting enzymes, and chromogen is only produced as required for oxidation. After the death of the plant the hydrolysis of the glucosides is rapidly increased, and the chromogen becomes entirely oxidised:

prochromogen (glucoside) + water → chromogen + sugar
chromogen + oxygen → 'respiration pigment'

Palladin included anthocyanin[2] among the respiration pigments, and

[1] Onslow, M. W., 'Oxidising Enzymes. II. The Nature of the Enzymes associated with certain Direct Oxidising Systems in Plants,' *Biochem. J.*, Cambridge, 1920, XIV, pp. 535–540.

[2] It must be clearly understood, however, that anthocyanins, apart from the fact that they are aromatic substances, have very little in common with the respiration pigments. The latter are formed only after death (unless we believe, with Palladin, that they are

explained its appearance in leaves fed on sugar, in young leaves and autumnal leaves, as due to excess of carbohydrates: "Diese Tatsache kann in der Weise gedeutet werden, dass durch Zuckerzugabe die Atmungsenergie so gesteigert wird, dass ein Teil des oxydierten Chromogens nicht momentan wieder reduziert werden kann."

It was first suggested in 1909 by the present writer (227) that anthocyanins may be derived from the flavone or flavonol pigments[1]. The flavones are a group of natural colouring matters, some of which have been artificially synthesised by Kostanecki[2]. As a group they are widely distributed, and a number have been isolated by Perkin[3] from plants used commercially for dyeing. They may be regarded as oxy-derivatives of β-phenyl-benzo-γ-pyrone:

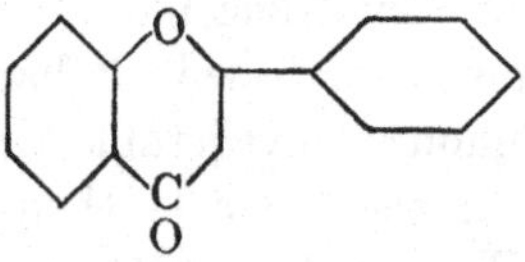

O
C
O

The flavones differ from each other in the number and position of their hydroxyl groups. They are all substances coloured yellow, and according to Witt the colour is due to the chromophore group:

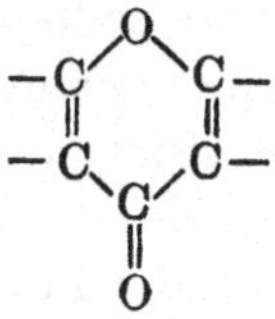

O
–C C–
‖ ‖
–C C–
C
‖
O

in combination with the auxochrome (hydroxyl), –OH, and the intensity of coloration is said to depend on the position of the hydroxyl groups.

The flavones, as a class, are yellow crystalline substances with high melting points (see accompanying table). They give either a deeper yellow, or an orange-yellow, coloration with alkalies, correspondingly coloured precipitates with lead acetate, and a green or brown coloration with iron salts. That their distribution in plants is practically universal can be readily demonstrated by the colour reaction with alkalies. This reaction is best shown by colourless parts of plants, such as white flowers.

reduced immediately in the living plant), the former only in living plants. There is no evidence at present that they play any direct part in respiration.

[1] In subsequent pages 'flavone' will be used to represent both flavone and flavonol pigments (see p. 111).

[2] See Abderhalden, E., *Biochemisches Handlexikon*, Berlin, 1911, Bd. VI.

[3] Perkin, A. G., various papers in *Chem. Soc. Trans.*, 1895–1904.

Table giving properties and characteristics of the commoner flavones and flavonols.

	Constitutional Formula	Melting point	Products of Decomposition from Alkali Melt	Distribution
		Flavonols.		
Quercetin	O OH HO OH OH C HO O	Sublimes above 250°	Phloroglucinol and protocatechuic acid	Free and as various glucosides in *Quercus* (bark), *Rhamnus* (berries), flowers of *Cheiranthus, Crataegus, Viola, Prunus, Hibiscus,* leaves of *Ailanthus, Rhus, Arctostaphylos, Calluna, Eucalyptus* and many others
Myricetin	OH O HO OH OH OH C HO O	357°	Phloroglucinol and gallic acid	*Myrica* (bark), leaves of *Rhus, Haematoxylon, Arctostaphylos*
Kaempferol	O HO OH OH C HO O	276°	Phloroglucinol and *p*-oxybenzoic acid	In flowers of *Prunus, Delphinium, Robinia,* leaves of *Polygonum, Indigofera*
Fisetin	OH O HO OH OH C O	Above 360°	Resorcinol and protocatechuic acid	*Rhus* (wood)
		Flavones.		
Chrysin	O HO C HO O	275°	Phloroglucinol and benzoic acid	*Populus* (buds)
Apigenin	O HO OH C HO O	347°	Phloroglucinol and *p*-oxybenzoic acid	Leaves of *Apium, Reseda*
Luteolin	OH O HO OH C HO O	327°	Phloroglucinol and protocatechuic acid	Leaves of *Reseda, Genista, Digitalis*

Placed in ammonia vapour, almost any white flower will turn bright yellow (with the exception of certain albinos of *Antirrhinum*, *Phlox Drummondii* and *Portulaca*, see pp. 150, 160, and 183). The same yellow reaction is given by green organs, though it can only be detected microscopically on account of the presence of chlorophyll. On reference to the table, p. 111, it will be seen that some flavones occur in genera of many natural orders, while others are limited to one or a few; this apparent limitation is probably only due to lack of knowledge of their occurrence in a number of plants. Further investigation will no doubt show a very much wider field of distribution for all the flavones.

There is little doubt that the flavones, as well as many other aromatic substances, are synthesised in the leaves from sugar. The actual steps of the process would be very difficult to demonstrate. On fusion with alkalies, the flavones split, as a rule, into phloroglucinol and an oxybenzoic acid. Conversely, they may be synthesised from oxybenzoic acids, or their derivatives, and phloroglucinol. As we know, somewhere in the plant and at some stage of plant metabolism, some aromatic nucleus must be synthesised, that is an aromatic substance such as phloroglucinol must be synthesised from an aliphatic substance such as glucose. Putting together what physiological and chemical evidence there is to hand, it seems most likely that the leaf is the organ where such a synthesis takes place.

The flavones, moreover, are usually present as glucosides in the plant, one or more hydroxyl groups being replaced by sugar; hence the crude alcohol or water extracts of many plants are only pale yellow (the auxochrome groups being replaced). On hydrolysis with dilute acids, the sugar is split off, and the colour deepens; at the same time a deposit of flavone is formed, as the free pigment is less soluble than the glucoside. According to Witt, the capacity for dyeing of the flavones depends on the presence of free hydroxyl groups. Thus, it comes about that aqueous or alcoholic extracts of most plants dye but slightly when boiled with mordanted cloth, but after hydrolysis of the glucoside with dilute acid, and neutralisation, the same extract dyes more deeply.

It was suggested by the author (241) that the flavones may, in many cases, be the chromogens from which anthocyanins are derived. The reactions involved would then be expressed in very general terms as follows:

$$\text{glucoside} + \text{water} \rightleftarrows \text{chromogen (flavone)} + \text{sugar}$$
$$x\,\text{(chromogen)} + \text{oxygen} \rightarrow \text{anthocyanin}$$

It was also suggested that the first reaction is controlled by a glucoside-splitting enzyme or enzymes, and the second reaction by an oxidising enzyme. Also that several of the hydroxyl groups of the flavones, as they actually exist in the cell-sap, are replaced by sugar. After hydrolysis of one or more, but not necessarily all, of these hydroxyl groups, oxidation of the flavone molecule, accompanied by condensation of either two flavone molecules, or of a flavone with other aromatic substances, may take place at these points. *Hence the final product—anthocyanin—would be itself a glucoside, and the reacting substances would at all times be in the glucosidal state.*

Exception has been taken by Everest (264) to the above hypothesis, which will in future be referred to as the glucoside hypothesis, and it is advisable to consider his criticisms at this point, since they do not affect the general evidence for the hypothesis to be considered later.

Everest makes the following statements which are to some extent the outcome of his investigations:

1. No known glucoside of a flavone has more than two hydroxyl groups replaced by sugar; most of the glucosides contain only one sugar molecule.
2. All anthocyanin pigments present in the natural state in plants are glucosides. No free anthocyanidins (non-glucosidal) have been detected (Willstätter & Everest, 260).
3. Artificial pigments identical with natural products can be prepared from flavones. From the flavone glucosides (except in the case of quercitrin, a glucoside of quercetin) anthocyanins are formed, and from the flavones themselves, anthocyanidins.
4. When anthocyanins are prepared artificially from flavone glucosides, no anthocyanidin stage is passed through. This may be shown by conducting the experiment under amyl alcohol, in which the anthocyanidins are soluble, should they be formed.
5. When flavone glucosides are hydrolysed, and the flavone converted artificially into anthocyanidins without removal of sugars, the latter do not again combine with anthocyanidins to form anthocyanins.

On the basis of the above statements Everest maintains that the glucoside hypothesis is untenable. For (1) precludes the possibility of the reacting substances being, as a rule, in the glucosidal state throughout the reaction; (2) makes it essential for the final product to be a glucoside, whereas (5) appears to make it impossible for such a recombination, i.e. between anthocyanidin and sugar, to take place. Finally (3) and (4) render hydrolysis unnecessary.

Let us now consider these points. As regards (1), it is by no means proved that flavones in the living plant have at most two hydroxyls replaced by sugar. It is quite conceivable that in the living cell more hydroxyls are replaced, and that only stable glucosides containing one or two molecules of sugar have been isolated, and this by virtue of their stability. No definite investigations have been made as to how many hydroxyls are replaced in the flavone in the plant, and in absence of further evidence, no conclusive statements can be made. With (2) the author is in agreement (Wheldale, 259). Again, it is not yet determined whether the artificial products mentioned in (3) and (4) are anthocyanins, and the matter is discussed later in the chapter. But, for the moment, grant them to be so. Then, with regard to statement (3), that it is not essential for hydrolysis to precede the formation of anthocyanin, it may be pointed out that nothing is known of the reactions giving rise to the artificial pigment; nor is there any reason for supposing the artificial and natural reactions to be the same. As regards (4), the glucoside hypothesis is still in agreement with a mode of formation which does not involve complete hydrolysis at any moment. Criticism (5) is of no value since it is well known that glucosides are not synthesised *in vitro* in the absence of an enzyme. Moreover, in the presence of an enzyme, synthesis only takes place to a small extent under special conditions which are certainly not realised in Everest's experiments. In this respect, the results obtained *in vitro* have no bearing upon the reactions taking place in the plant.

In the light of further research, the evidence in favour of the glucoside hypothesis which still holds good may be arranged under the following headings:

1. *Evidence from results obtained in the cross-breeding of* Antirrhinum. These results will be discussed in greater detail in Part II, but a short statement is necessary at this point, since the facts recorded are of value in this connection. The original type of *Antirrhinum* has magenta flowers, the colour being due to anthocyanin. During cultivation, two varieties, among others, have arisen as sports, viz. an ivory, and a white, both incapable of producing anthocyanin. Ivory, as the name suggests, is ivory-white in colour, and has on the palate a spot of yellow which is common to all varieties of *Antirrhinum*, except white. White is dead white, without the yellow spot on the palate. When a white[1] is crossed with an ivory, a plant having magenta flowers like the type is produced, and in these flowers, magenta anthocyanin is present in the epidermis of the corolla. Hence the original ivory and white varieties must between them contain the materials for the formation of anthocyanin. It has

[1] Of certain ancestry. See p. 163.

been shown by Wheldale & Bassett[1] that the pigment in the ivory variety is the flavone, apigenin:

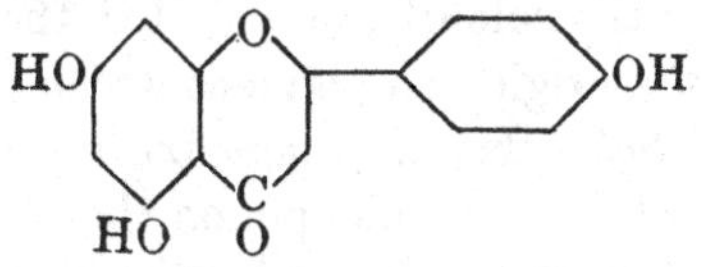

O
HO OH
C
HO O

which occurs in the plant as a glucoside.

No flavone is present in the white variety, but it must nevertheless contain a factor which, in some way, acts upon the chromogen, apigenin, with the production of magenta anthocyanin. The magenta flowers contain apigenin in the inner tissues of the corolla, and anthocyanin in the epidermis. This anthocyanin, isolated according to methods given elsewhere and purified from apigenin, gave on combustion the following percentages as compared with apigenin[2]:

	C	H	O
From anthocyanin	50·50 %	5·11 %	44·39 %
From apigenin	66·66 %	3·70 %	29·64 %

from which it will be seen that anthocyanin in *Antirrhinum* is a more highly oxidised substance than apigenin. Determinations of the molecular weight of this anthocyanin gave results of the order of 700 showing that the red pigment has a much larger molecule than apigenin, the molecular weight of the latter being 270. Hence it is possible that in the formation of anthocyanin, either two or three molecules of flavone condense with oxidation, or the flavone condenses with some other aromatic substance present in the plant.

2. *Evidence from analogous reactions.* It has been suggested above that sugar is first split off from certain hydroxyl groups of the flavones (which we know to occur as glucosides), and only then can changes, such as oxidation and condensation, take place at these points—the hydroxyl groups. Co-ordinated reactions of this kind are known to be common in plant metabolism, the most familiar case being that of indigo which may be represented:

$$C_{14}H_{17}O_6N \text{ (indican)} + H_2O = C_8H_7ON \text{ (indoxyl)} + C_6H_{12}O_6$$
$$2C_8H_7ON + O_2 = C_{16}H_{10}O_2N_2 \text{ (indigo)} + 2H_2O$$

[1] Wheldale, M., & Bassett, H. Ll., 'The Flower Pigments of *Antirrhinum majus*. II. The Pale Yellow or Ivory Pigment,' *Biochem. J.*, Cambridge, 1913, VII, pp. 441–444.

[2] It should be noted that in preparation, the glucosides of both the anthocyanin and apigenin are split up, and the pigments obtained for analysis free from sugar. Any sugar present in the molecule would naturally raise the percentage of oxygen to a considerable extent.

The first reaction is brought about by a glucoside-splitting enzyme, indimulsin, which hydrolyses the glucoside indican; the second by an oxidase which oxidises the colourless indoxyl to the pigment indigo.

The development of a bright red pigment which rapidly appears when flowers and leaves of *Schenckia blumenaviana* are placed in chloroform vapour has led Molisch[1] to form the opinion that the reactions involved are the hydrolysis of a glucoside and subsequent oxidation. Quite similar pigments have been described by Parkin, Bartlett, Tammes and others[2]. The reactions taking place in the formation of the respiration pigments of Palladin are probably of the same nature. An analogous case too is the development of blue pigment in the white flowers of an Orchid (*Phajus* sp.) on the death of the tissues. The change of colour has been used to determine the death-point of the protoplasm when subjected to freezing[3].

3. *Evidence from data as to the absorption of oxygen in gaseous exchange in red and green leaves.* By making analyses of gaseous exchange in red and green leaves respectively, Combes (466) has shown that the red absorb more oxygen during the reddening process than the normal leaves. Katić (441) also demonstrated that leaves kept in culture (sugar) solutions in the absence of oxygen failed, although they remained alive, to form any anthocyanin in contrast to similar leaves when the experiments were conducted in air. Molliard (463) also showed that oxygen is necessary for the formation of anthocyanin in radishes by totally submerging the roots in sugar solution, under which conditions the pigment failed to appear.

Viewing the above hypothesis in the light of more recent research, especially in connection with oxidases[4], evidence points to the fact that

[1] Molisch, H., 'Ueber ein neues, einen carminrothen Farbstoff erzeugendes Chromogen bei *Schenckia blumenaviana*,' *Ber. D. bot. Ges.*, Berlin, 1901, XIX, pp. 149–152.

[2] Parkin, J., 'On a brilliant Pigment appearing after Injury in species of *Jacobinia*,' *Ann. Bot.*, Oxford, 1905, XIX, pp. 167–168.

Bartlett, H. H., 'The Purpling Chromogen of a Hawaiian *Dioscorea*,' *U.S. Dept. of Agriculture, Bureau of Plant Industry, Bull.* 264, 1913.

Tammes, T., 'Dipsacan und Dipsacotin, ein neues Chromogen und ein neuer Farbstoff der Dipsaceae,' *Rec. Trav. Bot. Néerl.*, Nijmegen, 1908, V, pp. 51–90.

Palladin, W., 'Die Bildung roten Pigments an Wundstellen bei *Amaryllis vittata*,' *Ber. D. bot. Ges.*, Berlin, 1911, XXIX, pp. 132–137.

[3] Müller-Thurgau, H., 'Ueber das Gefrieren und Erfrieren der Pflanzen,' *Landw. Jahrb.*, Berlin, 1880, IX, pp. 133–189. Also:

Prillieux, Ed., 'Coloration en bleu des fleurs de quelques orchidées sous l'influence de la gelée,' *Bul. soc. bot.*, Paris, 1872, XIX, pp. 152–157.

Bommer, J. E., 'Étude sur le bleuissement des fleurs du *Phajus maculatus* Lindl.,' *Bull. soc. bot.*, Paris, 1873, XX, pp. xxvii–xxxiii.

[4] Onslow, *loc. cit.*

true oxidases play no direct part in the formation of anthocyanin. The oxidase system, though not universally distributed, is present in a large number of orders and genera; plants, however, without the system produce anthocyanin pigments equally well. The evidence, supplied by Mendelian researches on the white and ivory varieties of *Antirrhinum*, suggests that development of anthocyanin and flavone may depend on some common basal metabolism for production of aromatic substances, i.e. if flavone is not formed, neither is anthocyanin. The fact that the *Antirrhinum* anthocyanins isolated were non-crystalline casts doubts on the value of the analyses. Nevertheless, as will be seen later, apart from the intervention of direct oxidase action, the hypothesis still remains to be disproved.

That anthocyanin is produced by the action of an oxidase or peroxidase on a chromogen formed also the basis of the hypothesis of Keeble & Armstrong (245, 246). It would appear to be evident from the results obtained by these authors that there is an intimate relationship between the distribution of oxidising enzymes and the development of anthocyanin. There is reason to believe, on their evidence, that oxidation processes would readily take place in those tissues in which anthocyanin is found. These authors consider that the formation of anthocyanin is represented by the following equation:

$$\text{glucoside} + \text{enzyme} \rightarrow \text{sugar} + \text{chromogen}$$
$$\text{chromogen} + \text{enzyme} + \text{peroxide} \rightarrow \text{pigment}$$

Their evidence may be summed up under two headings: viz. evidence from (1) presence of oxidising enzyme, (2) presence of chromogen.

1. Presence of oxidising enzyme.

In order to demonstrate the existence of oxidising enzymes in the tissues, Keeble & Armstrong have adopted a microchemical method. The mode of procedure is to place the tissue to be examined in either a 1 % solution of benzidine in dilute alcohol, or in an equally dilute solution of α-naphthol and incubate at 37° C.[1] If no direct oxidase reaction results, the material is removed from the tube and washed with a dilute solution of hydrogen peroxide. The above method was employed on petals of *Primula sinensis*. It was found that α-naphthol gave a delicate lilac-blue colour with peroxidases, but only detected them in the bundle sheath of the veins, whereas benzidine gave a brown coloration

[1] Benzidine and α-naphthol are artificial acceptors. Whether the reaction obtained is direct or indirect depends on the presence of a natural peroxide suitable for acceptors used. Absence of direct action does not preclude a system peroxide-peroxidase in the plants.

and detected peroxidases both in the epidermis and in the bundle sheath. The results as regards *Primula sinensis* may be summed up as follows:

Flowers from all coloured varieties and recessive whites (see p. 186) gave a benzidine reaction in the epidermis and an α-naphthol reaction in the bundle sheath. These tissues being the chief seat of the pigment, it may be said that the distribution of enzyme and pigment in coloured varieties is practically coincident. Flowers of dominant white varieties gave no peroxidase reaction. In the case of a blue variety with inhibited white patches on the corolla segments, there was a more or less corresponding lack of peroxidase reaction in the inhibited patches. It was found that the inhibitor could be removed by treatment with hydrocyanic acid and other methods, and the above varieties then gave the usual benzidine and α-naphthol reactions. Flowers of certain flaked (magenta and white) varieties showed no peroxidase reaction in the white parts.

Other genera were also investigated by the same method. In Sweet William (*Dianthus barbatus*) flowers, it was found that the (in this case) oxidase reaction was entirely proportional to the amount of pigmentation. Of white varieties, some were found to contain oxidase; white parts of other varieties were found to contain no oxidase. In *Geranium sanguineum* the purple type contains epidermal oxidase, but the white variety gives no such reaction. In the Sweet Pea (*Lathyrus*) and Garden Pea (*Pisum*), no whites were found which did not contain peroxidase. Hence Keeble & Armstrong hold the view that in those cases where albinos give the oxidase or peroxidase reaction, albinism is due to lack of chromogen, but in the case of *Geranium sanguineum*, *Dianthus* and flaked *Primula*, lack of oxidase or peroxidase causes lack of pigmentation.

In the absence of isolation and analyses the assumption that albinism in *Primula*, *Lathyrus* and *Pisum* is due to loss of chromogen is unreasonable. In the author's experience the only instances of absence of chromogen (as determined by the flavone reaction) are the albinos of *Antirrhinum* and *Phlox Drummondii*. Flavones can be detected by the intense colour given with alkalies, and this reaction was given by all albinos of *Lathyrus* and *Pisum* examined. Nevertheless it must be admitted that this is no direct proof, since there may be several flavones from only one of which anthocyanin is derived.

The improbability of true oxidase being concerned with anthocyanin formation has already been dealt with (p. 117). That peroxidase also plays no part in anthocyanin formation is more difficult to prove. Keeble & Armstrong's results cannot, however, explain the case of

Antirrhinum, for, on their assumption, we must suppose the ivory variety to be an albino through lack of enzyme, since it contains the chromogen (apigenin). Yet all ivory individuals tested by the author have contained peroxidase. Granted that peroxidase is associated with pigment formation, a possible explanation for this universal presence of enzyme in ivories is that a peroxidase may be one of the many factors involved. Thus only individuals free from all factors (and these are extremely rare, as, for instance, 1 in 1024 when five factors are concerned) would show no enzyme action. Against this supposition is the fact that practically every living cell contains peroxidase.

Examining Keeble and Armstrong's results critically we find that their only evidence for associating anthocyanin formation with peroxidase activity is gained from *Primula sinensis* (for *Dianthus*, *Geranium*, etc., the data are too scanty). In this species, tissues which give a peroxidase reaction often, though not always, produce anthocyanin; in other tissues both anthocyanin formation and peroxidase reaction may be simultaneously inhibited.

Apart, then, from the case of *Antirrhinum* of which the flavone has been isolated, and there is good evidence that only one exists, little can be gained from speculations as to the presence or absence of chromogens and oxidising enzymes, without more definite isolation and analysis. Atkins (248, 261, 276), too, working with *Iris* flowers has found that there is no satisfactory evidence for the coincidence of anthocyanin formation with the presence of oxidases.

Keeble, Armstrong & Jones (254) have also maintained that the behaviour of anthocyanin in alcohol further confirms the view that the pigment is formed by oxidation from a colourless chromogen. According to these authors when coloured petals of stocks (*Matthiola*) are placed in absolute alcohol, some colour is extracted by the alcohol, but both petals and alcohol become colourless in the course of an hour or so. If now the colourless petals are removed from alcohol and placed in water, the colour returns, and more rapidly if the water be hot. Keeble, Armstrong & Jones explain these phenomena in the following way. Anthocyanin is formed from a colourless chromogen by oxidation, the agent being an oxidase which can only act in the presence of water; there is present also a reducing agent (which is not an enzyme), and this reduces anthocyanin to its colourless chromogen when the action of the oxidase is inhibited by absolute alcohol. The validity of this explanation has been questioned both by Wheldale & Bassett (271) and Tswett (268). The two former authors note that an extract of anthocyanin made in boiling absolute alcohol (which cannot according to Keeble & Armstrong contain oxidase) loses its colour on standing. The colour can be brought back by dilution with water. Hence obviously no oxidase is necessary. Moreover colour can be restored to the alcohol solution by dry hydriodic acid gas, a powerful reducing agent. Colour is also restored to petals, decolorised in alcohol, when they are placed in boiled water through which a stream of hydrogen has passed for some time and still continues to pass; a condition under which any leuco-compound should be stable. Similar criticisms to these are also offered by Tswett. The decolorisation is now known to be due to isomerisation (see p. 63).

We now turn to the second class of evidence. (2) Presence of chromogen.

Evidence for the presence or absence of chromogens is given by Jones (252). According to this author four classes of white flowers may be distinguished. (*a*) Those, such as *Lychnis coronaria*, *Anemone japonica*, *Chrysanthemum* sp., which produce a brown or brownish-red pigment when subjected to alcohol, chloroform, etc. Such flowers contain a direct oxidase; (*b*) those, such as varieties of *Dianthus Caryophyllus* and *D. barbatus*, which give similar pigments only on addition of hydrogen peroxide and contain peroxidase only; (*c*) those, such as white varieties of *Plumbago capensis* and *Swainsonia Tacsonia*, which do not produce a brown pigment even after addition of hydrogen peroxide but contain a peroxidase, and (*d*) those, such as varieties of Sweet William mentioned above under (1), which have no peroxidase and hence no oxidase. Reference to the account previously given of Palladin's 'respiration pigments' at once makes it clear that Jones's chromogens are identical with these pigments. Class (*a*) contains some substance which can act as an organic peroxide, class (*b*) no such substance, so that hydrogen peroxide must be added in order to give a pigment. Class (*c*) apparently contains no substance which can act as a respiration pigment. There is no evidence whatever that the chromogens of the respiration pigments are in any way the chromogens of anthocyanin, and hence their absence has no bearing on albinism with regard to anthocyanin (see footnote on p. 109).

More recently Combes (249, 250, 263) has brought forward evidence which he considers to be in complete contradiction to the oxidation hypothesis of the formation of anthocyanin. Combes's work may be summed up under three headings:

1. The isolation of two pigments from leaves of *Ampelopsis hederacea*: namely, a brownish-yellow pigment crystallising in rosettes of needles and having the properties of a flavone, and a purple pigment also

crystallising in rosettes of needles and having the properties of an anthocyanin.

2. The transformation of the flavone into a purple pigment identical with anthocyanin by reduction.

3. The transformation of anthocyanin into a flavone by oxidation.

The method employed by Combes of obtaining anthocyanin from flavone (from *Ampelopsis*) is to dissolve the flavone in alcohol, acidify with hydrochloric acid and add sodium amalgam. The solution thus treated with nascent hydrogen becomes violet-red and increases in intensity of coloration. After neutralisation and filtration, the solution obtained gives a purple substance on evaporation. The latter crystallises in needles grouped in rosettes like the natural anthocyanin and has the same melting point and properties as the natural product. Combes also obtained crystalline anthocyanin from a crystalline flavone isolated from leaves of the Privet (*Ligustrum vulgare*) and from a similar substance extracted from a variety of Vine which does not redden in autumn. From these results he concludes that the phenomena observed for *Ampelopsis* are not confined to that plant. And, moreover, though the variety of Vine (Chasselas doré) does not redden in autumn, yet its leaves produce a substance capable of being reduced to form an anthocyanin. In addition, he employed flowers of *Narcissus incomparabilis*. The genus *Narcissus*, as is well known, contains a readily crystallisable flavone, and from this substance he also prepared an anthocyanin-like product by reduction.

As a converse to the above experiments, Combes oxidised anthocyanin and obtained a flavone. The method employed consisted in taking *Ampelopsis* anthocyanin and purifying it by crystallisation, first from alcohol, and secondly from water. The pure product is dissolved in alcohol, and an equal volume of hydrogen peroxide is added. The purple colour of the anthocyanin gradually disappears, and a yellow solution is left which deposits crystals of a flavone identical in properties and melting point with the natural product derived from the plant. Hence he concludes that the anthocyanin has been oxidised with the formation of flavone.

A criticism of this last experiment and the deduction therefrom may be made at this point. It is not mentioned in Combes's paper whether the crystalline anthocyanin had been tested in order to ascertain if it were free from flavone. It is the experience of the author (Wheldale, 270) that when a mixture of flavone and anthocyanin obtained from *Antirrhinum* is allowed to crystallise from alcohol, both pigments crystallise out together in plates which, in spite of the fact that they may consist largely of flavone, are deep red in colour, and very careful purification is necessary before anthocyanin can be obtained entirely free from flavone. When purified anthocyanin from *Antirrhinum* is treated with hydrogen peroxide, the pigment is destroyed but no flavone is found. Hence it is possible that the flavone derived from Combes's anthocyanin may have been present as impurity. If, however, this were not the case, it still may be possible that the action of the hydrogen peroxide is not necessarily one of oxidation.

Before making any criticism upon the question of the action of sodium amalgam, the work of Keeble, Armstrong & Jones, Tswett and Everest on similar lines, must be considered.

A few months previous to Combes's publication, Keeble, Armstrong & Jones (255) had obtained red pigments, giving, in some cases, the anthocyanin reaction by treating alcoholic extracts of a number of plants with zinc dust and hydrochloric acid. Under these conditions, extracts from the following plants gave red pigments: pale yellow 'primrose' variety of *Cheiranthus*, yellow Daffodil, yellow *Crocus*, cream *Polyanthus*, and the dominant white variety, but not recessive white variety, of flowers of the Chinese Primrose (*Primula sinensis*). The red pigment which is first formed will, after continued treatment with nascent hydrogen, become colourless, the colour returning on exposure to air. This phenomenon led the authors to conclude that a preliminary reduction, followed by oxidation, is the sequence of events.

Tswett (258) also found that an 'artificial anthocyanin' could be prepared from apple juice by the action of strong mineral acids in the presence of formaldehyde or acetic aldehyde. The artificial anthocyanin had properties and reactions very like the natural pigment, but was soluble in ether, whereas natural anthocyanins are insoluble in this solvent. Tswett produced similar pigments, though he did not study them in detail, from extracts of white grapes, bananas, flesh of purple grapes, white petals of *Rosa* and *Cyclamen*. He failed however to get any formation of pigment on treatment of the following: leaves of white Cabbage, mesophyll of red Cabbage, leaves of *Pelargonium*, Orange skins, petals of white Pinks, white petals of buds of red Pinks, flowers of Lily of the Valley, Carrots, Potatoes, Kohlrabi and Barley seedlings.

Artificial anthocyanin resembles natural anthocyanin in the following respects. It is soluble in alcohol, gives a green reaction with alkalies, red with acids and a green precipitate with lead acetate; it is decolorised by sodium bisulphite, but the colour returns again on acidifying with sulphuric acid. It differs from the natural product in its solubility in ether and also its ready solubility in water, the natural pigments being only slightly soluble in water, except in the condition of glucosides.

The work of Everest (264, 265) in this direction has been more elaborate. Mention has been previously made of the fact that Everest has prepared from the flavones, substances which give some of the qualitative reactions of anthocyanin. Everest's methods are as follows: He reduces an alcoholic solution of flavones, to which is added 2 N acid solutions, by means of nascent hydrogen formed by (1) addition of granulated zinc, (2) sodium amalgam, (3) electrolysis, using sulphuric acid and lead electrodes, and (4) magnesium ribbon. The materials employed were quercetin, quercitrin and extracts from various flowers and tissues. In many cases a red pigment is obtained as reduction proceeds. When quercetin was employed, the product could be extracted quantitatively from the acid solution by amyl alcohol showing it to be an anthocyanidin. When quercitrin, a monoglucoside of quercetin, was used, anthocyanidin, contrary to expectation, was also obtained. Extracts, however, from flowers and other tissues, in which the flavones are presumably in the glucosidal state, gave anthocyanins in the cold and anthocyanidins on hydrolysis.

Everest also lays stress upon the fact first noted by Keeble, Armstrong & Jones (255), that the artificial anthocyanin is reduced to a colourless substance by vigorous reduction with nascent hydrogen. Such is also the case with the natural pigment (see p. 57). This similarity in behaviour, together with the solubility in amyl alcohol and the green reaction with alkalies, he considers to be a complete proof of the identity of the two pigments.

The author, in conjunction with Bassett (272), has prepared red products from both quercetin and apigenin. In the case of apigenin, the substance was analysed and found to be a reduction product, but its reaction with alkalies in no way resembled that of anthocyanin. On the other hand, there is the possibility that the anthocyanin of *Antirrhinum* may be derived from apigenin by oxidation and condensation. Hence, in the only instance where both artificial and natural anthocyanins have been obtained from a flavone and have been analysed, they appear to be by no means identical.

More recently Willstätter (274) has investigated the problem of the artificial formation of anthocyanin, and maintains that two substances are produced when quercetin is reduced with nascent hydrogen. In his experiments an alcoholic solution of quercetin acidified with hydrochloric acid is reduced with sodium amalgam or magnesium. A purplish-red product is obtained, and the bulk of this substance he terms allocyanidin; on the basis of analysis (by combustion) of the product he suggests the following constitution and reactions:

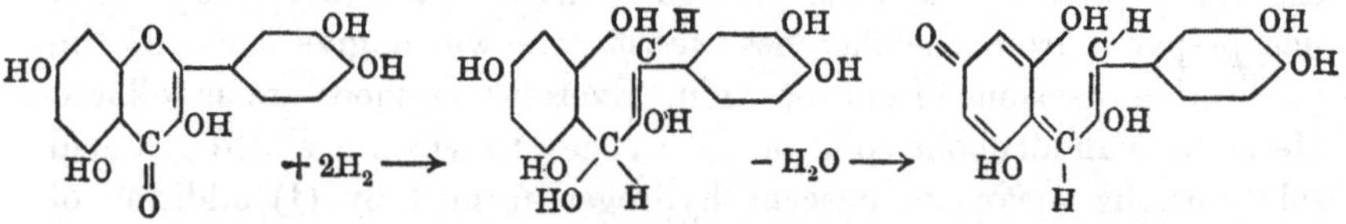

Quercetin $+ 2H_2 \longrightarrow$ $- H_2O \longrightarrow$ Allocyanidin

Allocyanidin forms a crystalline compound with hydrochloric acid, but is unstable on heating with the dilute acid. When, however, the crude solution obtained from the reduction of quercetin is diluted and heated, it does not entirely lose colour owing to the presence, in Willstätter's opinion, of a small amount of a second product. Isolated and analysed, this latter product was found to be identical with cyanidin chloride, and it is suggested that the reaction takes place as follows:

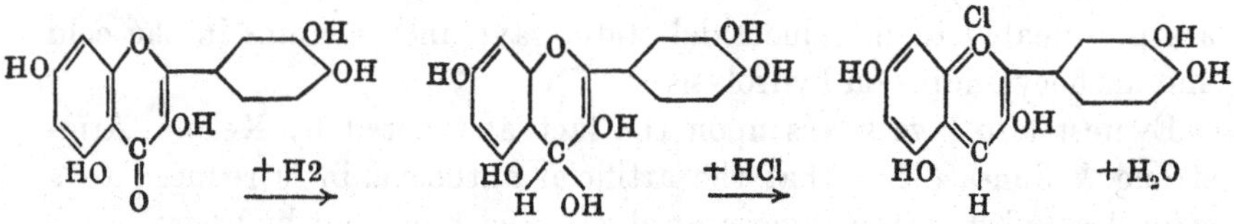

Quercetin $+ H2 \longrightarrow$ $+ HCl \longrightarrow$ Cyanidin chloride $+ H_2O$

From 33 gms. of quercetin, 0·165 gm. of the product was obtained. It seems to be open to criticism to assume that this product is formed by reduction; it is also doubtful whether the analysis (by combustion) forms an efficient proof of its identity with cyanidin, though its resemblance in chemical and physical properties certainly lends support to the statement.

Willstätter's preparation *in vitro* of cyanidin from quercetin, coupled with the fact that most of the flavones and flavonols will produce red pigments on reduction, led many investigators to the conclusion that *in the plant* flavone (or flavonol) glucosides are readily converted into anthocyanins by reduction and *vice versa*. This view has been exploited by Shibata (283), Shibata & Kishida (284), Shibata & Nagai (298), Shibata, Nagai & Kishida (299), Nagai (835) and Rosenheim (317) who extended the observations already made by other workers that crude extracts from many plants give reactions for flavones and, in such cases, also produce red pigments on reduction.

The evidence of anthocyanin formation *in the plant* by reduction is, however, slight, being chiefly that provided by Combes (249, 263), already mentioned. Combes prepared artificia anthocyanin by reduction of a flavone pigment from *Ampelopsis*; the artificial and natural products gave the same melting point. Conversely, the natural anthocyanin gave, on treatment with hydrogen peroxide, a flavone with the same melting point as the natural product. In no case, as already stated, was any product identified or analysed[1].

On the other hand, there exists evidence in opposition to this view. We must assume, if the above hypothesis be correct, that the anthocyanins, pelargonidin, cyanidin and delphinidin are likely to be accompanied in the plant by the corresponding flavonols from which they are formed by reduction, namely kaempferol, quercetin and myricetin. The only plants known where careful investigation has proved the identity of both flavone and anthocyanin present in the same plant are *Delphinium Consolida* and *Viola tricolor*. In the blue flowers of the former, A. G. Perkin was able to find only a glucoside of kaempferol; of the purple variety, Willstätter showed the anthocyanin to be a glucoside of delphinidin, not, as one would expect, of pelargonidin. Similarly, A. G. Perkin isolated only glucosides of quercetin from both the Pansy (*Viola tricolor*) and the Purple Violet (*V. odorata*), while Willstätter and Weil isolated from the blue-black Pansy, violanin, a glucoside of delphinidin —not of cyanidin. It is difficult, however, to draw conclusive evidence from the above, since it may be possible, though unlikely, that more than one flavone pigment is present and that one only is reduced to form anthocyanin.

Everest (314) has carried out an investigation designed to elucidate the above problem. He isolated the anthocyanin from Viola 'Black Knight,' and found it to be identical with Willstätter's violanin. He claims, too, by qualitative reactions to have shown that two flavones were present, one of which is myricetin (or, less likely, gossypetin). He did not however isolate either pigment, and the evidence is not by any means conclusive.

The hypothesis of anthocyanin formation by reduction has been further elaborated by Noack (315). This investigator maintains that in the shoots of *Polygonum compactum* there may be present an anthocyanin (glucoside), the pseudobase or colourless isomer of the corresponding anthocyanidin, and a yellow (presumably flavone) pigment. When the

[1] Later Combes (349) gives some figures for analyses. Anthocyanin, C 53·07: H 4·13: O 42·80. Flavone, C 52·42: H 4·06: O 43·52.

developing shoots lose colour, there is hydrolysis of the anthocyanin into anthocyanidin and sugar; the anthocyanidin then isomerises to its colourless pseudobase which is oxidised to a yellow pigment. The latter may be again reduced, and will give rise to anthocyanidin. The reduction of the yellow pigment is photochemical; in the dark, conversely, the anthocyanidin base is again oxidised. Noack's experimental evidence is based on the behaviour of the products withdrawn by amyl alcohol from acid extracts of the plant tissues. The anthocyanidin pseudobase and the yellow pigment are both soluble in amyl alcohol. When the extract of pseudobase in this solvent is heated with acid, a red pigment, said to be anthocyanidin, is formed and taken up by the alcohol. An amyl alcohol extract of the yellow pigment gives a slight red colour on reduction with nascent hydrogen (magnesium, zinc and acid) which is intensified on heating with acid; this is explained by formation of pseudobase on reduction, which, subsequently on heating with acid, gives anthocyanidin. Shoots which have been kept in darkness give no pseudobase, but only yellow pigment as a result of oxidation; shoots in light give both pseudobase and yellow pigment.

Combes (336) has criticised Noack's results, pointing out that the pseudobase is probably a phlobatannin which gives, characteristically, a red pigment (phlobaphene) on heating with acid. Combes has himself prepared the pseudobase from oenidin of grapes, and, also, the phlobatannin from the unripe fruit. Though the two substances have some properties in common (solubility in acidified water from which they are withdrawn by amyl alcohol; formation of red colour on heating with acid; change to green on adding alkali; precipitation on heating with hydrochloric acid and formol), they differ in other respects. The phlobatannin, for instance, is precipitated by potassium bichromate, alkaloids and bromine water and is coloured green by ferric chloride. The pseudobase has none of these properties.

It would appear, too, that Noack's yellow pigments are flavones which give, on reduction, the red pigments of questionable constitution previously mentioned (p. 123); it is also possible that the anthocyanidin pseudobases are phlobatannins which are dependent on light for formation. Absence, on the one hand, of free anthocyanidins, and, on the other hand, of the glucosides of the pseudobases from the *Polygonum* plant in which these interchanges are taking place, is remarkable, if Noack's hypothesis be correct.

In 1921 Jonesco (331) published results based on the following experiments. He made acid extracts of colourless flowers of *Cobaea scandens*, of Wheat seedlings and leaves of *Ampelopsis*. On withdrawal from acid solution by amyl alcohol, he obtained solutions of yellow pigments which gave a red colour on heating with sulphuric acid and manganese dioxide (or, in the case of *Ampelopsis*, with acid alone), but no red colour on reduction with sodium amalgam in presence of acid. Hence he concluded that the red pigments obtained by Combes were formed by the action of acid alone and not by reduction; and, consequently, that anthocyanins are formed from flavones by oxidation.

In criticism of Jonesco's results, Combes (337) states that no flavones are extracted by the strength of acid used by Jonesco, and that the substances extracted by that worker are phlobatannins which, on heating with acids, give red phlobaphenes. Combes states, moreover, that he has extracted the phlobatannin from *Ampelopsis* by the method previously used in the case of grapes (336) and that it gives the reactions attributed by Jonesco to flavones.

In 1922, Jonesco (341) published a further note. In this he maintains that the oxydase from the Basidiomycete, *Russula delica*, will oxidise a flavone extracted from young flowers of *Medicago falcata* with the formation of a purple pigment which he considers to be anthocyanin. He also gives a series of reactions by which he claims to prove that he is oxidising a flavone and not a tannin, both in this and the case of *Ampelopsis* aforementioned.

Closely following the above note is another by Jonesco (342) in reply to Combes's criticisms that phlobatannins are the source of red colour. Jonesco also states that Combes does not effect a separation of phlobatannins from the pseudobase, and that consequently he (Combes) is dealing with a mixture of these two substances. Jonesco proceeds to describe a method for separation of a flavone and a tannin from *Prunus Pissardi*, and, also, in addition, the pseudobase of the anthocyanin. Finally Combes (349) replies that the pigment in the case of *Medicago* has not been identified with anthocyanin; that his method for separating tannin from pseudobase in *Ampelopsis* has been misinterpreted by Jonesco; that, finally, the results obtained for *Prunus Pissardi* are erroneous since Jonesco assumes that tannins are soluble, and pseudobase insoluble in ether, whereas the contrary is the case.

Kozlowski (335) had also previously stated that from white beet-root a pale yellow substance can be obtained which will, on oxidation, with sulphuric acid and manganese dioxide (or with acid alone), produce a red pigment with reactions and spectrum similar to the natural pigment. Against this Combes (349) again brings the criticism of lack of analyses.

Finally, the subject has again been reopened by Mirande (344) who observed that the scales of the bulbs of *Lilium candidum* form anthocyanin on exposure to light. On microscopical examination it was found that the seat of pigment formation was, as a rule, the sub-epidermal tissue, and to this tissue, also, the oxidase reaction is confined.

The subject now awaits further contributions, accurate analyses being the form of investigation most needed.

CHAPTER VIII

THE SIGNIFICANCE OF ANTHOCYANINS

Biological significance.

THE function of anthocyanins in rendering the corolla, perianth, bracts and other parts of the inflorescence attractive to insects may be regarded as their biological significance; so also the function of making ripe fruits conspicuous to birds. Neither of these subjects is dealt with in the present volume, but references are given to the work of Knuth (527), Focke (524) and others, from which further information can be obtained.

Physiological significance.

Several physiological functions have been assigned to anthocyanins, but, broadly speaking, they fall into three classes. Though each hypothesis has been closely criticised by the various investigators of the problem, the final issue is far from being complete and satisfactory. The three main ideas are:

1. That of shielding the chloroplastids from too intense insolation; this is known as the 'light-screen' hypothesis. In slightly different forms it has been advocated by Kerner, Kny, Wiesner, Keeble and Ewart.

2. That of assisting the action of diastase by screening it from deleterious rays, and thereby facilitating the hydrolysis of starch and subsequently translocation. This view has been mainly supported by Pick and Koning & Heinsius.

3. That of absorbing certain light rays and converting them into heat. Though the conception of anthocyanin as a medium for raising the temperature of tissues was due to earlier writers, notably Comes and Kerner, yet it was Stahl who first regarded this property as a valuable asset in assisting transpiration; he also considered other processes of general metabolism, including translocation and fertilisation, to be furthered by the warming effect of the pigment.

The 'light-screen' hypothesis had no doubt a basis in the work of Pringsheim. About 1880 Pringsheim[1] published the results of various

[1] Pringsheim, N., 'Ueber Lichtwirkung und Chlorophyllfunction in der Pflanze,' *Jahrb. wiss. Bot.*, Leipzig, 1879–1881, XII; 1882, XIII. Also 'Pringsheim's Researches on Chlorophyll,' translated and condensed by Bayley Balfour, *Q. J. Microsc. Sci.*, London, 1882, XXII, pp. 76–112, 113–135.

experiments on carbon assimilation, and among these, the effect of light of different colours and intensities on chlorophyll. He found that white light of a high intensity decolorised chlorophyll, but in red light the effect was suppressed. In yellow, green, or blue light, on the other hand, it is easy to decolorise and kill the cells of many Algae, Characeae, Musci, Filices and Phanerogams. Thus in white light the result was attained in two or three minutes, in green and blue in five minutes, whereas in red light of the same intensity and after twice or four times as long exposure no changes took place. Hence he supposes the red rays to be photochemically inactive, or only very slightly active, on plant cells. That the injurious effects were not due to heat is shown by the fact that in green light the temperature did not rise sufficiently to be harmful to plant cells. For instance, even in red light the temperature of the water drop in which the tissue was placed was raised to over 45° C. within five minutes, and yet after 15–20 minutes' exposure there was neither decolorisation nor death of the cell. On the other hand, in green and blue light, cells were decolorised and killed in five minutes, although after 15–20 minutes' exposure the water scarcely reached 35–36°, a temperature quite harmless to the cells. The solutions used were for red light, iodine in carbon bisulphide, for yellow, potassium bichromate, for green, copper chloride, and for blue, ammoniacal copper sulphate.

It would seem, however, that Pringsheim may have overrated the destroying effect of sunlight on chlorophyll, for Reinke[1], in a paper published in 1883, mentions some results he obtained by exposing both the green and red (anthocyanin) parts of plants to light intensified by a lens, the heat rays being cut off by a screen of alum solution. He found that the chlorophyll in leaves of *Elodea* and *Impatiens* and the red pigment in the petals of *Papaver* and *Rosa* were not decolorised unless the light intensity became 800–1000 times greater than normal sunlight. When in light of only 200 times the intensity, there was no bleaching of *Elodea* even after two hours' exposure.

Kerner (498) appears to be one of the first, if not the first, to make suggestions as to the significance of anthocyanin, though in his *Pflanzenleben* (1885)[2], he by no means confines himself to one use of anthocyanin to explain its appearance in different places under various conditions. Thus, for instance, in stems and petioles, he says, it may be of use in

[1] Reinke, J., 'Untersuchungen über die Einwirkung des Lichtes auf die Sauerstoffausscheidung der Pflanzen,' *Bot. Ztg.*, Leipzig, 1883, XLI, pp. 697–707, 713–723, 732–738.

[2] The *Pflanzenleben* appears to have been issued in parts several years before the first edition in 1888.

keeping back light rays which would decompose the travelling materials. (This view may be original or may be taken from Pick, see p. 134.) In other cases it is produced temporarily for the same purpose when there is a transmission of metabolic substances on a large scale as in the seedlings of certain starchy seeds (polygonums, oraches, palms and grasses). Again, in spring, it appears in young leaves and shoots when supplies are travelling to them from their place of storage in the stem. In autumnal leaves, all materials of use to the plant are passing out for storage; hence the value of autumnal coloration. Anthocyanin developed on the under surfaces of leaves is, on the contrary, not protective but absorbs light and changes it into heat which is serviceable for growth, metabolism and translocation. When, in shrubs and herbs, anthocyanin appears on the under leaf-surfaces, this is only the case in the lowest leaves near the ground; the upper leaves remain green and so do not prevent light from passing through to the leaves beneath them. The development on the under surface of leaves of marsh plants and floating plants has similar uses. Thus, there is not only retention by anthocyanin of rays injurious to metabolism, but there is, at the same time, a useful transformation into heat. When, Kerner continues, the surrounding temperature is low, plants are often entirely reddened, as in some small annuals developing early in the spring (*Saxifraga, Hutchinsia, Androsace*); also in seedlings germinating at low temperatures. It is, for the same reason, common in the vegetative parts of many High Alpine plants. Of similar value is its distribution on the under surface of petals and perianth leaves of flowers which close by night.

Two further observations of interest are mentioned by Kerner. One, that plants which have leaves covered with a felt or wool of hairs rarely develop anthocyanin, since such leaves do not need so much protection from light.

The second observation is that made on *Satureja hortensis*, a plant growing wild in the Mediterranean region and in cultivation in gardens where it is known as the Summer Savory. In the shade it is green; exposed to the sun its stems and leaves are dark violet. Kerner sowed the seeds of this plant in his Alpine garden, 2195 metres above sea-level, in the Tyrol. There, as a result, it is supposed, of the intensity of the sun's rays, it produced anthocyanin in extraordinary abundance. Such an adaptation, as he points out, can only take place in plants which are able to form the pigment. Seeds of *Linum usitatissimum* were sown next to the Savory; the seedlings turned yellow and failed to survive owing to the fact, in Kerner's opinion, that they are unable to form anthocyanin to protect the chlorophyll and are also without any development of hairs.

In 1885 Reinke[1] published further results which, if accepted, make it difficult to uphold the 'light-screen' hypothesis. The problem he set out to investigate was the effect of the different parts of the spectrum on the destruction of chlorophyll, and he claims to have shown that the red rays have the maximum destructive effect, and that this power decreases in the other parts in the following order—orange, violet, yellow, blue, dark red and green: when this sequence is compared with the absorption spectrum of chlorophyll, it will be seen that the rays which are most absorbed by chlorophyll have the greatest destructive effect upon it, and those which are least absorbed do not have any effect upon it.

Further support was given by Hassack (493) in 1886 to the screen hypothesis, though his work is chiefly confined to the histological distribution of anthocyanin in leaves.

In 1887 a paper appeared by Engelmann (494) which has considerable bearing on the problem. By means of a microspectral photometer, Engelmann investigated the spectra of the pigments in red leaves, and found that the absorption of red pigment is, on the whole, complementary to that of chlorophyll. Hence we are confronted with this dilemma. Anthocyanin absorbs those rays which are not absorbed by chlorophyll; the rays least absorbed by chlorophyll are, according to Reinke, the least harmful to chlorophyll. Therefore anthocyanin absorbs the rays which are least, and not most, harmful to chlorophyll. How then can it be a protective screen? If we accept Reinke's results, a green screen would be the best protection as it would absorb the rays which are most absorbed by chlorophyll and are, at the same time, most injurious. If, however, we accept Pringsheim's results that the green portion of the spectrum is most harmful, then it is possible to consider the question as to whether anthocyanin is a screen or not.

The matter was again brought up by Kny (497) in 1892. To test the efficiency of anthocyanin as a light screen, Kny placed a solution of chlorophyll behind a parallel-walled glass vessel which was filled in one case with an extract of red, in another with an extract of white, Beetroot. Behind the red extract the chlorophyll retained its colour much longer than behind the white. This experiment is of little value since chlorophyll in a dull light (such as would be the case behind the red solution) invariably retains its colour much longer than in a bright light.

[1] Reinke, J., 'Die Zerstörung von Chlorophylllösungen durch das Licht und eine neue Methode zur Erzeugung des Normalspectrums,' *Bot. Ztg.*, Leipzig, 1885, XLIII, pp. 65–70, 81–89, 97–101, 113–117, 129–137.

Again, in 1894, Wiesner (500) supported the screen theory for the protection of chlorophyll, especially in young leaves.

In 1895 Keeble (503) published a paper of some significance on the hanging foliage of tropical trees. The characteristic pendent position and red coloration of young leaves of some of these trees have been mentioned in Chapter II. Keeble is of the opinion that, though the development of red pigment is not universal among them, yet it may serve as a protection against strong insolation, not only as a screen but also by protecting the leaf from too great heating. This view is based on an experiment carried out with red and green leaflets of *Amherstia nobilis*. These were laid side by side in the sun, and the temperature taken by thermometers placed upon the upper surfaces of the leaves; the thermometers were then placed under the leaves and the temperature again taken. His results and conclusions are expressed as follows: "...when young thin red and older tougher green leaves of *Amherstia* are exposed side by side to the direct rays of the sun, the temperature, as registered by the thermometer, is higher to the extent of 1° C. at the upper surface of the red leaves: conversely, of the temperatures registered at the lower surfaces of the leaves (i.e. behind them), that beneath the green is higher than that beneath the red, by a similar amount.

"Put in general terms, the surface-layers of the red leaf reflect more heat than those of the green leaf. The green leaf, conversely, absorbs more of the sun's thermal rays than the red. Now, the two sets of leaves differ in two respects: first, in colour; then in that the green has a thicker, more developed cuticle (and mesophyll) than the red. This might, therefore, be expected to oppose a more solid resistance to the heat-rays than the thin immature cuticle of the red leaf. That such is not the case must be regarded as due to the fact that the different colouring-matters have different powers of reflection and absorption of heat, and that this difference is of such a nature that the red colouring-matter more effectually cuts off heat-rays from the body of the leaf than does the green. That is to say, the red colouring-matter acts as a screen by which the thermal effects of the sun's rays are moderated." In conclusion Keeble says: "In addition, then, to its value as a screen, the red or reddish coloured saps of such trees as *Amherstia nobilis* have the capability of affording to the young leaf, if necessary, a protection against too great *heating* effects of the rays of a tropical sun."

Later, Keeble's conclusions have been adversely criticised by both Stahl (505) and Smith (520): the latter remarks, alluding to Keeble:

"From his somewhat rough experiments he drew the novel conclusion, which...is exactly the opposite of the conclusion logically to be drawn from his observations, that the red colour is a protection against too great heating up of the leaf."

In 1895 a long paper appeared by Ewart (501) on 'Assimilatory Inhibition in Plants.' Ewart maintains that leaves, especially of shade plants, when exposed for a long time to strong illumination, suffer from inhibition of photosynthetic power. He considers the development of anthocyanin, on exposure to light, to be an adaptation for protection against this light rigor, rather than a protection against the destruction of chlorophyll. We shall return later to further criticisms by Ewart.

We will now pass on to another suggestion as to the uses of anthocyanin, though it was the first in point of time. It is that made, in 1883, by Pick (490), who maintained that red light increases the rate of hydrolysis and translocation of starch, while it allows photosynthesis to go on unhindered. Hence its frequent occurrence in stems and petioles; also in young leaves where there is active metabolism and in autumnal leaves from which there is a migration of substances which may be useful for storage. Pick brings forward certain evidence, namely that in red leaves the palisade parenchyma contains less starch than the spongy, on account of the increased translocation under the influence of the red light. He, moreover, gives the results of an experiment with a view to proving the same point. Of the lobes of a large leaf of *Ricinus*, one was illuminated by light coming through orange glass, a second by light through ruby glass, a third by light passing through anthocyanin solution (of beet-root), while a fourth remained in bright insolation. All the lobes were illuminated thus for four hours. He then found that the insolated lobes had more starch in the palisade parenchyma, though there was a considerable amount in the spongy tissue too, whereas the lobes covered by red glass and anthocyanin solution had more starch in the spongy parenchyma; of the lobe covered by orange glass, there was nothing in particular to note. From these results Pick concludes that the lobes illuminated by ruby glass and anthocyanin solution had assimilated as much as the directly insolated lobe, but translocation had been more rapid.

Pick's conclusions were adversely criticised by Wortmann (492) who maintained that the observations on starch distribution in red leaves are valueless because, first, it was not noted whether *green* leaves behave differently under similar circumstances, and secondly, the effect of red light on diastase activity outside the cell should be investigated also. As regards the experiment on the *Ricinus* leaf, he points out that, according to Stahl, the palisade cells and chloroplastids are adapted to starch formation in bright light, while those of the spongy parenchyma are adapted to diffuse light. Hence the distribution of starch in the lobes exposed to red light may be explained on the ground that they receive less light than the directly insolated leaf. Since Wortmann's criticisms have not been refuted, the question still remains unsolved. Later, Ewart (506) also criticises Pick's deductions. Pick has stated at one point that young red leaves have little starch because the translocation is furthered, but as the leaves mature, and the pigment disappears, more starch develops. Ewart makes what appears to be a more natural suggestion, namely that photosynthetic activity increases as the leaf matures, and this is so even if the leaf should retain its red colour.

More recently Koning & Heinsius (516) have again reopened the question as to whether anthocyanin acts as a screen for diastatic activity. These authors quote the results of Brown & Morris to the effect that the diastase content of leaves decreases after a period of exposure to bright light, and that the enzyme is chiefly destroyed by the violet and ultra violet rays. Koning & Heinsius claim to have shown experimentally that the above-mentioned rays are absorbed, not only by a water extract of anthocyanin, but also by the pigment in the living leaves; also, by placing branches of *Quercus rubra* and other species under double-walled vessels filled with anthocyanin solution, that it is these same rays which cause the formation of anthocyanin. Finally, it was shown that red leaves always contain more diastase than green. From these results the authors conclude that anthocyanin has a protective action on diastase.

The third theory, which has been so largely supported by Stahl (505), is based on the power shown by anthocyanin of converting light rays into heat. Kerner, as previously mentioned, had the idea of this function of anthocyanin, but Stahl laid stress on a special aspect, viz. that of accelerating transpiration in red leaves.

Stahl, however, was by no means the first to suggest a connection between light absorption and transpiration. As early as 1879–80, Comes (487, 488, 489) published the results of work on the effect of light on transpiration. He used pot plants which were enclosed in a zinc case with a glass front, and the loss of water was determined by weighing. The plants were also exposed to differently coloured light obtained by using potassium bichromate, ammoniacal copper, and alcoholic chlorophyll solutions. The transpiration of various coloured corollas was observed, the absorption bands of the pigments being also ascertained. From his experiments Comes arrives at the following conclusions. Transpiration is affected by light as well as by physical conditions; other things being equal, a plant transpires more in the light than in the dark, and the effect of light is proportional to its intensity. Light favours transpiration only in so far as it is absorbed by the colouring matters, and thus, that organ transpires most which is most highly coloured. (This point was demonstrated by the coloured corollas; those of which the pigment showed absorption bands in the greatest number, breadth and intensity, transpired most strongly.) Moreover, the rays which are most favourable for transpiration in a coloured organ are those which are most absorbed by it. Hence the transpiration of an organ is slightest under the influence of light which is of the same colour as the organ itself, and strongest when under the rays of the complementary colour. We have here then in Comes's results the basis of Stahl's hypothesis, that the red pigment of leaves brings about additional absorption and heating of the leaf, and consequently greater transpiration.

In addition, Kny (497) claims to have shown, by means of a simple experiment, that red leaves do attain a higher temperature than green. Kny filled two parallel-sided vessels with green and red leaves respectively of the same plant, and then filled up the vessels with water. When these were exposed to light, the heat rays being cut off by alum solution, the temperature in the vessel containing red leaves rose higher than that in the one containing green.

Stahl's (505) experiments and his criticism of the work previously done in this direction are all included in his paper 'Ueber bunte Laubblätter.' An account is first given of his experimental work on the method of finding the respective temperatures of red and green leaves. The apparatus used was a thermo-junction with spathulate electrodes which could be buried in the substance of the leaf. The source of light was a gas burner, and sometimes a Leslie cube was used by which dark heat only was obtained. Owing to the size of the electrodes, thick and fleshy leaves were most suitable for use. One of the species employed was an epiphytic orchid, *Sarcanthus rostratus*. On taking the temperature of both green and red leaves, it was found that the red had a temperature of 1·5° above the green in one experiment, in another of 1·82°. Similar results were obtained with the red and green parts of one and the same leaf of *Sempervivum tectorum*. Though more difficult to manipulate, some thin-leaved plants were used with similar results; for example, a difference of temperature of 1·35° for *Begonia heracleifolia* var. *nigricans*, 0·22 for *Pelargonium peltatum* and 0·14 for *Tulipa Greigi*. A rise in temperature of red leaves over green was also obtained when the Leslie cube was used. In addition, other investigations were made using coco-butter instead of the thermopile. The melting point of the butter was raised by mixture with beeswax, and was then spread in a liquid condition as thinly as possible on the under surface of the leaf. When the butter was set, the upper leaf-surface was exposed either to the sun's rays or to the heat of the Leslie cube. The butter was found to melt more rapidly, and to a greater degree, on red leaves, or parts of leaves, than on green.

Stahl then proceeded to apply these results to cases which have been discussed by previous investigators. With regard to Kerner's experiments on *Linum* and *Satureja* in the Tyrol, which we have already noted, Stahl suggests that it is not from lack of protection of chlorophyll that the flax suffers, but from the low night temperature, against which it would be protected were it able to form anthocyanin with the resultant raising of temperature. Stahl proposes that the additional experiment should be conducted of covering up the *Linum* by night. If then the plant is still unable to survive, Kerner's view is more feasible. In general, also, with Alpine vegetation, Stahl is of the opinion that the red leaves and stems take a higher temperature than the green, and thus in a wider and more general sense than Pick he says "in dem wärmeabsorbierenden Blattrot besitzt die Pflanze ein Mittel, die Stoff- und Kraftwechselprocesse zu beschleunigen." Stahl suggests that at such low night temperature, *Linum* is unable to translocate starch and thus the plant becomes 'starch-sick,' and synthetic reactions cannot proceed; but he admits that the experiments require further attention. He has himself made observations upon cultures of *Linum* and *Satureja* at Pontresina and found, after a night with temperature about 0° C., in the morning the *Linum* leaves were still full of starch, whereas *Satureja* leaves were starch-free, although exposed during the previous day to intense sunlight. Stahl's view on this function of anthocyanin was also applied to autumnal leaves.

Again, the reddening of stigmas of anemophilous flowers is considered by Stahl to be another illustration of the value of the heating properties. As examples of trees and plants possessing such a characteristic, he quotes species of *Populus*, *Salix*, *Platanus*, *Ulmus*, *Ostrya*, *Carpinus*, *Corylus*, *Alnus*, *Acer*, *Fraxinus*, *Poterium sanguisorba* and *Rumex scutatus*, and of these, the trees, he points out, flower early in the year.

Hence the presence of anthocyanin is a special adaptation for raising the temperature and thereby furthering the growth of the pollen-tube.

Stahl next deals with his own suggestion as to a special significance of anthocyanin which it possesses by virtue of its temperature-raising properties. He maintains that anthocyanin is frequently found in leaves of water-plants inhabiting marshy places; also to a considerable extent in the leaves of shade-loving plants in damp tropical regions. The presence of anthocyanin in these leaves leads to a rise of temperature, and thereby accelerates transpiration which is rendered difficult by the conditions of such habitats. He quotes as examples the red under surfaces of leaves of *Nymphaea, Villarsia* and the frond of *Lemna*; of marsh plants, *Orchis maculata, O. latifolia, Ranunculus acris*; of shady wood plants, *Arum maculatum, Phyteuma spicatum, Hypochaeris maculata.* These are, however, insignificant as compared with tropical plants of which he quotes many examples from Borneo, Java and Mexico, of the orders Begoniaceae, Orchidaceae, Acanthaceae, Gesneriaceae, Marantaceae, Araceae and Melastomaceae; Stahl even goes so far as to say that in the case of leaves which develop blotches of anthocyanin, the lower leaves near the moist ground are more strongly marked than the upper leaves which may not be marked at all (*Polygonum Persicaria*). Also in Java he has noted that shaded pitchers of *Nepenthes* lying on the ground amongst ferns and mosses were deep red, while those borne high up above the vegetation had only a slight reddening.

One item of special interest he also notes, namely, that red blotches on leaves give off less water (tested by cobalt paper method) than green; this he found to be due to the fact that the red parts contained fewer stomata. Stahl, however, seems to think this compatible with his views, as too greatly increased transpiration might otherwise occur. Branches of the red-leaved Beech and Hazel he finds to transpire more strongly than branches of the ordinary species when both were placed in the shade, but in the sun and a dry atmosphere the opposite was the case.

As regards the coloration of young leaves, Stahl considers that reddening in temperate regions assists metabolic processes at low temperatures. In the tropics, on the other hand, the most intense colouring, he says, frequently occurs in the densest shade-forests, and is then an adaptation for promoting transpiration by heating up of the leaf; a view completely opposed to that held by Keeble (503).

Ewart's (506) work and criticisms are so much connected with both the screen hypothesis and that of Stahl, that it will be more convenient to consider the arguments in the order which he follows in his papers. He first concludes, from a number of observations and experiments, that strong insolation may bring about inhibition of photosynthesis, and hence too bright light may be injurious in this way rather than in the destruction of chlorophyll, and this is especially so in the case of shade plants. He is of the opinion, moreover, that anthocyanin does act as a protection in such cases, for he apparently trusts Pringsheim's observations, in preference to Reinke's, that concentrated blue or green sunlight will kill and bleach chlorophyll grains in five minutes, whereas concentrated red rays (the heat rays being eliminated) will only cause bleaching after twenty minutes. Anthocyanin absorbs 70–90 % of the green and 50 % of the blue, that is just those rays which are most harmful to chloroplast activity, while rejecting those useful for photosynthesis. If, according to Stahl's view, its function were that of a heat absorber it would, instead of absorbing the green and yellow rays (of which the heating effect is comparatively slight), show a marked absorption of the dark heat rays, and this is not the case. But he admits that Stahl has obtained a rise in temperature of one or two degrees in red leaves by exposure to dark heat rays. This higher temperature he considers to be a disadvantage, and since the red leaves might thus be liable to transpire too much, excess of transpiration is prevented by development of fewer stomata. Stahl has, as we have seen, shown that red leaves, or areas, have fewer stomata than green, and that green leaves also give off more vapour in sunlight and dry air than red ones.

The development of anthocyanin in young leaves and many tropical shade-loving plants is not considered by Ewart to be an adaptation for promoting transpiration. Its purpose, on the other hand, is to guard against assimilatory inhibition to which shade plants may be especially liable if accidentally exposed to intense light. He points out that in trees and shrubs with young red foliage, individuals growing in shady positions are less red than those in the sun, whereas, according to Stahl, it should be the reverse. Further, though it is true that certain plants growing in the shade form a considerable amount of pigment, yet of these, again, individuals growing in the more exposed positions have more colour than those in the deepest shade. Ewart infers that such plants are extremely sensitive to, and are injured by, light of marked intensity. He further points out that in many plants which have anthocyanin on the under surfaces, the ventral surface of the young leaves is exposed to light; in one variety of *Musa*, for instance, anthocyanin develops on the under surface when the young leaves are vertical and rolled up, but it disappears as the leaf unrolls and expands horizontally. In *Uncaria sclerophylla* the young leaves are so folded that the ventral surface is most exposed, and this develops anthocyanin which disappears as the leaf expands. Several other instances are also quoted, one of the most interesting being *Mimosa pudica*, in which red coloration is developed on the parts of the under surfaces of the leaflets which are exposed when folded. Ewart, however, does admit that, in the case of certain Begonias which have horizontally expanded leaves with red under surfaces, Stahl's view of the function of anthocyanin may be the correct one, and there are other cases where it is difficult, he says, to find any use at all for the pigment.

That the distribution of anthocyanin does not conform to Stahl's hypothesis is emphasised in the following passages from Ewart's paper. "If the primary function of the red dye in the tropics were to increase the amount of transpiration, then it would be only natural to expect that it would be formed in greatest abundance where the temperature is lower and the air more nearly saturated with water-vapour. The very opposite is however the case. Thus at the foot and sides of the volcanic mountain of Gedeh, and in the valleys around, very many plants have a reddish colour, especially in the young leaves. As one ascends this becomes less marked, until at Tjibodas and in the forests above it (4500 ft. to 6500 ft.) the number of plants showing a red colouration, and the intensity of the latter when present, reach a minimum. The vegetation at this elevation is almost entirely green, a few plants only, especially if growing in open clefts or glades in the forest, having more or less reddish young leaves. Yet it is just here where a power of stimulating transpiration is apparently most needed; for at this elevation the air is, during the greater part of the time, at or near saturation-point. On the other hand if the red pigment acts as a protection against sunlight, it is easy to understand why here, where the sun rarely shines for more than a few hours daily and then generally through a haze of clouds, the protective red pigment should almost entirely disappear; for it is just the more refrangible photochemical rays which the air saturated with water-vapour absorbs in greatest amount."

And again, "In Java at the commencement of the wet S.W. monsoon and in Ceylon at the rainy commencement of both monsoons, the vegetation acquires a more marked reddish tinge than the dry periods between the monsoons. This is, however, simply due to the fact that the young foliage, which in most tropical plants is more or less tinged with red, is very much more abundantly formed at this period than during the dry season. Even during the wet season in West Java, there is almost always bright sunlight until mid-day, lasting often till 3 or 4 p.m., and occasionally all day; so that the young foliage which the rain has caused to be produced in such abundance is exposed for six hours on the average to very bright illumination, the sunlight from 9–12 being the brightest of the day. Hence the protective red colouration is perhaps quite as necessary during the wet season as during the dry."

The question as to the significance of the non-development of pigment in the stomata in leaves with red epidermal cells is also considered by Ewart. Stahl's view is that the stomata by this means transpire less, and so are able to keep open longer for purposes of transpiration and photosynthesis. Ewart, on the contrary, regards the absence of anthocyanin as being due to the fact that the stomata are organs which react to light, and it is important that they should be exposed to the same intensity of light as that falling on the rest of the leaf even at the risk of injury.

Again, Stahl, as we mentioned previously, looks upon the development of anthocyanin in stigmas of anemophilous plants as an adaptation for increasing the temperature and aiding the growth of the pollen-grain. Ewart, however, notes that it has been shown that the pollen-tube growth is retarded by light and hence is protected by the pigment. Similarly, Stahl considers the reddening of young shoots in temperate regions in the spring to be protective against cold, whereas Ewart considers the function to be protective for the chloroplast, since the latter is more sensitive to light when the temperature is low. Reddening of submerged plants in sugar-cultures Ewart explains as being due to an unhealthy condition of the plant in which state it requires more protection against light.

Ewart concludes thus: "There can be little doubt that, both in the tropics and in temperate climes, the main and primary function of the red dye, when present in exposed parts, is to act as a protection against light of too great intensity; though in all cases its presence at the same time confers upon the plant a slightly increased power of absorbing heat. For calling attention to this latter possibility Stahl deserves full credit from both the physiologist and the biologist: in a few cases, such as in the horizontal leaves of shade plants having the red colouration present on the under surfaces only, the relatively slight heat-absorbing power of the dye may, by secondary adaptation, have become its most important function."

The latest work on the physiological significance of anthocyanin, and undoubtedly the most accurate, is that published by Smith (520) in 1909. This author determined the temperature of leaves in tropical insolation in Ceylon, using a thermo-electric apparatus. The latter was of the improved pattern which had been employed by Blackman & Matthaei, and had the advantage that it could be used even in the lamina of thin leaves and also for the internal temperature of leaves in natural illumination.

In the first of a series of experiments connected with this point young leaves of *Amherstia nobilis* and *Saraca indica* were compared. Both trees have young foliage of the pendent type we have previously described. The leaves of *Amherstia* are of a deep brownish-red colour due to the presence of anthocyanin, and after this pigment is removed by a solvent, there is seen to be but a slight development of chlorophyll. The leaves of *Saraca* are greenish-white and have, in the same way, a small development of chlorophyll. Both leaves are also thin and flaccid and hence form good objects for comparison. On exposing the leaves and testing the temperature, it was found that the coloured leaf of *Amherstia* reached a temperature of 2° C. higher than the leaf of *Saraca*. Another experiment was conducted with a young red leaf of *Mesua ferrea* as compared with a young leaf of *Saraca indica*. The red *Mesua* leaf registered a temperature of 2·8° C. higher than the *Saraca* leaf.

It was next found by comparing green and white leaves, that a green leaf also reaches a higher temperature than a white one. Hence it seems clear that any pigment, either chlorophyll or anthocyanin, raises the temperature of the leaf, and that the simultaneous presence of anthocyanin and chlorophyll in the leaf will raise the temperature considerably above that of a leaf without either pigment. The latter point was demonstrated in a comparison between a green and white leaf of *Caladium* sp. and a red and green leaf of the same genus, the electrodes being placed in the green and white leaf where there was little chlorophyll. The red and green leaf then showed a most striking rise of 3·9° C. above the green and white leaf temperature.

Attention was finally given to the differences between the temperature of young flaccid and coloured leaves as compared with older green leaves from the same tree. With *Saraca indica*, a mature green leaf showed a difference of 4° C. above a young colourless leaf. When, however, a young leaf of *Theobroma Cacao*, which is red, was compared with a mature green leaf, the young leaf showed a difference of 3·5° C. above the mature leaf, and this one would suppose to be due to the presence of the anthocyanin. On the other hand, comparisons of young red leaves with mature green leaves of *Amherstia nobilis* resulted in a decidedly higher temperature in the mature leaf.

"Thus," Smith concludes, "it seems, that we have a series beginning with *Saraca indica*, in which anthocyan is almost absent, and in which the mature leaf is always higher in temperature than the young leaf. Then comes *Amherstia nobilis* with a brownish-red colour, in which the mature leaf is, as a rule, only slightly higher in temperature than the young leaf. Lastly, we have *Theobroma Cacao* with the young leaf an intense pinkish-red and the mature leaf lower in temperature than the young leaf.

"No doubt the relative temperatures of mature and young leaves are to be correlated with the amount of anthocyan in the young leaf. The young leaves, without this pigment, would be always cooler than the mature leaf, as is the case in *Saraca indica*. The presence of more or less anthocyan produces a temperature in the young leaf which almost reaches (*Amherstia*) or exceeds (*Theobroma*) the temperature of the mature leaf. Thus the general tendency of these results is to confirm and extend Stahl's conclusion that the presence of anthocyan tends to raise the internal temperature of the leaf. What biological advantage, if any, is gained by the plant in this way is quite another question, but it is well to have this physical effect definitely established."

When we review the work on the subject of the uses of anthocyanin, we find it singularly unconvincing in any direction. In favour of the light-screen hypothesis, there is undoubtedly a greater development of anthocyanin in sun-exposed leaves than in leaves of the same plant in the shade; and, conversely, in red-leaved plants like the Blood Hazel and Copper Beech, a lack of colour in leaves which happen to be shaded. But, on the other hand, Kerner's experiments on *Satureja* and *Linum* are not very convincing, since so many factors might account for the unhealthiness of the Flax plants. And again, even if Pringsheim's red screen were protective to chlorophyll, it consisted of iodine in carbon disulphide which has a different absorption spectrum from anthocyanin. Moreover, Engelmann's analyses seem to prove that the anthocyanin only absorbs those rays which pass through chlorophyll, and an effective screen would be one with a similar absorption spectrum to chlorophyll itself. As regards Stahl's hypothesis, it is certainly supported by a definite piece of evidence, viz. that the temperature of red leaves is higher than that of green. Also, there are without doubt a number of shade plants which do possess anthocyanin to a considerable extent.

Yet it is difficult to find a hypothesis which would fit all cases of anthocyanin distribution without reduction to absurdity. The pigment is produced, of necessity, in tissues where the conditions are such that the chemical reactions leading to anthocyanin formation are bound to take place. For the time being we may safely say that it has not been satisfactorily determined in any one case whether its development is either an advantage or a disadvantage to the plant.

PART II

ANTHOCYANINS AND GENETICS

PART II
ANTHOCYANINS AND GENETICS

Classes of Variation.

As pointed out in Chapter II, practically all flowering plants produce anthocyanin; moreover, when this pigment is developed in the flower, we are able to see that each specific type forms anthocyanin of a certain characteristic colour. In nature to some extent, but under cultivation very commonly, colour-varieties arise from the type. The underlying cause of these variations is still a matter for conjecture. Colour-varieties have afforded plentiful material for Mendelian research, and it is to the cases of inheritance involving anthocyanin as a character that the following pages are devoted.

Among the colour-varieties of different genera and species there is a certain correspondence in the series of varieties produced, that is, we may find one series in a number of plants not necessarily related, and another series in a number of other plants, and so forth[1]. The kinds of variation which may occur we are able to classify as follows:

1. The loss of power to form anthocyanin pigments which results in albinism. The albino may be white or yellow; if yellow, the pigment may be either plastid or soluble.

2. The loss of power to produce blueness. The type has blue, or purple, anthocyanin: the variety has red anthocyanin.

3. The loss of power to produce redness. The type has red anthocyanin: the variety blue, or purple, anthocyanin.

4. The loss of power to augment the formation of anthocyanin and hence to intensify its colour. The type is fully coloured: the variety tinged only.

5. The loss of power to inhibit the formation of anthocyanin and hence to diminish its intensity. The type may be pale in colour and the variety deep: or the type may be tinged only and the variety fully coloured.

The following variations are independent of variation in the anthocyanin pigments, and may (with a certain exception in 9) take place simultaneously with variation in the anthocyanins.

[1] See Vavilov (861) on the law of homologous series in variation.

6. The loss of power to produce yellow pigment in the plastids. The type has yellow plastids: the variety colourless plastids.

7. The loss of power to inhibit the formation of pigment in the plastids. The type has colourless plastids: the variety yellow plastids.

8. The loss of power to inhibit the formation of yellow soluble pigments. There is no yellow in the type: the variety is yellow.

9. The loss of power to produce yellow soluble pigment. This variation is of rare occurrence. The type may contain both yellow soluble pigment and anthocyanin, and the variety may be without one or both: or the type may be without yellow colour, but may produce a soluble yellow variety, which, in turn, may lose its power of forming yellow pigment and is then also unable apparently to form anthocyanin.

It should be understood that the above is only a classification on the broadest basis; when we come to the details of variation it is almost necessary to consider each case separately. It is difficult, for various reasons, to make any kind of comprehensive classification, apart from the fact that no two cases are exactly alike; in many species, for instance, the inheritance of colour has not been systematically worked out, and relationships between variations cannot be correctly judged; new 'breaks,' moreover, are continually occurring in horticultural plants; there are also the complexities introduced by species-crossing, and in these circumstances it is often difficult to identify the type from which the variety has arisen. In the following paragraphs an attempt is made to indicate some of the main series or ranges of variation. It has only been possible to include a selection from the great mass of material provided by observations on plants under cultivation, but it is hoped that the lines suggested may provide a basis for further classification, as our knowledge of the inter-relationships of varieties increases.

A point to be emphasised is that one cannot judge correctly of the inter-relationships of varieties from appearances. For true knowledge it is not only necessary, as we have said, to find out the behaviour of the pigments in heredity, but we must examine their properties, and above all we should know their chemical composition. The colour series in *Dahlia* and *Tropaeolum*, for instance, are similar to the eye, but are really fundamentally different. Hence, in the following series, from lack of knowledge, instances may be grouped together which are not in reality of the same nature.

We may now consider the ranges of variation in greater detail. The simplest series is that which involves variation to albinism only, and, in the case of some genera grown as horticultural plants, this is the only important colour-variation known: the type produces anthocyanin, and the albino is without that pigment. As examples we may quote: *Datura Stramonium* (Thorn-apple), *Dictamnus Fraxinella* (Rue), *Geranium sanguineum*, *Epilobium angustifolium*, *Lavatera trimestris*, *Linaria Cymbalaria*, *Malope trifida*, *Malva moschata* and *Polemonium caeruleum* (Jacob's Ladder); in some of these species there may also be different intensities of type colour due to heterozygous forms (see p. 194).

If it should happen that the type produces both anthocyanin and plastid pigment, then loss of anthocyanin will not give a white variety but a yellow. Examples of such a case are the yellow varieties of *Abutilon* spp., *Fritillaria imperialis* and the variety *lutea* of *Atropa Belladonna*; in all these species the type is yellow suffused with anthocyanin.

Variation to redness, in addition to variation to albinism, is characteristic of another series. Variation to redness is a more complex phenomenon than albinism, and the series requires analysis. In the first place the type may be one of two kinds; it may be either some shade of magenta, purple, or purplish-red, or it may be blue. As examples of the first group we may suggest *Anemone Pulsatilla* which has a variety *rubra*; *Clarkia elegans* (Bateson, 627) which has a red (pink) variety, the type being magenta. *Linaria alpina* (Saunders, 691) and *Pisum sativum* (Garden Pea) (Lock, 621) have definite red varieties; *Salvia Horminum* (Saunders, 590) is violet in type with a red variety, and of *Viola odorata* there is a variety redder than the type. With the exception of the last, all the above mentioned also give true albinos. Examples of the second group (type blue) are *Centaurea Cyanus* (Cornflower) of which purplish-red and pink varieties are known; *Lobelia Erinus* and *Vinca minor* (Periwinkle) produce a purplish-red variety and *Myosotis sylvatica* (Forget-me-not) a pink; *Delphinium Ajacis* (Larkspur) and *Campanula medium* have both mauve and pink varieties; in *Aquilegia vulgaris* (Columbine) the type is violet-blue, and there are several purplish-red and pink varieties. This second group also varies to white, though it is doubtful in some cases whether there is complete albinism. In other species there is more than one grade of variation to redness. In the garden Stock (*Matthiola*) (Saunders, 590) it is believed that the type was some shade of purple; there is a crimson (blue-red) variety and also a true red 'Terra-cotta.' Of *Dianthus barbatus* (Sweet William) the type was doubtless of a magenta shade and there is variation to crimson and also to a true red, 'Scarlet.' Similarly *Primula sinensis* (Chinese Primrose) had in all probability a magenta type (Hill, 682) and varies to crimson and true red, 'Orange King.' In *Antirrhinum majus* (Wheldale, 638, 651) the type is magenta and the red variety, 'Rose Doré'; the case of *Antirrhinum* (Snapdragon) is further complicated by the existence of a yellow variety (see below). Variation to redness also occurs in *Cheiranthus* (Wall-flower), *Hyacinthus* and others, but these are dealt with later in connection with more complex series.

Another variation-series is that which includes a yellow variety. Colour in the yellow variety may be due either to plastid or soluble pigment, and the two series have very different characteristics. It should be emphasised that the soluble-yellow series is essentially different from the plastid-yellow series. In the latter, as we have already pointed out, the yellow is really the albino as regards anthocyanin pigment, whereas a soluble yellow variety is formed, as a rule, by the loss of a factor from an albino. In the case of the plastid-yellow series, further complexities, in addition to the simple loss of anthocyanin, are introduced by variation in the plastid colours, or by total loss of these pigments.

To deal first with the soluble-yellow series: *Antirrhinum majus* forms a typical example. The type is magenta; loss of anthocyanin gives an ivory-white variety. From the ivory a soluble-yellow variety is derived; mixture of yellow with the magenta anthocyanin of the type, i.e. simultaneous presence of both pigments, produces another variety, crimson. A red variety, 'Rose Doré,' has arisen from the magenta, and a mixture again of rose doré with yellow results in another variety, bronze. Finally there is a white variety which contains neither anthocyanin nor yellow pigment. Hence the series can be expressed: magenta, crimson, rose doré, bronze, ivory[1], yellow and white. Though we have no exact knowledge of the colour relationships in *Althaea rosea* (Hollyhock), *Dahlia variabilis* and *Dianthus Caryophyllus* (Carnation), yet these species in the main exhibit a similar series to *Antirrhinum*, for they produce magenta, crimson, yellow and ivory-white varieties; *D. Caryophyllus*, however, is characterised by many other shades. It cannot at present be stated whether *Dahlia* and *Althaea* have a red corresponding to rose doré. In many respects *Phlox Drummondii* shows the same series as *Antirrhinum*, though violet is included in addition. It is interesting to compare the series given by *Primula sinensis* (Gregory, 660) and *Dianthus barbatus* with *Antirrhinum*. In the two former there is a crimson similar in appearance to the crimson of *Antirrhinum*, but it is not due to mixture

[1] In the soluble-yellow series ivory or ivory-white is used for the albino, which is without anthocyanin, in contrast to the true white which is without both anthocyanin and soluble yellow pigment. At present the only two species in which both these varieties are known are *Antirrhinum majus* and *Phlox Drummondii*. I am indebted to Miss Killby for information about *Phlox*.

with yellow, since a soluble yellow variety is unknown in *Primula* and *D. barbatus*. But the 'Scarlet' of *D. barbatus* and the 'Orange King' of *P. sinensis* appear to be truly comparable to the rose doré of *Antirrhinum*.

The Marvel of Peru, *Mirabilis Jalapa* (Marryat, 636), produces a variety with soluble yellow pigment, and although this species must obviously be included in the present series, yet it is, in a sense, fundamentally different from *Antirrhinum*. From our knowledge of the colour-inheritance of *Mirabilis*, one is led to believe that the original type was crimson, and this apparently contains a mixture of magenta anthocyanin and soluble yellow pigment. Loss of anthocyanin gives a yellow variety, and further loss of yellow pigment, a white variety. Anthocyanin may also be present on a pale yellow ground, and then we have a magenta variety. Thus the series is an inversion, so to speak, of *Antirrhinum* and runs: crimson, magenta, yellow and white. It is further complicated by the existence of heterozygous forms, but these will be considered later (see p. 175).

Cheiranthus Cheiri (Wall-flower), on the other hand, is typical of the plastid-yellow series. The original wild type has deep yellow flowers tinged with brown, from which has arisen in cultivation the ordinary brown variety (see p. 154). The brown colour is due to the simultaneous presence of purple anthocyanin and deep yellow plastids. Loss of anthocyanin from the brown gives a yellow variety, which, strictly speaking, is the albino as regards anthocyanin. Some loss from, or change in, the deep yellow plastids results in a lemon-yellow variety. When the purple anthocyanin is present with pale yellow plastids, the latter are masked, and we get the purple variety now commonly grown. During recent years further varieties have appeared, some of which are practically cream in colour. The purple anthocyanin, too, has produced a red variation comparable to the red group previously mentioned. This red, on a background of lemon or pale yellow plastids, gives us the varieties 'Eastern Queen' and 'Ruby': on a background of deep yellow plastids, the new 'Scarlet.' Thus the *Cheiranthus* series runs: crimson (brown), purple, scarlet, ruby, yellow, lemon and cream. In main outline a similar series, brown or crimson, purple or magenta, deep yellow and pale yellow is shown by *Zinnia elegans*, and also by the new varieties of the Sunflower, *Helianthus annuus* (Cockerell, 709, 710, 722). Variation in the Garden Nasturtium (*Tropaeolum majus*) is on the same lines as in *Cheiranthus*, though it differs in one respect, namely that in *Cheiranthus* the type anthocyanin is purple and gives rise to a red variety, while in *Tropaeolum* the type anthocyanin is scarlet, or carmine red, and gives rise to a purple variety.

The colour representing the original type of *Tropaeolum* is the orange-red due to carmine anthocyanin on yellow plastids. A deep yellow variety is the albino after the loss of anthocyanin and, as in the case of *Cheiranthus*, it produces several pale yellow varieties approaching to cream as a result of the loss of some factor from the plastids. On this pale yellow ground the red anthocyanin appears as carmine or ruby. The red anthocyanin also varies to a blue-red or purple, which, on the deep yellow plastids, appears as maroon, and on the pale yellows as purple. There is additional variation caused by anthocyanin blotches at the base of the petals; these may be carmine or purple, and may be retained when the main part of the flower is free from anthocyanin; they may, however, be entirely lost, so that the flower is wholly yellow. Variation in *Salpiglossis sinuata* is also in all probability on similar lines to *Cheiranthus*.

We may next consider an additional range in the plastid-yellow series due to complete loss of colour from the plastid. In this way a white arises and it may exist as such, or anthocyanin may be present on the white ground. Typical of this series is *Chrysanthemum*; the type was in this case yellow, probably slightly tinged with anthocyanin. Loss of an inhibiting factor (see p. 154) would give rise to a crimson variety, i.e. purple anthocyanin on yellow plastids. Loss of yellow plastids, without loss of anthocyanin, gives purple, purplish-red or magenta; loss of anthocyanin gives either yellow or white. The series is then: crimson, magenta, yellow and white. *Helianthemum vulgare* (Rockrose), *Viola tricolor* (Pansy) and the garden Tulips probably come into the *Cheiranthus-Chrysanthemum* class, but our knowledge of these species is not very systematic. *Viola tricolor* possibly contains species-crosses. We can readily distinguish two groups of the garden Pansy: one includes white, reddish-purple, purple and purple-blue; the other yellow, brown and crimson. But we have no evidence as to how the groups are related.

We can thus differentiate two groups or series, soluble-yellow and plastid-yellow: in the former the type is generally magenta and gives the series crimson, red, orange, yellow and ivory-white: in the latter the type is, as a rule, crimson and gives the series magenta, orange, red, yellow, pale yellow and sometimes cream or white.

There is another series we may best consider at this point, and that is one in which a particular variety of plastid pigmentation occurs known as 'cream.' The plastids contain an orange-yellow pigment, but only in sufficient quantity to give the petals a cream appearance, and the variety is further characterised by being recessive to white containing colourless plastids. Such a 'cream' variety is found in the Sweet Pea, *Lathyrus odoratus* (Bateson, 627) and *Matthiola* (Saunders, 578); in *Matthiola* the series runs: purple, crimson, terra-cotta, white and cream. The cream may underlie any of the anthocyanin pigments, but it is too pale to affect the resultant colour, and such individuals are indistinguishable from those having anthocyanin on a white ground, except at the 'eye' of the corolla where the cream or white ground, as the case may be, is shown. In *Lathyrus* there is a very great range of colour, and the exact inter-relationships of many varieties are still unknown; the series includes blue, purple, mauve, crimson, pink, salmon, white and cream. In some varieties (Thoday, 650) the underlying cream modifies the effect of the anthocyanin. Cream plastids are also found in *Hyacinthus orientalis* in which they produce a fairly deep yellow; in this species the range of colour is also very great, since it includes several shades of blue, purple, magenta, pink, as well as white and cream, but, as there have been no systematic breeding experiments, their relationships to each other cannot be stated. It is not known whether the cream plastids in varieties of *Rosa* are of the same nature as those of *Lathyrus* and *Matthiola*.

Finally there is the series which includes a blue variety. Variation to redness when the type is blue or purple is, as we have seen, a common phenomenon, but variation to blueness is much less frequent. In *Lathyrus odoratus* varieties have appeared which are bluer than the type, 'Purple Invincible,' and the same is true for *Phlox Drummondii*. In *Primula sinensis*, true blue-flowered varieties occur, the type in all probability having had pale magenta flowers (Hill, 682); so also from the original *Cineraria* with flowers of a pale magenta, blue-flowered varieties have arisen, though probably under the influence of species-crossing. A striking instance of a blue variation from a scarlet is the variety, *coerulea*, of *Anagallis arvensis* (Scarlet Pimpernel)[1].

We may next consider those variations which affect the intensity of colour. If a colour series such as we have described for *Lathyrus* or *Antirrhinum* be examined, a number of definite varieties will be found of a paler or deeper shade of the same colour. Without Mendelian analyses it is not as a rule possible to arrive at the underlying significance of the shade differences. Thus, a paler shade than the type may signify either loss of a factor for full-colour, or the existence of a heterozygous form between type and albino. Hence, in dealing with this kind of variation, we can only quote as satisfactory instances those cases of which we have knowledge from experimental breeding. The main expressions of intensity-variation are tinged and deep varieties. The tinged variety

[1] Kajanus (684 states that blue is recessive to red in *Trifolium* (Clover).

forms a case in which a deepening, or full-colour factor, affecting anthocyanin formation has disappeared from the type, leaving the flower flushed or tinged with colour only. Examples are the 'tinged ivory' variety of *Antirrhinum* (Wheldale, 638) and the 'tinged white' and 'picotee' of *Lathyrus* (Bateson & Punnett, 603). The variation of tingeing is common to both red and blue classes.

The deep variety can be defined as one having a deeper shade of anthocyanin, i.e. more pigment, than the type from which it is derived, as the result of the loss of an inhibiting factor. Since it is often difficult, and sometimes impossible, to ascertain the original type in many horticultural plants, this definition cannot be rigidly applied. Like tingeing, the variation of deepening is common to both red and blue classes. Examples of such varieties are the deep shades of magenta, crimson, rose doré and bronze of *Antirrhinum* (Wheldale, 638, 651), purple-winged 'Purple Invincible' and 'Miss Hunt' varieties of *Lathyrus odoratus* (Bateson & Punnett, 603); deep shades of crimson and magenta in *Primula sinensis* (Gregory, 660) and deep purple and crimson in *Matthiola* (Saunders, 590); also deep varieties of *Cyclamen*. The following species, among many others, have varieties obviously deeper than the type: *Althaea rosea, Dahlia variabilis, Dianthus Caryophyllus, D. barbatus, Hyacinthus orientalis* and *Phlox Drummondii*.

In the instances quoted above, the type, from which the deeper variety is derived, is itself fully pigmented. There are, however, other cases in which the type is either unpigmented, or only slightly pigmented, and, on the loss of an inhibiting factor, a coloured variety is produced. As examples may be quoted *Anemone japonica* in which the type has a white petaloid calyx tinged with anthocyanin on the under surface, whereas the variety is purple-flowered; the wild *Bellis perennis* (Daisy) has white ray florets tinged with anthocyanin and the garden variety is red-flowered. *Cheiranthus Cheiri*, also, in the wild state is only tinged with brown (anthocyanin), and from it in cultivation have appeared varieties fully coloured with anthocyanin of deep shades of brown and purple. In *Crataegus Oxyacantha* (Hawthorn) the white-flowered type has produced a red-flowered variety. In *Cyclamen persicum* the type has white flowers with a basal magenta spot on the petals and has given rise to varieties with fully-coloured magenta and crimson flowers. In *Helianthemum vulgare, Primula acaulis* (Primrose) and *P. veris* (Cowslip), the types are yellow and the varieties both crimson and magenta. Further examples are *Freesia* and *Achillea Millefolium* (de Vries, 601) with coloured flowers. De Vries (601) also notes this phenomenon. He says:

"a pink-flowered variety of the 'Silverchain' or 'Bastard-Acacia' (*Robinia Pseud-Acacia*) is not rarely cultivated. The 'Crown' variety of rice, oats and barley are also to be considered as positive color-variations, the black being due in the latter cases to a very great amount of the red pigment." And again: "The best known instance is that of the ever-flowering begonia, *Begonia semperflorens*, which has green leaves and white flowers, but which has produced garden varieties with a brown foliage and pink flowers." Cockerell (709, 710, 722) has also drawn attention to a red-flowered variety of *Helianthus* of which the type is yellow. Again, the red colour may be developed in the carpels as in the varieties of *Primula* with red stigma (Gregory, 660) and the purple-podded forms of *Pisum* and *Phaseolus*. The variety appears in other fruits too (the 'Blood Orange' and the red Banana); also in the Gooseberry, *Ribes Grossularia* (de Vries, 601).

Another phenomenon which sometimes appears in variation is that which may be termed partial albinism; the type has coloured flowers but the variety has white flowers, though all the vegetative organs may still form anthocyanin. This is the case in the blue *Vinca minor* with its white-flowered variety of which the corolla is slightly tinged with purple. Of *Geranium Robertianum* (Herb-Robert), also, there is a white-flowered variety with red stems and leaves. A similar variation is sometimes found in the native flora, i.e. *Polygala vulgaris*, *Jasione montana*, and others, but it is not really possible, in the absence of Mendelian evidence, to distinguish these varieties from the tinged varieties already mentioned. Partial albinos, moreover, are similar in appearance to types in which anthocyanin is inhibited; but an important distinction lies in the fact that partial albinos cannot give rise to coloured varieties though inhibited types may.

A case difficult to place is that of the dominant or inhibited white of *Primula sinensis* (Gregory, 660). This variety has apparently the power to form pigment in the flower, but colour is prevented from appearing by an inhibiting substance. Hence the case most closely resembles the inhibited types of *Anemone*, *Crataegus*, *Primula veris*, etc., mentioned above, and must, for the present, be included in this class.

Up to this point we have been solely concerned with variation in flower-colour. But it should be borne in mind that the colour-varieties already described may be characteristic of the fruit and seed and also of the vegetative organs; in fact, the whole plant may either be a total albino, or pigment may be lost from various organs or parts independently of each other. These phenomena will be dealt with later in greater detail (see p. 190). A few instances of colour-variation in different organs may however be mentioned. Of albinism in fruits we have the white-fruited varieties of *Atropa Belladonna*, *Daphne Mezereum*, *Fragaria vesca*, *Ribes rubrum*, *Rubus Idaeus*, *Solanum nigrum*, *Vitis vinifera* and many others. De Vries (669) also notes a red-berried variety of *Empetrum nigrum*. In seeds there are white-seeded varieties of *Pisum* and *Phaseolus*. Of coloured roots also, white varieties may occur as in *Beta vulgaris*. Variation to redness may also affect the vegetative organs, as for instance in the red-flowered varieties, rose doré of *Antirrhinum* and 'Orange King' of *Primula sinensis*, in which the leaves, stems and petioles develop red anthocyanin instead of purple. In the tubers, moreover, of the Potato, *Solanum tuberosum* (Salaman, 647), one may have both red and purple varieties (pigment due to anthocyanin) as well as dominant and recessive white tubers.

Loss of an inhibiting factor, which in the case of the flower gives a coloured from a white variety, is quite common in the vegetative organs. This coloration of vegetative organs may be connected with deepening of flower-colour, for example, deep varieties of *Antirrhinum* and *Primula sinensis* which have pigment in leaves and petioles. Or it may occur more obviously as a red-leaved variety, as in the following: *Atriplex hortensis*, *Berberis vulgaris*, *Beta vulgaris*, *Brassica oleracea*, *Canna indica*, *Corylus Avellana*, *Fagus sylvatica*, *Lactuca sativa*, *Perilla nankinensis*, *Plantago major*; de Vries (601) also notes *Tetragonia expansa* and the brown-leaved *Trifolium*[1]. Lastly a very interesting case of the occurrence of a red variety in roots is recorded by Wittmack (598), that is the variety of the Carrot, *Daucus Carota Boissieri Schweinfurth*, which has anthocyanin in the root, in addition to the orange plastid pigment (carotin).

As to variation in nature, that of albinism is most frequent. It is found in the wild state among the British flora in *Armeria vulgaris*, *Calluna vulgaris*, *Campanula rotundifolia*, *Carduus nutans*, *C. palustris*, *Digitalis purpurea*, *Erica cinerea*, *Lamium purpureum*, *Lychnis Flos-cuculi*, *Ononis arvensis*, *Pedicularis sylvatica*, *Symphytum officinale* and *Scilla nutans*. Bentham & Hooker[2] also mention varieties of *Ajuga*

[1] With regard to red-leaved varieties, de Vries (601, 669) notes that they may produce green-leaved branches (red *Corylus*, *Fagus*, *Betula*). Also red bananas have produced a green variety with yellow fruits. Whether this phenomenon is bud variation, or whether it may be in some instances due to the production of the true albino and not the inhibited type, is uncertain. An analogous example in flower-coloration would be *Cheiranthus Cheiri*; the type is yellow tinged with anthocyanin: this varies to deep brown, and this again gives a yellow (the albino) free from anthocyanin.

[2] *Handbook of the British Flora*, London, 1896.

reptans, Campanula latifolia, Centranthus ruber, Delphinium Ajacis, Iris foetidissima and *Malva moschata* without anthocyanin in the flowers, but it cannot be deduced from the text in each of the above cases whether there is complete absence of the pigment from the plant, or loss from the flower only. As regards variation to redness, among native species *Polygala vulgaris* has a red-flowered variety. Bentham & Hooker[1] record in addition red-flowered varieties of *Delphinium Ajacis, Veronica spicata, V. officinalis, V. Beccabunga* and *V. Chamaedrys*. Of species in which the flowers are normally white though they may produce red varieties, or red pigmentation under certain conditions, Gillot (560) gives a considerable list. Guinier (559) also notes a variety of strawberry (*Fragaria*) with purple flowers, and Chabert (564) records red flowers of certain species of *Galium* which are normally white-flowered. A variation, extremely rare in nature, if not altogether unknown, is that of white from a type with plastid yellow pigment, though as a species differential character it is quite common, for instance in the Ranunculaceae and Compositae.

Details of Cases of Mendelian Inheritance in Colour-varieties.

Many of the varieties mentioned in the last section have provided material for cross-breeding work on Mendelian lines. Since most of the species investigated differ more or less in their colour series and behaviour, a separate account will be given of the more important cases in turn. The genera and species which have been employed in these investigations are the following:

Agrostemma Githago (Corn Cockle), de Vries (577, 601).
Amaranthus caudatus, de Vries (601).
Anagallis arvensis (Scarlet Pimpernel), Heribert-Nilsson (681), Weiss (670).
Andropogon sorghum (Juar Plant), Graham (742).
Antirrhinum majus (Snapdragon), Bateson (583), Baur (620, 639), Wheldale (617, 638, 651), de Vries (669).
Aquilegia vulgaris (Columbine), Kristofferson (850).
Arum maculatum, Colgan (655).
Aster Tripolium, de Vries (577, 601).
Atropa Belladonna (Deadly Nightshade), Saunders (578).
Beta (Beet-root), Kajanus (662, 685, 704, 759), Lindhard & Iversen (794), Rasmuson (797, 874), Roemer (766).
Brassica (Cabbage, Turnip), Hallqvist (731), Kajanus (686, 704, 758), Kristofferson (832).
Campanula carpatica, Pellew (764).
C. medium, Lathouwers (851).

[1] *Loc. cit.*

Canavalia ensiformis, Lock (607).
Canna indica, Honing (715, 734).
Cattleya, Hurst (605, 634, 643, 702).
Clarkia, Bateson & Punnett (627), Rasmuson (814), de Vries (601).
Collinsia bicolor, Rasmuson (813).
Corchorus capsularis (Jute), Finlow & Burkill (677).
Coreopsis tinctoria, de Vries (577).
Cypripedium (*Paphiopedilum*), Hurst (605, 634, 643, 702).
Datura Stramonium (Thorn-apple), Saunders (578, 590), de Vries (577).
Dendrobium, Hurst (702).
Digitalis purpurea (Foxglove), Jones (683), Keeble, Pellew & Jones (645), Miyake & Imai (795, 811), Saunders (666).
Dolichos lablab (Bonavist or Hyacinth Bean), Harland (807).
Fragaria vesca (Strawberry), Richardson (719, 780, 815).
Geum (Avens), de Vries (601).
Godetia, Rasmuson (796, 836).
Gossypium (Cotton), Balls (618, 626), Fletcher (611), Martin Leake (664).
Helianthus (Sunflower), Cockerell (709, 710, 722, 724, 753, 772), Shull (623).
Hepatica (Anemone), Hildebrand (576).
Hieracium (Hawkweed), Ostenfeld (608, 646).
Hordeum (Barley), Biffen (604).
Hyoscyamus (Henbane), Correns (593), de Vries (577).
Impatiens (Balsam), Rasmuson (737, 814).
Ipomoea (Morning Glory), Barker (752), Hagiwara (864), Miyazawa (777, 834).
Lactuca (Lettuce), Dahlgren (773).
Lathyrus odoratus (Sweet Pea), Bateson (602), Bateson & Punnett (590, 599, 603, 619), Thoday (650), Punnett (765, 854, 873).
Linaria alpina, Saunders (691).
Linum usitatissimum (Flax), Fruwirth (846), Tammes (668, 721, 739, 748, 749, 858, 875), Vargas Eyre & Smith (751).
Lupinus (Lupin), Burlingame (824), Hallqvist (828), Kajanus (717).
Lychnis dioica, Correns (588), Saunders (578), Shull (623, 649, 693), de Vries (577).
Malope trifida, Rasmuson (745).
Matthiola (Stock), Correns (575), Saunders (578, 590, 599, 603, 609, 665), Tschermak (597, 695).
Mirabilis Jalapa (Marvel of Peru), Baur (620), Correns (585, 600, 640), Marryat (636).
Nicotiana (Tobacco), Allard (785), Clausen & Goodspeed (825), Haig Thomas (700), Lock (635), Sachs-Skalińska (840), Setchell, Goodspeed & Clausen (856).
Oenothera (Evening Primrose), Cockerell (723), Davis (697, 711), Gates (642, 658, 678, 713, 727, 728), Heribert-Nilsson (680, 732, 733), Renner (838), Shull (720).
Oryza sativa (Rice), Hector (743), Parnell Ayyangar & Ramiah (763), Yamaguchi (843).
Oxalis, Nohara (736).
Papaver Rhoeas (Field Poppy), Becker (769), Kajanus (790), Rasmuson (812), Shull (693).
Papaver somniferum (Opium Poppy), Hurst (605), Kajanus (791), Martin Leake & Ram Pershad (809, 852), Miyake & Imai (872), Przyborowski (853), de Vries (577, 601).

Petunia, Malinowski & Sachsowa (744), Rasmuson (779), Sachs-Skalińska (839).

Phaseolus multiflorus (Scarlet Runner) and *P. vulgaris* (French Bean), Emerson (586, 595, 631, 632), Kajanus (716), Kristofferson (879), Lock (607), Lundberg & Aakerman (762), Shaw & Norton (782), Shull (616, 624), Sirks (817, 857), Tjebbes & Kooiman (801, 841, 860), Tschermak (582, 597, 695).

Phlox, Gilbert (729), Kelly (808).

Phyteuma (Rampion), Correns (589).

Pisum sativum (Edible Pea), Bach (786), Bateson (590), Hammarlund (865), Kajanus (867, 878), Kajanus & Berg (792), Kappert (869), Lock (615, 621), Meunissier (810), Tedin (819, 876), Tschermak (582, 597, 695), Vilmorin (842), White (767, 768).

Polemonium (Jacob's Ladder), Correns (589), Dahlgren (773), de Vries (601).

Portulaca grandiflora, Ikeno (830), Yasui (820).

Primula sinensis (Chinese Primula), Gregory (660, 714), Gregory, de Winton & Bateson (863), Keeble (687), Keeble & Pellew (644).

Raphanus (Radish), Riolle (816).

Reseda (Mignonette), Compton (674).

Ricinus (Castor-oil Plant), Harland (806, 847), White (783, 784).

Rudbeckia, Blakeslee (823).

Salvia Horminum, Saunders (590).

Senecio vulgaris (Groundsel), Trow (694).

Silene Armeria (Lobel's Catchfly), de Vries (577, 601).

Sisyrinchium angustifolium, Miyake & Imai (833).

Solanum nigrum, de Vries (577).

S. tuberosum (Potato), East (641), Salaman (647, 690).

Trifolium (Clover), Gmelin (730), Kajanus (684), Raum (798, 837).

Triticum (Wheat), St Clair Caporn (781).

Tropaeolum (Nasturtium), Rasmuson (778, 814), Weiss (637).

Veronica longifolia, de Vries (577, 601).

Vicia Faba (Broad Bean), Heuser (866).

Vigna sinensis (Cow Pea), Harland (788, 805, 848), Mann (718), Spillman (667, 708).

Viola, Kristofferson (870), de Vries (577, 601).

Vitis vinifera (Vine), Rasmuson (746), Stuckey (800).

Zea Mays (Maize), Anderson (821), Anderson & Emerson (862), Bregger (770), Burtt-Davy (672), Collins (673, 787), Correns (579), Coulter (803), East (675), East & Hayes (656), Emerson (657, 676, 699, 712, 725, 755, 774, 826, 827), Fugii & Kuwada (741), Hayes (757), Hutchison (829, 849), Jones & Gallastegui (789), Kempton (760, 793, 831), Lindstrom (761, 871), Lock (596, 607, 688), Remy (855).

Agrostemma Githago. Pigmented type dominant to albino. De Vries (577, 601).

Amaranthus caudatus. Red-leaved type dominant to green-leaved variety. De Vries (601).

Anagallis arvensis. Weiss (670) crossed the scarlet-flowered type with the blue-flowered variety and obtained the red type in F_1. In F_2 there was segregation into red and blue in the proportion 3:1. Hence the blue arises from the red by the loss of a single dominant factor.

Heribert-Nilsson (681) crossed a variety which had pink, almost white, flowers, which bred true, with the red type. F_1 was red like the type. In F_2 there was segregation into red and pink in the proportion 3:1.

Andropogon sorghum. Graham (742) found red grain (anthocyanin in pericarp) dominant to both yellow and white. Red-grained plants have anthocyanin also in leaves. Colour depends on presence of two factors which segregate normally.

Antirrhinum majus. De Vries (669) made the first experiments on this species using the following varieties:

Rot ....tube and lips of the corolla red, the lips deeper.

Fleischfarbig......tube and lips pale red.

Delila ....tube pale or white, lips fairly deep red.

Weiss...............white, often with a distinct very pale red tinge.

De Vries regarded red as made up of fleischfarbig, F, and delila, D, two dominant characters, and white as carrying two characters, w and w′, recessive to F and D respectively. Then F_1 from red × white would be FDww′, and F_2 would give red, flesh, delila and white in the proportion 9:3:3:1, which de Vries maintains agrees with his experimental results. Various F_1 reds, delilas, and flesh colours were selfed, and the F_2 was in accordance with the above scheme.

Bateson (583) suggested that a better interpretation of de Vries's results would be to suppose that the F_1 plants produced, in each sex, in equal numbers, gametes having the characters, R, F, D and W, and these, on fortuitous mating, would give 9R:3F:3D:1W. In view of more recent work, both suggestions have to be revised.

In 1907 the author (617) published results of work on a number of colour-varieties; those used were

1. White—lips and tube of the corolla pure white.
2. Yellow—lips yellow; tube ivory.
3. Ivory—lips and tube ivory.
4. Crimson—lips crimson; tube magenta.
5. Magenta—lips and tube magenta.

With the exception of white, all varieties have a constant orange-yellow palate. Further varieties occur in which the lips are magenta or crimson, and the tube is ivory; the term 'delila' was used by de Vries for these varieties and has been retained.

The inheritance of colour in the varieties employed was represented by the following factors:

Y. A factor representing yellow colour in the lips associated with ivory tube-colour.

I. A factor representing ivory colour in the lips.

L. A factor representing magenta colour in the lips.

T. A factor representing magenta colour in the tube.

It was found also that certain relationships existed between the factors, i.e.

1. All zygotes, from which Y is absent, are white, though they may contain any of the factors I, L and T.

2. The effects of the factor T are not manifested unless L is also present in the zygote; that is, no magenta colour appears in the tube unless there is magenta colour in the lips.

3. All zygotes containing Y are coloured, the actual colour being modified and determined by the presence of one or more of the remaining factors. A zygote containing Y only, or Y and T is yellow.

4. Ivory is dominant to yellow; a zygote containing Y and I, or Y, I and T is ivory.

5. Since magenta superposed upon yellow gives crimson, a zygote containing Y, L and T is crimson, Y and L only, crimson delila.

6. Magenta upon ivory gives magenta. A zygote containing Y, I, L and T is magenta, Y, I and L only, magenta delila.

A further variety termed 'rose' was employed, but full details of the results of crossings with this variety are not given, though it was shown to be recessive to magenta.

Rose, which has the lips of the corolla tinged with magenta and a pale magenta tube, was identified with de Vries's fleischfarbig; magenta with rot, delila with magenta delila, and weiss with ivory.

Hence de Vries's results may be expressed:

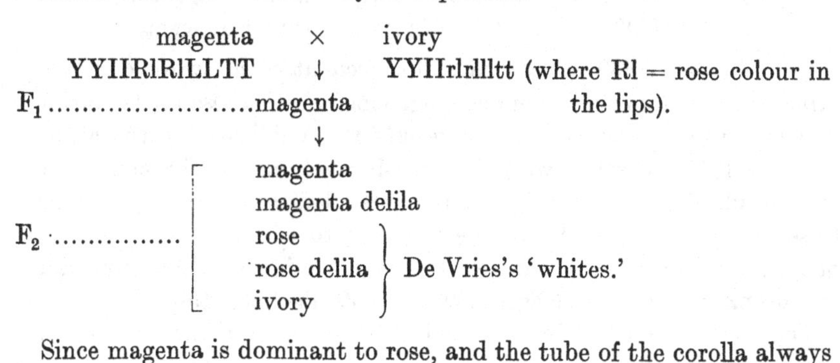

Since magenta is dominant to rose, and the tube of the corolla always takes the same depth of colour as the lips, rose × magenta delila would give magenta which would also corroborate de Vries's suggestion that rot consists of fleischfarbig and delila.

In 1908 Baur (620) published some results of his work on the same species. He worked with magenta (rot), ivory (elfenbein), yellow and white, and his results are entirely in agreement with those published by the author in 1907.

In 1909 the author (638) published further work and certain additional results of which the most important are as follows:

'Rose' is now termed 'tinged ivory,' and the factors L and T represent tingeing in lip and tube respectively.

D is used to denote a deepening, or full-colour, factor which converts tinged ivory into magenta.

It was found that zygotes heterozygous for L in the presence of D are pale magenta in colour, whereas those homozygous for L are intermediate magenta. The colour remains unaltered whether the plant be homo- or heterozygous for D.

Sometimes the deeper colour occurs only in streaks and blotchings when S may be substituted for D. When the zygote is heterozygous for L, the streaks are pale magenta; when homozygous for L, intermediate magenta. In non-striped forms, as mentioned above, the flower-colour is the same whether the zygote is homo- or heterozygous in D. But striped forms, heterozygous in S, have magenta stripings on a ground tinged with magenta, whereas, when they are homozygous in S, the ground colour between the stripings is ivory. The four zygotic types may be represented as follows:

1. YYIILLTTSS......Ivory striped with I magenta.
2. YYIILlTTSS ......Ivory striped with P magenta.
I. YYIILLTTSs......T. ivory striped with I magenta.
A. YYIILlTTSs ......T. ivory striped with P magenta.

In 1910 Baur (639) published a full account of his work on *Antirrhinum*; his results are in complete agreement with those of the author. A comparison of the two sets of results was published by the author (651) in 1910, together with further observations on the crossing of certain additional, true red, varieties, rose doré (on ivory) and bronze (rose doré on yellow). Rose doré is shown to be recessive to blue-red or magenta; similarly bronze is recessive to crimson. Both rose doré and bronze occur in the tinged, pale and intermediate states.

The work, published in 1913 and 1914 by Bassett and the author[1]

[1] Wheldale, M., & Bassett, H. Ll., 'The Flower Pigments of *Antirrhinum majus*. II. The Pale Yellow or Ivory Pigment,' *Biochem. Journ.*, Cambridge, 1913, VII, pp. 441–444. 'The Chemical Interpretation of some Mendelian Factors for Flower-colour,' *Proc. R. Soc.*, London, 1914, LXXXVII B, pp. 300–311.

on the pigments of *Antirrhinum*, has shown yellow to be identical with the known substance luteolin, and ivory with the known substance apigenin. The pigment in rose doré was shown to be a red anthocyanin, and the same pigment mixed with yellow gives the colour, bronze; similarly magenta contains a magenta anthocyanin, and this mixed with yellow gives crimson colour.

All the varieties may now be expressed in terms of the following factors:

Y. A factor leading to the production of luteolin in the lips and apigenin in the tube.

I. A factor leading to the suppression of luteolin in the lips, apigenin being formed instead.

L. A factor leading to the production of a tingeing of red pigment in the lips.

T. A factor leading to the production of a tingeing of red pigment in the tube.

D. A factor leading to the production of more anthocyanin pigment, i.e. a deepening, or full-colour, factor.

B. A factor converting red into magenta anthocyanin.

The different varieties may be represented as:

yyI(i)I(i)L(l)L(l)T(t)T(t)D(d)D(d)B(b)B(b)—white.
YY(y)iillT(t)T(t)D(d)D(d)B(b)B(b)—yellow.
YY(y)II(i)llT(t)T(t)D(d)D(d)B(b)B(b)—ivory.
YY(y)iiLL(l)ttddbb—yellow tinged bronze delila.
YY(y)iiLL(l)TT(t)ddbb—yellow tinged bronze.
YY(y)II(i)LL(l)ttddbb—ivory tinged rose doré delila.
YY(y)II(i)LL(l)TT(t)ddbb—ivory tinged rose doré.
YY(y)iiLL(l)ttDD(d)bb—bronze delila.
YY(y)iiLL(l)TT(t)DD(d)bb—bronze.
YY(y)II(i)LL(l)ttDD(d)bb—rose doré delila.
YY(y)II(i)LL(l)TT(t)DD(d)bb—rose doré.
YY(y)iiLL(l)ttDD(d)BB(b)—crimson delila.
YY(y)iiLL(l)TT(t)DD(d)BB(b)—crimson.
YY(y)II(i)LL(l)ttDD(d)BB(b)—magenta delila.
YY(y)II(i)LL(l)TT(t)DD(d)BB(b)—magenta.

Certain varieties deeper than intermediate magenta are known to exist; they are recessive to intermediate magenta, but otherwise they have been very little worked with. The heterozygous forms have already been mentioned above. The original wild type is thought to be intermediate magenta and would have the constitution YYIILLTTDDBB.

As would be gathered from the above statements, ivory × white will give magenta; hence two albinos, as regards anthocyanin, will give colour. Whether or no this case is strictly comparable to the results of crossing two whites in *Lathyrus* and *Matthiola* will be discussed later.

Aquilegia vulgaris. Kristofferson (850) found factor, B, gave light-blue, R red flower colour. If B and R both present, flower dark blue; if both absent, flower white. C, a factor which produced colour all over corolla; if absent, petals white-margined. When R is present, leaves have anthocyanin on both sides; if R absent, only on under-side.

Arum maculatum. A. maculatum occurs in nature in two forms; one has black (anthocyanin) spots on the leaves, the other is without spots. Colgan (655) obtained seeds from a spotted plant, and found, out of 11 seedlings, 5 were spotted and 6 unspotted. Colgan suggests the female parent was heterozygous for spotted character and was fertilised by the unspotted form, but there was no evidence in support of the suggestion beyond the fact of equality of forms among the offspring.

Aster Tripolium. De Vries (601) notes that pigmented type is dominant to albino.

Atropa Belladonna. Saunders (578) made crosses between *A. Belladonna typica* and *A. Belladonna* var. *lutea*. The type has brown (anthocyanin on plastid) flowers, black (anthocyanin) fruits, and stem tinged with anthocyanin: the variety, which is without anthocyanin, has yellow flowers and fruits, and a green stem. F_1 was found to have the fruits of the type, but the intensity of colour in flowers and stems was sometimes diminished. Further results clearly indicated that there was normal Mendelian segregation in F_2.

Beta. Kajanus (662, 685) has experimented with a large number of varieties. As far as anthocyanin pigmentation in the root is concerned three classes can be differentiated:

(*a*) Red	With anthocyanin.
(*b*) Rose	
(*c*) White, yellow	Without anthocyanin.

The results from crossing are at present rather obscure. Red was obtained in F_1 from crossing:

Red × red, rose, white and yellow.
Rose × yellow.
White × white and yellow.

Rose was obtained in F_1 from crossing:

Rose × rose.
White × yellow.

In F_2 red gave red, rose, white and yellow, and also every selection and combination of these varieties except red only.

Rose, on the other hand, gave red, rose, white and yellow, and again every combination and selection except red only, and red, white and yellow.

White gave:

Red, rose, white and yellow.
Red, rose and white.
Rose and white.

Kajanus suggests a number of factors, but in the absence of further data, very little light is thrown on the results.

Kajanus notes that the pigment is not always confined to the skin of the root. In the salad beet, the flesh is completely violet-red; in the common red beet it is red, reddish or colourless; in the rose and in the white it is colourless.

The leaves on the whole are green, but in red beets they are sometimes red. Quite red leaves only occur in the case of red-fleshed beets. In red-fleshed beets there are also varieties in which the petioles and larger veins only are red, the leaf tissue being green. The results of crossing red- by green-leaved varieties seem to show that in some cases the green colour is due to inhibition of red, in other cases, not.

Later, Kajanus (704, 759) considered red colour to be due to two factors. GGrr (yellow) × ggRR (white) gave F_1 (red), and in F_2, 9 red, 3 yellow and 4 white. Experimental numbers do not fit well, owing, probably, to insufficient isolation.

Roemer (766). Occurrence of red beets among progeny of white.

Rasmuson (797, 874). Some red sugar beets threw red, yellow and white in proportion 9:3:4. Therefore, probably, hybrids between sugar beet and fodder beet.

Lindhard & Iversen (794). Crosses have been made between red, yellow and white beets. Suggested RG (red), G (yellow) and Rg, rg (white). Ratio, however, of 9:3:4 in F_2 does not hold very well. Linkage between R and G suggested, but not proved.

Brassica (Turnip). Kajanus (685, 686, 704, 758) has worked with two species *B. napus* and *B. rapa*.

B. napus. As regards the colour of the root, distinction must be made between upper and lower parts. Anthocyanin is only found in the upper part, which may be violet-red, intermediate, or green. When the root is intense violet, the neck (the lower basal portion of the stem) is also violet-red. If the root-colour is reddish only, the neck is usually green. Hence there are three classes:

1. Red with red neck.
2. Red with green neck.
3. Green with green neck.

It has been deduced from crossing the varieties that there are two factors involved in anthocyanin pigmentation, i.e. P_1 which gives pale violet-red, and P_2 which gives deep violet-red colour. P_2 is dominant to P_1. When both P_1 and P_2 are absent, the root is green. The differentiation between the classes is often not sharp.

B. rapa. Here also anthocyanin is only found in the upper part of the root, which may be violet-red (deeper or paler), or if pigment be absent, green. The violet-red colour may be continuous or blotched. It has been deduced, from crossing varieties, that anthocyanin is due to the presence of one dominant factor, P, and when this is absent, the root is green.

(Cabbage). Kristofferson (832). Cross made between white cabbage and one with green mid-rib (red in autumn). F_1 dark red-violet. F_2 9 dark red-violet : 3 light red-violet : 4 green.

Campanula carpatica. Pellew (764). Blue type dominant to white variety. Normal segregation of blue and white factors occurs in ovules, but 97 % of pollen-grains carry white factor, and only 3 % blue factor.

C. medium. Lathouwers (851) found four factors for flower-colour. VABR deep violet, VAR lilac, ABR violet and AR rose. If A or R be absent, flowers white.

Canavalia ensiformis (Leguminosae). Lock (607) showed pink colour in the flower to be dominant, or nearly so, to white, the latter reappearing in F_2. Absence of red pigment from the testa (the author does not state whether this is anthocyanin) is dominant to red. In F_2 red reappeared, but in nothing like its former intensity. Some of the plants of F_2 bore mottled grains, but in these also the pigmented patches were of a very faint reddish colour. In F_2, plants with no red pigment in the testa were more numerous, probably three times so, than those with reddish and mottled testas taken together.

Canna indica. Inheritance of anthocyanin when red- are crossed with green-leaved plants. Honing (715, 734).

Cattleya. Hurst (605, 634, 643, 702) notes that purple pigment is dominant to its absence in the flower.

Clarkia elegans. Bateson (627) states that the magenta-flowered type is dominant to the salmon-pink variety.

Rasmuson (814). Preliminary experiments only. *C. elegans.* Crosses with purple, purplish-red, salmon-red and white. *C. pulchella.* Purple, purplish-red, white and purple with white edge.

Collinsia. Rasmuson (813). White-flowered variety was crossed with lilac. F_1 lilac. F_2 segregated into 9 lilac : 7 white. Green stem recessive to red. In F_2, 9 red : 3 tinged red : 4 green. A, factor representing white flowers and tinged stem; B, white flowers, green stem. AB gives lilac flowers and red stem. Spots on upper lip dominant and segregate normally.

Corchorus capsularis. Finlow & Burkill (677) investigated the inheritance of anthocyanin in the Indian Jute plant. As regards pigmentation 33 races were broadly classified into the following types:

(*a*) Deep red stem, petioles and fruits; the teeth of the leaves also tipped with red.

(*b*) Brownish red stems, petioles and fruits, with no distinct red borders to the leaves.

(*c*) Green stems with reddish petioles and fruits.

(*d*) Pure green stems, petioles and fruits.

The red pigment is found:

1. Chiefly in the parenchyma cells which lie immediately under the epidermis of the stems and petioles.
2. In the parenchyma of the petioles, sporadically, even as deep as the phloem.
3. In sub-epidermal cells near the margin of the leaf.
4. In small multicellular hairs on the leaf and on the stipules.

The intensity of coloration is due to the general distribution of the pigment. Conversely, the fewer the pigment cells, the less red in the stem. The authors point out that, in fact, there are only two real colour types, red and green, since the classes (*a*), (*b*) and (*c*) are without definite boundary lines.

The results of crossing red races with green were as follows. When pure green is crossed with a fixed red, red is dominant. In F_1 the hybrids appear to consist entirely of plants of one tint of redness, which is less dense than the colour of the red parent. The red plants of the F_2 generation vary widely in the amount of red colour they contain. F_3, from red F_2, though fixed as regards red colour, shows the same variation in intensity as F_2. As a result of the experiments examples were produced, either fixed or unfixed, of all intermediate colour types of jute hitherto met with, including a pure fixed culture of one of the commonest of these.

Coreopsis tinctoria. De Vries (577) has shown that the yellow type is dominant to the variety, '*brunnea*,' in which the brown colour is due to the development of anthocyanin. The type has evidently an inhibitor of anthocyanin.

Cypripedium (*Paphiopedilum*). Pigmented (anthocyanin) flower is dominant to albino. Hurst (605, 634, 643, 702).

Datura Stramonium. Saunders (578, 590) used two types, *D. Tatula* having reddish stems and violet flowers, and *D. Stramonium* with green stems and white flowers. The F_1 had red stems and violet flowers, though the intensity of colour varied. In F_2 there was evidence of typical Mendelian segregation.

Dendrobium. Pigmented (anthocyanin) flower dominant to albino. Hurst (702).

Digitalis purpurea. Keeble, Pellew & Jones (645) made certain observations on the inheritance of anthocyanin in flowers of this species. The plants used were:

1. White with yellow spots (mmddww).
2. White with red spots (MmddWw).
3. White with red spots (MMddWw).
4. Purple with red spots (MmDdww).
5. White with purple flush and red spots (???).

Colour is due to the factor, M, producing magenta sap. Absence of M gives a recessive white. A deepening factor, D, is dominant to M and changes it to purple. The colour may be inhibited by a dominant factor, W, so that the corolla is white except for red spots. The spots on the corolla are always present. In recessive whites they are brown or yellow: in magenta and dominant whites they are red. They depend on the presence of the factor M and they are not inhibited by W.

Saunders (666) confirms the results of Keeble, Pellew & Jones for the inheritance of spot colour. It is stated that white-flowered plants with red spots may either breed true, or give a mixture of whites with red spots (dominants) and white with greenish-yellow spots (recessives) according as they are of pure-bred or cross-bred parentage. Coloured flowers may vary from deep purplish-red to white with a faint flush. The white-flowered plants with red spots frequently become tinged as they get older.

Miyake & Imai (795, 811) found that plants with coloured (anthocyanin) stems have either purple flowers or white flowers with red spots, and are dominant to white-flowered. A factor, C, represents purple stem, corolla spots and dots on anther. An additional factor, P, gives purple corolla.

Dolichos lablab. Harland (807). Two white-flowered varieties gave a purple-flowered. One variety contains a factor, C, which produces purplish-brown anthocyanin in testa and a minute amount of anthocyanin in stipular hairs. The other variety contains R (which has no effect except in presence of C), is entirely devoid of anthocyanin and has cream seeds. The purple-flowered F_1 has black seeds and anthocyanin in vegetative parts (purple nodes). The factors segregate normally.

Fragaria vesca. Richardson (719, 780, 815). Pink flowers dominant to white or very pale pinks. Red fruit dominant to white.

Geum. De Vries (601) mentions the inheritance of anthocyanin in hybrid from the cross of type (yellow plastids plus anthocyanin) by yellow (plastids) variety.

Godetia. Rasmuson (796, 836). Varietal crosses of *G. Whitneyi* made. The following factors for flower-colour identified. aa gives yellow-margined petals. Aa, AA non-yellow. B gives pale lilac. C gives rose, heterozygotes nearly white. D, together with B or C, gives lilac. E gives red. F, with E, gives red with light-margined petals. G gives red spot in middle of petal; Gg smaller spot. H enlarges the spot (alone no effect). I, with B (probably), gives rose-lilac. *G. amoena.* Variety with large spot and another with small basal spot. All with both spots are heterozygous and gave 1 large spot, 2 double spots and 1 small basal spot. Both gave 3:1 ratio with unspotted. Linkage between B and E, E and G, C and F.

Gossypium. Balls (618, 626) working on Egyptian cotton recognised the following pairs of Mendelian characters which are connected with anthocyanin:

Full red spot on the leaf and faint spot.

Large purple spot on the petal and no spot.

The red spot on the leaf is due to the development of anthocyanin in the epidermal and sub-epidermal cells of the petiole at the point where it divides into the leaf-veins. Crosses of spotted with spotless give spotted F_1, but the intensity of colour in the spot is less than in the spotted parent. In F_2 the heterozygote spot can be distinguished from the homozygote. In the case of flower spot, the homozygote has a large purple spot, the heterozygote a small purple spot.

Martin Leake (664) worked with Indian cotton. Four varieties connected with anthocyanin pigmentation were used, i.e.

1. A variety in which the petal is entirely red having a darker spot at the base.
2. A variety in which the petal is yellow with a deep red spot at the base.

3. A variety in which the petal is pale yellow with a basal spot.
4. A variety in which the petal is white with a basal spot.

Leake identifies the following factors. A factor for yellow which is dominant to white, and a factor for red which is dominant to white and yellow. Individuals heterozygous for the reddening factor have petals only partially coloured; this is very obvious in red on yellow, but less so when on white. A scheme of the factors may be represented as:

YRr.........red on yellow.
YRR ......red.
RR(r) ......red on white.
Y ....yellow.
– ....white.

In some of the strains used there was also anthocyanin in the vegetative parts, that is in the young leaves, and in the ribs and veins of the mature leaf; in other strains these were quite green. In the F_1 from a cross between these red-foliaged and green-foliaged strains the red colour was dominant, though diminished in amount. In F_2 there was segregation into red and green in the proportion 3:1. Among the individuals with red colour there was a considerable range in intensity, though the DD individuals had more pigment in the leaf than the DR, and by this means they could be separated with a fair degree of certainty. There is an association also between anthocyanin in the vegetative parts and the complete redness of the flowers.

Helianthus. Shull (623) has made some experiments with this genus. The wild *Helianthus annuus* of the Prairie region has a purple disk, the colour being found in the tips of the paleae which are a deep metallic purple, the margin of the corolla which is brownish-purple and the style and stigmas which are reddish-purple. The 'Russian Sunflower' (*Helianthus annuus*, var.) has the tips of the paleae yellowish-green, the corolla a clear lemon yellow, and the styles and stigmas usually have the same colour as the corolla. Shull concluded, on the results of crossing, that the purple disk is a strict Mendelian character and is dominant to the yellow disk.

The red sunflower mentioned by Cockerell (709, 710, 722) appears to be dominant to the yellow, though the F_1 may vary in the amount of anthocyanin it produces. The red variety is no doubt formed owing to the loss of an inhibiting factor, and the F_1 plants would only receive the inhibitor from one parent. When anthocyanin appears in the primrose-coloured variety of *Helianthus*, the result is purple and not chestnut-red.

Later Cockerell (724, 753, 772) showed that chestnut-red, wine-red, orange and primrose segregate out in F_2 in the ratio of 9:3:3:1. Other observations concerning anthocyanin recorded.

Hordeum. Biffen (604) mentions that in *Hordeum* the paleae may be white, black, brown or purple. The grain also may be white, bluish-grey or purple. Apparently the purple colour is due to anthocyanin. According to Biffen, purple paleae as contrasted with white, and dark grain with light grain, form pairs of Mendelian characters.

Impatiens balsamina. Rasmuson (737, 814). Blue, blue-red, red, pink, white and white variegated. Factors suggested for flower-colour are BA blue, RA red and BRA blue-red. Rose is brA. White flecking on coloured flowers, due to F, is dominant.

Ipomoea (Japanese). Miyazawa (777). A dark-red variety is dominant to white and gives a magenta (heterozygous) form in F_1. A variety, in which the coloured flowers have a white margin (hukurin), is dominant. Further results (834) showed that flower-colour may be magenta, scarlet and deep red. There are three shades, light, medium and deep, of each colour. The colours depend on the following factors: G, factor for green colour in leaf. D produces dark-red flowers if G homozygous; scarlet with Gg or gg if B is also absent. B is factor for blue colour; in presence of D it has no effect, if G homozygous, but produces magenta with Gg or gg. M is a factor which modifies shades in connection with D as follows: DdM light, DDM medium, DDmm or Ddmm deep. White flowers have green, coloured flowers, red stems. Both scarlet and red flowers may show magenta streaking.

Hagiwara (864), though difficult to follow, finds in corolla four groups, blue BPRK, purple Bb(b)Pp(p)RK, scarlet pbRK and dark red Bb(b)Pp(p)rK; also yellow and white. Blue colour regarded as wild type.

(American.) Barker (752). Two factors, C and R, required for colour (pink). X modifies pink to mauve (CRX). I deepens mauve to magenta (CRXI). B modifies pink to light blue (CRB), mauve to dark blue (CRBX) and magenta to dark purple (CRBXI). Various intermediate heterozygous forms exist. Flaking is a dominant character.

Lactuca muralis. Dahlgren (773) made cross *L. muralis normalis* × *atropurpurea.* Latter variety is dark brown-red with anthocyanin. F_1 was green and segregated normally.

Lathyrus odoratus. Bateson & Punnett (590, 599, 603, 619) have carried out extensive work on the colour inheritance in the Sweet Pea.

The original wild type is probably most nearly represented by the variety now known as 'Purple Invincible' with chocolate standard and bluish-purple wings.

Loss of a factor for light wing produces a variety with deeper wings, 'Purple-winged Purple Invincible.' Loss of a full-colour factor gives a tinged variety, 'Picotee.'

When the blueing factor is absent, a series of red varieties appears comparable to the above: 'Painted Lady,' with a deeper variety, 'Miss Hunt,' and a tinged variety, 'Tinged White.'

It was shown early in the experiments with Sweet Peas that two white varieties, indistinguishable except that one has long, the other short pollen grains, gave a 'Purple Invincible' hybrid, and from this result the fact was deduced that colour production is dependent on two factors, or that the two factors taking part in its formation can be inherited independently. As in *Matthiola*, we must suppose that two factors produce the most hypostatic colour, i.e. 'Tinged White,' and that 'Purple Invincible' resulted in the original cross because the white varieties employed carried both B and a full-colour factor.

These facts may be represented in tabular form as follows (L = factor for light wing, in absence of which, wing is dark, De = full-colour factor):

CRBDeL......Purple Invincible.
CRBDe ......Purple-winged P.I.
CRBL ....Picotee (see Bateson & Punnett, 590).
CRB ....Picotee.
CRDeL ......Painted Lady.
CRDe ....Miss Hunt.
CRL ....Tinged White.
CR ....Tinged White.

Any plant without C or R is white. Hence whites can carry any of the other factors in any combination or arrangement, but never C and R simultaneously. There is also a certain connection between colour and the hooded form of the standard. In hooded varieties the standard always approaches in colour to the wings, and has never the bicolor appearance of the varieties with erect standard.

M. G. and D. Thoday (650) have given an account of experiments with certain varieties of *Lathyrus*, and the crosses form a very complex series in F_2. The chief point of interest is that they find a scarlet anthocyanin, distinct from, and recessive to, the bluish-pink of the 'Painted Lady' variety.

Later Punnett (854) described relationship between additional varieties. If in Purple Invincible and Purple-winged P.I., the flowers become hooded by loss of factor E, the colour becomes more uniform, and these varieties are known as Duke of Westminster and Duke of Sutherland respectively. Further, a factor J may be lost from each of the above varieties giving a second, 'red-purple' series. Similarly, loss of a factor, D, from the original series gives a 'blue' series. Finally there is a 'red-blue' series corresponding to the 'red-purple.'

Purple Invincible (P.I.)	ELDJ	Red P.I.	ELDj
Deep Purple (Ppw)	ElDJ	Red Ppw	ElDj
D. of Westminster (D.W.)	eLDJ	Red D.W.	eLDj
D. of Sutherland (D.S.)	elDJ	Red D.S.	elDj
Blue bicolor	ELdJ	Violet bicolor	ELdj
Deep blue	EldJ	Deep Violet bicolor	Eldj
Blue hood	eLdJ	Violet Duke	eLdj / eldj
Lord Nelson (L.N.)	eldJ		

All above series contain B and none of the forms should be confused with 'true' reds. Linkages detected by Punnett (765, 873) involve colour factors, C, R and B. Also anthocyanin in leaf-axil (D), the factor, F, the loss of which causes 'marbling' of pigment and, finally, the above-mentioned factors D and J.

Linaria alpina. Saunders (691) made crosses between the type and a variety. The type has purplish-blue flowers and the variety pink flowers. Both pigments are anthocyanins. It was found that blue is dominant to pink.

Linum usitatissimum. Tammes (668, 721, 739, 748, 749, 858) has investigated a complex of factors concerning flower-colour and other characters. For flower-colour six factors A, B′, C′, D, E and F co-operate. B′ and C′, when combined, produce colour (extremely light pink). A and E are intensification-factors, and act accumulatively in each other's presence. E is stronger than A. The intensifying action of A is different for various complexes of factors. When B′, C′, D and F are present, A dominates a, when B′, C′ and F are present, a dominates A. D modifies the pink colour caused by B′ and C′ into lilac, and acts also as intensification-factor. F does not appreciably modify the pink, but it changes the lilac into blue. In both cases it acts simultaneously as a diluting factor. Blue colour of anthers depends on co-operation of B′, D and H. If one of these is absent, anthers are yellow. Consequently blue and lilac flowers have either light or dark anthers, white flowers may have blue or yellow, but pink flowers only yellow.

A few observations on colour also by Vargas Eyre & Smith (751) and by Fruwirth (846).

Lupinus angustifolius. Hallqvist (828) found four factors concerned with flower-colour. R gives pure red; if R absent, flowers white. B is blueing factor and gives bluish-red flowers (RB). V transforms red into violet (RV). B and V do not show in absence of R. R and V required for full blue (RVB). F is deepening factor, without which all the colours are tinged. Flower-colour is closely associated with seed-colour. Nature of seed-pigment not described, though apparently anthocyanin in unripe stages. Coupling between B and V, B and F, V and F.

Lychnis dioica. Saunders (578) used this species in certain crosses. *L. dioica* was crossed with a glabrous form of *L. vespertina*, and a glabrous variety of *L. dioica* was crossed with *L. vespertina*. F_1 and their offspring showed pink, the depth of colour varying according as the wild hairy species or de Vries's glabrous strains (see original paper) were employed in the cross.

Shull (623) first mentions the results of crosses between white- and purple-flowered *Lychnis dioica*, the purple being dominant.

Later, Shull (649) published further results, in which he states that two varieties of coloured flowers can be detected in *Lychnis*, i.e. a reddish- and a bluish-purple. The former is dominant to the latter, this being the reverse of what is found in other plants. Two factors are necessary for formation of the bluish-purple, and a third factor modifies this to reddish-purple.

In a still later paper Shull (693) again states that the type has reddish-purple flowers, and that there is a bluish-purple variety recessive to the type. The albino is without anthocyanin. No albinos mated together produced colour. Two new German strains were introduced, *Melandrium album* and *M. rubrum*. A certain individual of *M. album* × white *dioica* gave reddish-purple offspring. *M. rubrum* × *M. album* gave a mixture of both purple- and white-flowered offspring in the proportion of 4:23.

Malope trifida. Rasmuson (745) found white- by red-flowered (corolla, anthers and pollen red) gave F_1 all red. One factor necessary for pigment in flower and vegetative parts. Segregates normally.

Matthiola. Our knowledge as regards the inheritance of colour in this genus is due to the work of Saunders (578, 590, 599, 603, 609, 665).

It is not known what variety represents most nearly the original type, but if it be assumed that each variety arises by loss of some factor from the type, then the latter would be represented by a plant with pale purple flowers. Loss of a diluting factor gives rise to deep purple; loss of another factor from the deep purple gives a duller shade of purple termed 'plum.'

Loss of the blueing factor—B—from each of the above varieties gives rise to the corresponding blue-red series; rose, a dilute variety; carmine and crimson, deep varieties; and 'copper,' a dull red variety represented by plum in the bluer series.

Loss of a further factor from the blue-red class reveals a true, less blue, red class containing a dilute variety, 'flesh,' and a recessive deeper variety, 'terra-cotta.'

Early in the experiments with *Matthiola* it was ascertained that two factors are necessary for the production of colour, and that certain white varieties crossed together produce coloured offspring, purple in the original experiment, since one at least of the original whites used contained the blueing factor, B.

The varieties may be represented in the following scheme. C and R are the factors for colour, B modifies the blue-red class to the purple class, D causes dilution in colour and X the difference between the pure and dull, or impure, colour:

CRBDX......pale purple.
CRBX ......deep purple.
CRB ....plum.
CRDX ......pale red (rose).
CRX ....deep red (carmine, crimson).
CR ....copper.
C..............white.
R..............white.

In some families other varieties, such as terra-cotta, flesh and lilac, appear in addition, but, apart from the fact that they are recessive to crimson and purple, their relationship to each other and to the other colours is not clear at present. As we see from the table, individuals without either C or R are white; thus white can carry every factor and combination of factors, except C and R simultaneously.

Many of the above results have been confirmed by Tschermak (695).

Mirabilis Jalapa. Correns (585, 600, 640) published the first work on the colour-inheritance in this genus. He used several varieties, white, yellow and 'red.' In F_1 he obtained a 'rose' by crossing white × yellow, but as he did not realise the existence of heterozygous forms (see below), he was unable to solve successfully his results in the F_2 generation.

Baur (620) made crosses with two varieties, and obtained a heterozygous form in F_1, but his results are similar to those of Marryat (636) to whom we owe the bulk of our knowledge on the inheritance of flower-colour in *Mirabilis*. The work of the latter author leads to the following conclusions:

We must suppose that the original type had crimson flowers. Loss of a factor for anthocyanin production from crimson gives a variety with yellow flowers free from anthocyanin. Further loss of a factor produces white, an albino both as regards red and yellow pigment. The peculiar interest of *Mirabilis* (among other points) is centred in the occurrence of heterozygous forms which may be best represented by the following scheme:

CCMM......crimson.
CcMM......magenta.
CcMm......magenta rose.
CCMm......orange red.
CCmm......yellow.
Ccmm......pale yellow.
ccMM......white.
ccMm......white.
ccmm......white.

It was also found that a certain two white individuals crossed together gave coloured (anthocyanin) F_1, as in *Matthiola* and *Lathyrus*. Hence the factors for anthocyanin production can be separated into two components, of which one is M, and the other is not represented in the above scheme. Absence of colour may be due to loss of either of these components, or to loss of the yellow pigment.

Nicotiana. Inheritance of anthocyanin in species-crosses. Lock (635), Haig Thomas (700).

Clausen & Goodspeed (825, 856). Red of *macrophylla* recessive to light pink of *angustifolia*; also red *calycina* to light pink *virginica*. Factors suggested for flower-colours are WRP carmine, WR light pink, W red and wR white. In cross between white and *macrophylla*, F_1 was pink. In F_2, 9 pink, 3 red and 4 white. In the case of *purpurea*, however, flower-colour (carmine) is darker than *macrophylla* and is dominant to pink.

Allard (785) made crosses with white, carmine and pink varieties of *tabacum* and white-flowered *sylvestris*. In this case carmine found to be dominant to pink, and segregates normally. White is recessive to both. Sachs-Skalińska (840). Various observations on flower-colour factors.

Oenothera. Since *Oenothera* is a plant which forms anthocyanin in its stems, petioles, buds, etc., the complex inheritance of characters among the numerous strains which have been employed experimentally involves also the inheritance of this pigment. Mention will only be made of one or two cases in which anthocyanin pigmentation has been considered an important character.

The first case to be dealt with is that introduced by Gates (642, 658, 678), and which concerns the appearance of *O. rubricalyx*. This variant was found among the offspring of self-fertilised *rubrinervis* plants, the latter being characterised by having the calyx of the buds streaked with anthocyanin, whereas *O. rubricalyx* has a completely red calyx. The segregation of the offspring from self-fertilised *rubricalyx* plants in two generations, into *rubricalyx* and *rubrinervis*, led Gates to consider the case purely Mendelian. The original *rubricalyx* plant was regarded as a heterozygote which has acquired a dominant Mendelian character, the character being purely quantitative, i.e. as causing an increased formation of anthocyanin.

In a later paper (713) Gates again states that the red pigmentation character, R (which originated by a mutation and distinguishes *rubricalyx* from *rubrinervis*), is more or less completely dominant in F_1 from the cross *grandiflora* × *rubricalyx* and its reciprocal. In F_2, 3:1 ratios were obtained, and also ratios of 5:1 and 10:1 as well as 3:1. Gates has no satisfactory explanation of these facts.

Shull (720) makes further experiments on *rubricalyx* by selfing this strain and crossing it with *rubrinervis* and with *Lamarckiana*; he obtains what he calls a series of negative correlations in the distribution of the red pigment, the pigmented buds of *rubricalyx* being invariably associated with a low degree of pigmentation in stems and rosettes. Moreover, the segregation of the *rubricalyx* character was not found to be a simple Mendelian case. Shull maintains also that certain of Gates's conclusions are erroneous, viz. that the *rubricalyx* character represents a quantitative difference, and that it can be expressed by one factor.

The other case of interest in connection with anthocyanin is that considered by Davis (697, 711) in connection with the stem coloration (the formation of papillae or glands coloured with anthocyanin at the base of long hairs) in parents and hybrids of crosses, *O. grandiflora* × *biennis*, *O. franciscana* × *biennis* and their reciprocals. Davis considers this character to be dominant to the green stem, but it has not been shown to segregate in a Mendelian way.

Gates (727, 728). Shull's criticisms based on error, his '*rubricalyx*' being hybrid between *rubricalyx* and *grandiflora*.

Cockerell (723) and Heribert-Nilsson (732, 733). Red veining.

Oryza sativa. Hector (743). Presence of anthocyanin in leaf-stalk, glume apex and stigma due to interaction of several factors.

Parnell, Ayyangar & Ramiah (763). Presence of anthocyanin in whole plant dominant to absence and segregates normally. Same true for characters purple pulvinus and auricle and for purple colour in leaf-sheath. Purple lining of internode, purple glumes, purple stigmas, purple axil (inside of sheath) all dominants and segregate normally. Purple lining associated with purple glumes, purple stigma with purple axil. Green internode and glumes with purple stigma and axil.

Oxalis. Nohara (736) worked with varieties of the so-called *Oxalis corniculata* L. which differed from each other in the presence or absence of purple (anthocyanin) in the eye of the corolla and in the leaves. The presence of anthocyanin was found to be dominant to its absence, and the intensity of pigmentation in F_1 from pigmented by unpigmented was found to be intermediate. The purple colour in eye and leaf is due to one factor so that eye- and leaf-purples are associated, but the leaf-purple can appear without the eye-purple.

Papaver Rhoeas. Shull has published work (693) on this genus. It was found that varieties with a white margin were dominant to varieties without the margin, i.e. the type. Certain whites crossed with some reds gave white or striated offspring, but the same whites were found to be recessive to pink or orange. Some red-flowered varieties crossed together gave whitish offspring. Two suggestions are made: (*a*) that only spectrum red in inhibited; (*b*) that two factors are necessary for inhibition.

Rasmuson (812) crossed *Rhoeas* with *laevigatum.* Dark eye-spot (S) dominant to white (W) and segregates. If both missing, petals unspotted.

Becker (769). Markings on base of petals may be dark fleck and white spot (+ s + w) or the following variations: (+ s − w), (− s + w) and (− s − w). The combination + s + w appears only on outer petals if also on inner, and may be lacking. Two inhibitory factors, H_1 and H_2, postulated, H_1 acting only on inner petals. If heterozygous for s and w, H_1 is dominant, if homozygous, H_1 and H_2 dominant.

Kajanus (790) also crossed varieties of *Rhoeas* and *glaucum* and others. Dark eye-spot dominant to white and segregates.

Papaver somniferum. De Vries (577) and Hurst (605) have shown that the basal patch on the petals is dominant to its absence. Hurst (605) also states that colour in the rest of the petal is dominant to albinism, and that purple is dominant to red.

(Indian.) Martin Leake & Ram Pershad (809). Impossible in short summary to do credit to the complexity of the case. Factors for flower-colour as follows: P, factor for pink margin and white eye. R, an intensifier of P, and, in its presence, produces a scarlet or crimson margin separated from a white eye by a blue or purple band. M, factor which develops a mauve-purple in the margin and a deeper eye. This factor shows complete dominance. L, an intensifier of M, converting the mauve-purple into a rich magenta-purple. Considerable variation of intensity of colour is produced by heterozygous forms. Five groups arise from the combination of these factors: I. White-eyed group. P and R present, M absent. It includes light pink, full pink, purple white eye, red-purple white eye. II. Purple group. M and L present, P absent. Includes light mauve-purple when L is absent, or true mauve-purple to full magenta-purple when present. III. Dilute colour group. Marginal colour is dilute and eye is coloured. P and M present, L absent. R, if present, has no visible effect in presence of M. Includes pink-purple eye, red-blue eye and others not readily described. IV. Full colour group. Marginal colour full and eye coloured. Factorially the same as III, but L also present. Includes red, crimson and blue shades of crimson-purple. V. White-flowered group. P and M are absent. Pure white. The seed may be coloured pink, deep or bright blue or brown, straw or white. The nature of the pigment is not stated. White-eyed varieties have white seeds; when eye is coloured, seeds are coloured.

Later (852), extension of previous data on seed-colour, but nature of pigment still not stated.

Miyake & Imai (872) give, in connection with flower-colour, a series of factors, but account difficult to follow.

Kajanus (791) distinguishes a red and a violet group, having in each a set of intensifying factors. The factor for violet makes the petals coloured both on upper and under surfaces; that for red on upper surface only. Both combined produce red on upper and violet on under surface. In absence of red and violet factors, petals are white. There is a factor for striping of the petals.

Przyborowski (853) investigated white-flowered variety with deep violet spot, and found one factor for pale spot and a deepening factor for same.

Petunia. Rasmuson (779). Deep flower-colour is dominant to pale. Blue colour in anthers (throat blue) is dominant to yellow (throat yellow). In both cases there is normal and independent segregation. Sachs-Skalińska (839). Polymorphic *P. violacea* crossed by *P. grandiflora*; colour of latter dominant.

Phaseolus multiflorus. The characters which have received most attention in this genus are those concerned with the pigmentation of the seed-coat.

Lock (607) made some preliminary experiments by crossing a dark purple-seeded bean with a dark yellow-seeded bean. The F_1 was dark purple. The F_2 could be subdivided into two groups: (*A*) containing beans of various shades of purple; (*B*) of various shades of yellow. It was found that members of (*A*) might throw (*B*) but not *vice versa*.

Shull (616) gives an account of the results obtained by crossing several varieties, i.e.

'Prolific black wax'—purple-black seeds (anthocyanin in testa).
'Ne plus ultra'—yellow-brown seeds.
'Long yellow six-weeks'—light greenish-yellow seeds.
'White flageolet'—seed-coats white.

The results in F_1 were as follows:

Purple × yellow-brown = purple.
Purple × yellow = purple.
White × purple
White × yellow-brown } All gave similar F_1 with testa mottled with purple.
White × yellow

Shull then postulates the following factors:

P = pigment.
B = modifier which changes pigment to purple.
M = mottling factor.

Then the constitution of the different beans is

Brown and yellow—Pbm.
Black—PBm.
White—pBM.

Shull concludes that the mottling factor is carried by the white bean, whereas Bateson and Tschermak had regarded the mottling factor as latent in the pigmented bean.

Shull in a later paper (624) gives the proportions of the varieties in F_2 from the above cross. They were found to be

Purple mottled—18.
Purple self-colour—18.
Brown or yellow mottled—6.
Brown or yellow self-colour—6.
White—16.

Shull's explanation for this result is that beans containing PB and heterozygous in the M factor are mottled, whereas those beans homozygous in M are self-coloured. Hence it is possible for purple to carry the mottling factor, and Tschermak, Bateson and Lock are also correct. Mottled beans of the above constitution are heterozygous and should never breed true, and this was found to be the case. Shull points out that there are however other strains of mottled beans which do breed true.

Emerson (631) publishes a long paper on pigmentation in bean seeds. He gives first a list of the crosses made by Tschermak, Shull and himself. He points out, as Shull has done, that there are two kinds of mottled beans, viz., strains which breed true, and heterozygous forms not breeding true. Emerson then suggests a scheme to explain the existence of two sorts of mottling, namely by postulating two factors for mottling: M the sort of mottling which breeds true, and X, the sort which is visible only in the heterozygous condition. The results he obtained experimentally can be explained on the basis of this scheme. In a later paper (632) Emerson gives further results of the inheritance of total and eyed (round hilum) pigmentation, but since the kind of pigment is not described, it is not clear how far it concerns anthocyanin. Emerson also mentions another hypothesis suggested to him by Spillman to explain the two kinds of mottling mentioned above. Spillman supposes that the mottled races which breed true have in them two correlated factors, and that there are three types of non-mottled beans resulting from the loss of one, or the other, or of both of the correlated factors. On Spillman's plan, just as on Emerson's two factor hypothesis, a definite formula can be assigned to all the races used in crossing, and in this way all the results, with one or two exceptions, can be accounted for. Emerson himself inclines to the coupled-factor, rather than the independent-factor, hypothesis.

P. vulgaris. Striping of the testa is also another character described by Tschermak and others. Facts about actual colours (usually yellows, browns and blacks) are in a state of confusion which is increased by absence of statements as to nature of pigments. Tschermak used violet beans, and Shaw & Norton (782) red. Tjebbes & Kooiman (801, 841) deal with inheritance of violet striping and also factor for blue. Sirks (817) gives V for violet and B, changing violet to blue; also S which restricts blue and violet into stripes. Sirks (857) used violet-marbled. Tjebbes & Kooiman (860) showed connection between colour of flower and seed-coat. Factor F, together with ground factor for colour and a third factor for specific colour, changes seed-coat into black or dark blue; also flower-colour from white into pale purple. In addition factor for pigment round hilum can give flower-colour. Other factors affect flower- but not seed-colour. Kristofferson (879) gives some summary of results.

P. multiflorus. As regards testa colour, violet used by Tschermak (695) and Tjebbes & Kooiman (801). Sirks (857) used both light and dark purple marbled with black (latter dominant); also light purple spotted with black, this being dominant to light purple marbled.

Phlox Drummondii. Gilbert (729) investigated varieties of flower-colour. Factor for dark-eye dominant to its absence. Also factors for blue and red and intensification of colour. Yellow factor (due to chromoplastids) acts only in presence of eye-colour. Results incomplete. Kelly (808). Two white-flowered strains gave full-coloured type, as also did white by pink. Three factors necessary for full colour, P, E and A. E and A with D give stippled shade. E and A with R give another, different, full-colour. There is also a blueing factor and one for dark-eye. Other colours due to further combinations of factors.

Phyteuma. Inheritance of anthocyanin in the cross *Ph. Halleri* × *Ph. spicatum.* Correns (589).

Pisum sativum. Here again as regards pigmentation, the colour of the testa has received a great deal of attention. Bateson & Killby (590) noted that greys and browns in the seed-coat are associated with coloured flowers, and crossed with whites they occasionally give 'reversionary' F_1 with purple (anthocyanin) spots, though spots were absent from the parents.

Lock (615, 621) sums up many of the results on pea colour. He states that the albino variety has no anthocyanin. The presence of a factor C produces grey (chromogen) in the testa, and red anthocyanin in the leaf-axils and flowers. The presence of S produces spotting of a reddish shade on the testa. The presence of P modifies red anthocyanin of the axils, flowers and spots on the testa to purple.

White (767, 768) gives a very useful summary of factors including those in which anthocyanin concerned; testa, purple or purple-spotted, violet eye, coloured leaf-axil, purple or red flower-colour, purple pod.

Bach (786). Red-flowered heterozygote from pink or red by white-flowered. Homozygous reds appear to contain same amount of anthocyanin in same concentration. Suggests red anthocyanin not only in greater concentration but of a different nature from pink.

Tedin (819). Three factors A, B and C concerned with flower-colour and give: light purple AAbbcc, rose AABBcc, violet AAbbCC, purple AABBCC, and white aaB(b)C(c). No heterozygous forms.

Meunissier (810). Black hilum dominant to white and segregates normally. Exceptions mentioned. Also some observations on purple seed-coat colour.

Vilmorin (842). Variety of purple-flowered pea has purple (or partly purple) pods; on yellow, the effect is red. Some white-flowered plants have traces of purple in young pods. When purple-flowered with green pod is crossed with white-flowered with tinged pod, F_1 all purple. F_2 green, purple, tinged, yellow and red pods and white, purple and pink flowers. Seed, garnet marbled, plain garnet and (in white flowers) white or faintly marbled.

Kajanus & Berg (792) used varieties of *P. sativum* and *arvense*. Kajanus's factors R, G, P, S and M correspond to Lock's C, P, S, D and M. R gives rose flowers, light-brown testa and dark hilum. G and R together dark red flowers, grey-green testa, sometimes somewhat violet spotted. O hinders the development of blood-red to a red-brown testa. P produces dark violet spotting of testa which can extend to flecking and violet colour. M causes dark brown marbling of testa. S causes black colour of hilum. Effect of P, as well as that of M, is dependent on R and G.

Later Kajanus (878) suggests further factors. Z gives, with R, red colour to testa. RGZ blood-red, RgZ red-brown; without R, white. O suppresses red ground colour to light brown or grey-green. If Z and O absent, ground colour pale brown. If Z absent and O present, brownish. P produces, if R and Z present, a violet spotting; blue-violet if RGZO present, red-violet if RgZO. In red seeds, spots dark, with RGZo black. M produces a brown marbling, if R present. Flecks with RGZO dark brown, with RgZO rust-brown; in red seeds, especially RGZo, indistinct, and if Z absent, very pale. S causes black hilum. If absent, and R present, light brown, though in red seeds it is red. If S and R absent, hilum white. With red-violet flowers, leaf bases deep red, paler with rose flowers and green with white flowers.

Hammarlund (865) notes linkage between violet flower and green pod, white flower and yellow pod.

Kappert (869) is concerned with the development of a ring of pigment at base of stipules.

Polemonium. Inheritance of blue anthocyanin in F_1 from cross of *P. caeruleum* by *P. flavum.* Correns (589).

Dahlgren (773) crossed white variety by blue. Blue dominant and segregates normally.

Portulaca grandiflora. Yasui (820). Pale yellow variety has yellow pigment, probably flavone, represented by C. White variety (c) is without this pigment, and flowers do not react with ammonia. When mated together, they produce magenta and segregate normally. CR magenta, C yellow, R white and r white.

Ikeno (830) deals with orange, yellow, red, magenta and white. Four factors are involved C, G, R and B. CC or Cc is orange, G lightens C and gives yellow. R causes C to become red, and B produces magenta if C and R are present. White may carry R and B, but not C. R and B said to be linked. White flowers have green foliage, coloured flowers red foliage.

Primula sinensis. The bulk of the work on this genus is due to Gregory & Bateson. The following is a summary of the results which have been published by Gregory (660).

All colour in *P. sinensis* is due to anthocyanin except the yellow of the eye which is plastid pigment.

Varieties of flower-colour. Albinos: The flowers contain no anthocyanin; they are differentiated into dominant and recessive whites, but the flowers of both look alike.

Full colours: Salmon-pink. True red—'Orange King.' Blue reds—Light (dilute) and deep shades of magenta and crimson. Crimson is less blue than magenta, but is more blue than 'Orange King.' Blues—various shades.

Pale colours: Shades of pink, of which 'Reading Pink' is the deepest; of these apparently some correspond to magentas, and others to crimsons, but the difference cannot be detected except by crossing.

Distribution or pattern of colour. 'Sirdar.' The pigment of the petals occurs in separate minute dots, and the edges of the petals are white. The pigment itself may be either magenta, crimson or blue.

'Duchess.' The pigment occurs as a flush round the eye of the corolla, and may be either magenta or crimson.

Colour in the spot external to the eye. In certain varieties there are spots of deep colour on the petals just external to the eye. Deep spots are not fully developed unless the stigma is coloured; nor even if the stigma is coloured are they developed in plants which have no yellow (plastid) eye. Spots are deeply coloured only in deeply coloured flowers; on light flowers they are similar to those in the flowers with a green stigma. They also depend on the base colour, since they are not visible in pale coloured flowers, nor in flaked flowers unless the stripe occurs on the area occupied by the spot.

Colour in the ovary, style and stigma. By loss of an inhibiting factor, colour may be formed in these organs.

Varieties of stem-colour. Anthocyanin may be entirely absent when the stem is green. This condition is associated with recessive white flower-colour (occasionally dominant white, i.e. 'Pearl'), or with pale colours of which the deepest is 'Reading Pink,' and no flower-colour deeper than

'Reading Pink' is ever found on plants devoid of anthocyanin in the stem.

There may be faint colour in young leaves and petioles associated with pale flower-colour.

There may be pigment at the bases only of petioles and pedicels, associated with 'Sirdar' flower-colour.

The stem may be fully coloured with purplish-red sap, either light (dilute), or deep, associated with full flower-colours, i.e. salmon-pink, crimsons and magentas; but whereas light magentas and crimsons can be borne on both light- and deep-stemmed plants, the deepest magentas and crimsons can only be borne on deep-stemmed plants.

The stem may be coloured with pure red sap associated with 'Orange King' flowers.

The stem may be coloured with blue sap associated with blue flowers.

Hence the direct relationship between stem- and flower-colour may be summed up as

Recessive whites and green stem.

Pale flower-colours and green, or faintly coloured, stem.

Full flower-colours and full-coloured stem (and of these deep flower-colour and deep stem).

True red ('Orange King') flower-colour and true red stem.

Blue flower-colour and blue stem.

Factors (*stems*). Colour in both flowers and stems can be produced by two or more complementary factors. Keeble & Pellew (644) obtained an F_1 with magenta flowers from two whites. As regards stem, there is no case of colour from mating two greens, but plants heterozygous for colour in the stem have given 9 coloured : 7 green stems.

Slight colour in the stem is due to the factor Q, which is dominant to complete absence of colour.

The difference between full colour and faint colour is due to a dominant factor, R.

The difference between 'Sirdar' and full colour is due to a factor, F, which regulates distribution of colour. Thus an individual heterozygous for all three factors will give

9 FRQ......full colour (magenta or crimson).
3 RQ ......Sirdar (magenta or crimson).
3 Q ....faint colour (pinks).
1 – ....green (whites).

Purplish-red in the stem is dominant to the true red of 'Orange King,' and one factor only is necessary to produce the change.

Blue in the stem is recessive to all colours.

No case is known of inhibition in the stem, that is, there is no dominant green stem.

Light shades in the stem-colour (flowers light) are dominant to deep shades (flowers light or deep). Probably one factor is concerned with heterozygous forms, but there may be more.

In some cases there is one factor between crimson and magenta; other cases indicate two factors.

It is doubtful whether there exists one or more diluting factors. There is one factor diluting stem-colour and another flower-colour, and these are inherited independently.

Factors (flowers). Colour in flowers may be produced by two or more complementary factors.

The difference of one factor regulates the distribution of colour between 'Sirdar' and full colour, and also between pale and full colour as in stems.

Full colours are dominant to pale colours.

Magenta is dominant to crimson.

Magenta and crimson are dominant to full red ('Orange King').

Blue is recessive to all magentas and reds.

Whites may be dominant or recessive to colours.

Dominant whites are generally red-stemmed. A green-stemmed white variety, 'Pearl,' is a dominant white (Keeble & Pellew, 644); the same authors also report that a full red-stemmed white, 'Snow King,' may be recessive. The flower is only white in dominant whites if homozygous for the inhibiting factor. It has also been suggested that suppression of colour is due to two inhibiting factors, central and peripheral, one of which is represented in 'Duchess.'

Keeble & Pellew (644) provide further results on dominant whites. Hitherto all whites with red stems have been regarded as dominant whites. Keeble & Pellew give evidence for regarding 'Snow King,' which is a red-stemmed variety with white flowers, as a recessive white. Crosses between various plants of 'Snow King' and fully coloured varieties showed that 'Snow King' may be homo- or heterozygous for the dominant white factor, or may be entirely without it.

'Snow King' × 'Snow Drift' gave magenta F_1. Hence these two white-flowered varieties can give colour.

Further observations. Gregory, de Winton & Bateson (863). Blue flowers. Two varieties, one (blue) contains B and has soluble anthocyanin in petals; the other (slaty) has solid or crystalline anthocyanin. Thus the relations are: Magenta BR, red (crimson) bR, blue Br and slaty br.

The above-described blue and slaty are associated with light reddish (due to a factor now called L) stems. When this is absent and the stems are deep coloured, blue and slaty flowers become curiously mottled and slightly redder. In absence of a factor, G, for green stigma, the stigma becomes red. Fully red and dark blue flowers are not found unless G is absent. A variety 'Harlequin' is described in which one or two petals are fully coloured, the others pale. Also mosaics of red and blue or pale and deep magenta. The factors B, G and L are involved in one linkage group.

Raphanus. Riolle (816). Hybrids of *Raphanistrum* and *sativus*. Root-colour classified as rose, violet, black, grey or white giving, *inter se*, violet. Every possible combination of characters in F_2.

Reseda odorata. Compton (674) has shown that orange-red colour (anthocyanin) in the pollen is dominant to bright yellow; self-fertilised heterozygotes throw about three reds to one yellow.

Ricinus communis. White (783, 784). Varieties of stem and leaf coloration described. Harland (806, 847) found later that two factors, M and G, were involved giving 9 MG rose : 3 Mg mahogany : 3 mG green : 1 mg tinged. Factors M and G are linked.

Rudbeckia hirta. Blakeslee (823). Purple disk crossed with yellow disk, purple dominant and segregates normally. Two yellows identified, one giving reddish, the other blackish colour with alkalies. These two forms crossed together give purple F_1 and in F_2 9 purple : 7 yellow. These yellows could again be segregated by alkalies.

Salvia Horminum. Saunders (590) has worked with the type and varieties of this species. The type has violet flowers, and the bracts of the inflorescence are also coloured violet. A red variety was used in which the flowers and bracts were red. Both pigments are anthocyanins. Loss of pigment gives an albino from which anthocyanin is completely absent.

The inheritance can be represented by the following scheme:

BR.........purple.
R ....red.
B ....white.
– ....white.

Senecio vulgaris. Trow (694) has published results of crossing a number of elementary species. It was found that red colour, anthocyanin, in the stem is dominant in some elementary species. It is suggested that one factor is involved, or possibly two.

Silene Armeria. Colour in flower dominant to albinism (de Vries, 601).

Sisyrinchium angustifolium. Miyake & Imai (833). Strain having bluish-purple flowers crossed with strain having white flowers with coloured centre. Dominance of white, but not complete. Normal segregation in F_2.

Solanum tuberosum. East (641) has published some results. He states that presence of anthocyanin in the stem is dominant to its absence, and segregates on Mendelian lines. Purple in the flower is probably dominant to its absence. The colour of the tubers is either red or purple, and purple is dominant to red.

Salaman (647) has published more extensive results. Black tubers of the variety, 'Congo,' have purple anthocyanin in the skin. The red-tubered variety has red anthocyanin. White tubers have no anthocyanin. Two factors are required for colour (red), and a third dominant factor gives purple. *S. etuberosum* has a dominant white factor inhibiting the purple.

In the flowers all pigment is anthocyanin. Heliotrope colour is due to two factors, and purple to a third factor. In varieties of *tuberosum*, colour is confined to the upper surface. In *S. etuberosum*, pigment is developed on the lower, and is inhibited on the upper surface. In *S. verrucosum* both surfaces are probably free from inhibitors.

Trifolium. Inheritance of anthocyanin in flowers and seeds. Red in flower-colour is dominant to both white and blue (Kajanus, 684).

Raum (798, 837). A few facts about inheritance of violet seed-coat colour and flower-colour.

Triticum. St Clair Caporn (781). Purple grains (anthocyanin in pericarp) are dominant to unpigmented. Grains, with purple streak only, appeared in F_2.

Tropaeolum majus. Rasmuson (778, 814). Red anthocyanin is dominant to its absence and segregates normally, the heterozygotes being paler. There are also two classes of red according to whether it is present with a pale or deep plastid pigment. Variegated flowers dominant to self colour.

Veronica longifolia. Colour (anthocyanin) dominant to its absence (de Vries, 601).

Vicia Faba. Heuser (866). Both violet and black dominant over normal colour.

Vigna sinensis. Mann (718). Account of blue and red pigments in testa. Early work by Spillman (667, 708).

Harland (788). Three varieties of flower-colour were used, deep reddish violet, pale violet and white. The relationship on crossing is normal,

LLDD dark, LLdd pale, llDD white and lldd white. Seed-coat pigmentation was also investigated, but author does not state nature of pigment. A factor B producing black seed-pigment causes anthocyanin to appear in tip of young pod and in calyx and peduncle, but only has effect when factor for flower-colour and anthocyanin in stem and leaves is present. Later work (805). The presence of anthocyanin in stem and leaf-stalk is due to factor X, dominant to its absence, which segregates normally. The purple colour of pod is due to one main factor, P. There is an additional variety of flower-colour, tinged, due to factor G, dominant to its absence, but recessive to D.

Viola cornuta. Colour (anthocyanin) dominant to its absence (de Vries, 601).

Kristofferson (870). Cross between violet-flowered *V. tricolor* and light yellow *V. arvensis* is violet. In F_2 there was segregation into 9 violet, 3 blue, 3 red and 1 light yellow. Blue is due to factor B, red to R. When both are present (BR), the petals are violet; when neither, light yellow. The F_1 generation between above forms and *V. Munbyana* was also obtained.

Vitis vinifera. Rasmuson (746). Autumnal colouring either red or yellow. White and red-berried varieties of *vinifera* and all varieties of *Riparia* and *Rupestris* form yellow autumnal tints; blue-berried varieties form red. Red dominant to yellow and segregates normally. Berry colour due to two factors. One alone gives white, the other red, both together, black. F_2 9 black : 3 red : 4 white.

Zea Mays. Correns (579) made the first investigations on this species. As regards characters connected with anthocyanin he states that two such characters behave in a Mendelian way, i.e. colour in the pericarp (red or absence of red), and colour in the aleurone layer (blue or white). The degree of dominance was found to vary.

Lock (596, 607) has published further results. He states that blue, black or purple pigments are confined to cells of the aleurone layer. Red is confined to the pericarp. The pigment is situated in the cells, and to some extent also in the cell-walls of the pericarp. He points out that in cross-bred cobs, blue will occur mixed since the character is not maternal; the pericarp colour, on the other hand, appears in either all or none of the grains, as it is a maternal character. The work includes a large number of results in connection with the cross, blue × white and the reciprocal. The results show that there is irregularity in the dominance of blue. Red colour in the pericarp is dominant to its absence, and forms a simple Mendelian case.

East & Hayes (656) have published extensive results on Maize. They show that Lock's irregular results were due to the fact that the whites he employed in crossing were carrying different factors. They themselves worked with a variety having red (anthocyanin) pigment in the aleurone layer, and the formation of this pigment was found to be due to two factors, C and R. A third factor, P, modifies the red to purple. There is also a fourth factor, I, which inhibits the red and purple colour. Red pigment in the pericarp, cobs and silks, is dominant to its absence. It may be present in each of these parts separately and independently of the others. No plant has been obtained which has red glumes and yet shows no red colour in other parts of the plant. One, however, has been found that is pure for red glumes, and shows no red in other parts with the exception of the silks.

Emerson (774). In addition to C and R, a third factor, A, must be present before any endosperm colour can be developed. Emerson (826). As regards anthocyanin coloration four types are described; purple, sun-red, dilute purple, dilute sun-red, and four sub-types, weak purple, weak sun-red, green-anthered purple and green-anthered red. Sun-red and dilute sun-red are dependent on daylight. Purple and dilute purple develop in the dark. Infertile soil and accumulation of carbohydrates intensify colour. Two factors are involved and a series of multiple allelomorphs. Some of the factors also involved in development of aleurone colour. Bregger (770). C factor for aleurone colour linked with factor for waxy endosperm. Hutchison (849). Linkage between factors for anthocyanin in endosperm and shrunken endosperm. Lindstrom (761). R factor for aleurone colour linked with G factor for chlorophyll development. Anderson (821). Salmon silk (possibly anthocyanin) recessive to green and linked with factor, Pl. Kempton (793). Genetics of spotting in aleurone colour. Lindstrom (871). Summary of characters in Maize.

Connection of Flower-colour with the Presence of Anthocyanin in Vegetative Organs, Fruits and Seeds.

There are two main classes of phenomena which may come under this heading. First: the relationship between anthocyanin pigmentation in all the different organs of the plant. In many cases, however, as, for instance, in Maize and Rice, the pigment will be inherited independently in various organs. Secondly: the relationship of anthocyanin pigmentation to characters of quite different nature.

The causes of these associations may be various, and are usually not definitely known. The connection may be due to the effect of one factor, to linkage, to the interaction of various factors, etc.

As examples of the first class the following may be taken:

Antirrhinum majus. The white, ivory and yellow-flowered varieties never produce anthocyanin in any part of the plant. The author has noticed that the stems and leaves of deep magenta- and crimson-flowered varieties produce anthocyanin to a considerable extent, whereas in intermediate and pale varieties these organs are practically free from anthocyanin.

Aquilegia. The factor, R, for red flower-colour causes the leaves to produce anthocyanin on both surfaces. In its absence, pigment only on the under surface. (Kristofferson, 850.)

Atropa Belladonna. The type has anthocyanin in the flowers (brown), fruits (black) and stems (tinged with red). The albino has yellow flowers and fruits, and a green stem (Saunders, 578).

Beta. Some varieties have leaves coloured quite red with anthocyanin. This only happens in red-fleshed varieties, i.e. when anthocyanin is formed in the inner tissues, as well as in the skin, of the root. In varieties having colourless flesh, the leaves are green (Kajanus, 662).

Corchorus capsularis. Varieties with deep red (anthocyanin) stems, petioles and fruits have the teeth of the leaves also tipped with red. Other strains with less pigment in the stem have no colour in the leaf teeth (Finlow & Burkill, 677).

Datura. D. Tatula has violet flowers and red stems (anthocyanin). *D. Stramonium* has white flowers and green stems (Saunders, 578, 590).

Digitalis purpurea. The presence of a factor, M, for magenta pigment always brings with it colour in the spots on the lower inner surface of the corolla. Even in the presence of M the general colour in the corolla may be inhibited by another factor I, but the spots remain red (var. white spotted with magenta). If M is absent, the spots are colourless (Keeble, Pellew & Jones, 645; Saunders, 666).

Gossypium. In the Indian cotton, strains having yellow flowers, or white flowers with only a basal spot on the petals, have green foliage. In strains having red pigment, the varieties having red flowers (forms homozygous for the reddening factor, see p. 170) have the veins and lamina of their leaves suffused with red. In leaves of the plants which have flowers red on yellow, or red on white (forms heterozygous for the reddening factor) the anthocyanin is confined to the veins (Martin Leake, 664).

Helianthus. In the wild *H. annuus* the disk is purple, the colour (anthocyanin) being in the tips of the paleae, the margin of the corolla and the styles and stigmas. The 'Russian Sunflower' has the tips of the paleae yellowish-green, the corolla, styles and stigmas yellow (Shull, 623).

The chestnut-red-flowered Sunflower, which has anthocyanin on yellow in the ray florets, has dark reddish-purple stems. The purple-flowered variety, anthocyanin on primrose, has no reddish-purple in the stems (Cockerell, 722).

Lathyrus odoratus. The albinos always have green leaf-axils. When a certain factor for production of colour in the axils is present and the flowers are coloured, there is a reddish-purple spot in the axil; when the axil colour factor is absent, the axils are green (Bateson, 627). Red tendrils are (?) always associated with red in the axils of leaves (Bateson, 602). In the red-purple series there is more development of anthocyanin throughout the vegetative parts than in the normal, giving the plant a dusky appearance (Punnett, 854).

Oenothera. In the variant *Oenothera rubricalyx* red buds and hypanthia are said to be associated with red stems (Gates, 658). In the F_1, however, from reciprocal crosses between *O. rubricalyx* × *O. Lamarckiana*, plants appeared in which red buds were associated with a low degree of red pigmentation in stems and rosettes, whereas pale red buds and green hypanthia were associated with brilliant red stems, and buds entirely free from anthocyanin with dark stems (Shull, 720).

Pisum sativum. White-flowered plants have no colour in the leaf-axils and no red or purple (anthocyanin) colour in the testas. Red-flowered plants have red colour in the leaf-axils and red spots on the testa. Purple-flowered plants have purple colour in the leaf-axils and purple spots on the testa (Lock, 621).

Primula sinensis. Recessive white-flowered varieties may have green or red stems (Keeble & Pellew, 644). Pale-flowered varieties may be on green or faintly coloured stems; deep-flowered varieties appear on deep red stems only; whereas light-flowered varieties may be borne on either deep- or light-stemmed plants. The 'Sirdar' variety, in which the pigment is in minute dots and the edges of the petal are white, has stems with a red 'collar,' i.e. base of stem. If the flower-colour is crimson or magenta, the stem-colour is purplish-red; if the flower is true red ('Orange King') the stem-pigment is also true red. Blue flowers have blue stem-pigment.

The deep spots of colour external to the eye are not fully developed unless the stigma is coloured. They are deeply-coloured only in deeply-coloured flowers: in light flowers they are similar to those in flowers with a green stigma. They also depend on base colour as they are not present in pale flowers, nor in striped flowers, unless the stripe occurs in the area occupied by the spot (Gregory, 660).

The deepest-flowered varieties are only found with coloured stigma. Blue and slaty flowers become mottled when on deeply-coloured stems (Gregory, de Winton & Bateson, 863).

Salvia Horminum. The bracts of the inflorescence are purple, red, or green according to whether the flowers are purple, red, or white (Saunders, 590).

Zea Mays. Red pigment may appear in the pericarp, cobs and silks (stigmas) separately and independently in each part. If the glumes are red, red is present in some other part of the plant, though it may only be in the silks (East & Hayes, 656; Emerson, 657).

Other examples of association of coloured flowers or fruit with anthocyanin in the vegetative organs, white flowers or fruits being associated with its absence, are: *Andropogon* (Graham, 742), *Collinsia* (Rasmuson, 813), *Dolichos* (Harland, 807), *Ipomoea* (Miyazawa, 834), *Malope* (Rasmuson, 745), *Portulaca* (Ikeno, 830), *Triticum* (St Clair Caporn, 781), and *Vitis* (Rasmuson, 746).

In the following cases there is an intimate relationship between seed-colour and anthocyanin in flower or foliage. (Since information is not given as to nature of seed-pigment, these need not necessarily be derived from anthocyanin, though they may be in some cases, for example, *Lupinus.*) *Dolichos* (Harland, 807), *Linum* (Tammes, 858), *Lupinus* (Hallqvist, 828), *Papaver* (Martin Leake & Ram Pershad, 809), *Vigna* (Harland, 788, 805).

Secondly, we may consider those cases where a factor for colour is associated with characters of a different nature.

Antirrhinum. The true albino is associated with poor vegetative development.

Aquilegia. The factor, R, for red flower-colour is associated with higher stature, shiny (as contrasted with dull) seed-coat and greater weight of seed (Kristofferson, 850).

Lathyrus. In hooded varieties, the colour of standard and wings is more nearly alike than in the case of erect standard. In red-purple series, the plants are shorter than normal (Punnett, 854).

Linum. Some colour factors affect breadth and crimping of petals and also power of germination of seeds (Tammes, 858).

Matthiola. In certain strains of stocks, hoariness of leaves depends on two factors, H and K, but it is only manifested when the two colour factors C and R are present (Saunders, 692).

Papaver. Length of vegetative period is associated with certain flower-colour factors (Martin Leake & Ram Pershad, 809).

Pisum. In presence of A and C for flower-colour there is an abnormal structure of the hilum (Tedin, 819).

There are many other cases, less definite, where flavour is correlated with colour in fruit and vegetables (Goff, 547).

Heterozygous Forms.

Just as in the case of factors for other characters, so also in the case of factors concerned with pigmentation we find heterozygous forms. The one of most frequent occurrence is that due to the inheritance in the zygote, from one parent only, of the factor which actually produces colour (anthocyanin). We find it, for instance, in the following genera:

In *Atropa Belladonna*: when the red-stemmed, brown-flowered type is crossed with the green-stemmed, yellow-flowered variety, the colour in stems and flowers of F_1 is less intense than the parent (Saunders, 578).

In *Corchorus capsularis*: when deep red-stemmed varieties are crossed with green-stemmed varieties, the colour in stems of F_1 is less intense (Finlow & Burkill, 677).

In the cross *Datura Tatula*, with red stems and purple flowers, by *D. Stramonium*, with green stems and white flowers, the colour in flowers and stems in F_1 is less intense (Saunders, 578).

In Egyptian cotton, *Gossypium*: when the variety with a spot on the leaf is crossed with the non-spotted variety, the spot in F_1 is paler in colour (Balls, 618, 626).

In Indian cotton: when strains with anthocyanin in the vegetative parts are crossed with green-leaved strains, the colour in leaves and stems in F_1 is paler (Leake, 664).

In *Linum usitatissimum*: a cross between a blue-flowered and a white-flowered variety gave an F_1 with paler blue flowers than the parent (Tammes, 668).

On the other hand, in many cases where there are factors producing colour (anthocyanin) no heterozygous forms at all exist, as in *Lathyrus* and *Matthiola*.

Cases, fundamentally different, though giving a similar result to those above, are those in which the type has anthocyanin inhibition by an inhibiting factor, and the variety has lost the inhibitor and produces anthocyanin. When the type is crossed with variety, the F_1 is heterozygous for the inhibitor and is less intensely coloured than the parent. Examples of such are the F_1 from *Coreopsis tinctoria* with yellow (plastid pigment) flowers and the variety *brunnea* (anthocyanin on plastids) (de Vries, 577); the F_1 has paler flowers than *brunnea*; also the F_1, probably, from the yellow Sunflower by its chestnut-red variety.

Somewhat similar, though not strictly comparable, is the case in *Primula sinensis* where whites tinged with anthocyanin are obtained in F_1 from the cross—dominant white × fully-coloured variety (Gregory, 660).

In connection with the class mentioned above which is heterozygous for the factor for colour, *Antirrhinum* offers an interesting illustration. Colour, i.e. anthocyanin, in *Antirrhinum* is produced by the action of one factor, L, which gives ivory tinged with magenta, but there is no apparent difference in the flower-colour between individuals of the composition LL and Ll. In the presence of an additional factor D, a deepening factor, the zygote develops more pigment and is a deeper colour, but the ultimate colour depends on whether the zygote is homozygous or heterozygous for L, its condition as regards D having no effect; thus LlDD(d) is pale magenta, and LLDD(d) is intermediate magenta.

A second heterozygous form in *Antirrhinum* is also interesting. Individuals heterozygous for the striping factor have magenta stripes on a tinged ground, individuals homozygous for the striping factor, magenta stripes on an ivory ground (Wheldale, 638).

Mirabilis Jalapa is of special interest in this connection since it is a species in which individuals heterozygous for any of the factors so far identified have a heterozygous form, and this applies, not only to the factor converting yellow chromogen into anthocyanin, but also to the factor for the presence of chromogen (Marryat, 636).

In Indian cotton, individuals heterozygous for the factor which produces anthocyanin in the flower have this pigment diffused through part only of the petal, whereas homozygotes have anthocyanin throughout the flower (Leake, 664).

In some strains of *Phaseolus*, plants which are heterozygous for mottling with anthocyanin in the testa have mottled seeds, whereas homozygotes in this factor have self-coloured purple seeds (Shull, 624). It is at the same time well known that homozygous true-breeding mottled races do exist, and Emerson has attempted to explain this anomaly by postulating two factors concerned in mottling.

In *Godetia amoena*, Rasmuson (836) notes there are two varieties, one with large spot on petal, the other with small spot. On crossing, the heterozygotes have two spots, and these give, on selfing, 1 large spot : 2 double spot : 1 small spot.

In *Ipomoea*, according to Miyazawa (834), a factor, G, modifies the green colour of the leaves. If G is homozygous, then D produces a dark-red flower, and B has no effect. If G is heterozygous (or absent), D produces a scarlet flower, and DB produces a magenta flower.

Pattern in Colour-variation.

Most of the following cases have been mentioned in the previous accounts, but they are enumerated again here for the convenience of reference. Pattern generally implies the localisation of pigment in definite areas.

Antirrhinum majus. The 'delila' variety may perhaps be regarded as an instance of pattern. For every variety with anthocyanin there is a corresponding 'delila form,' that is one in which the lips are coloured but the tube is ivory; there is always a sharp line of demarcation at the point of union of the tube and lips (Wheldale, 638).

Aquilegia vulgaris. Petals are white-margined in absence of a factor C (Kristofferson, 850).

Arum maculatum. The black spots (anthocyanin) on the leaf are probably due to a Mendelian factor which is dominant to the lack of spots (Colgan, 655).

Collinsia bicolor. Spots on upper lip are dominant and segregate normally (Rasmuson, 813).

Cypripedium (*Paphiopedilum*). Patterns of spots or stripes (anthocyanin) may be present. On crossing, there is segregation into striped and spotted; the former appears to be dominant (Hurst, 605).

Digitalis purpurea. The presence of spots on the inner lower surface of the corolla is bound up with the presence of pigment in the corolla generally. The ground colour of the corolla can be inhibited by an inhibiting factor, but not the spot colour (Keeble, Pellew & Jones, 645).

Erodium cicutarium. A variety occurs without patches at the base of the petals (de Vries, 669).

Gentiana punctata. There is a variety in which the dark patches in the flower are absent (de Vries, 669).

Godetia amoena. One variety with large petal spot, another with small. Both dominant to unspotted and segregate normally (Rasmuson, 836).

Gossypium. The red anthocyanin 'spot' on the leaf is a Mendelian character, and is less intense in the heterozygous forms. In Indian cotton a red spot on the flower is characteristic of the heterozygote, but an almost completely red flower is characteristic of the homozygote (Balls, 618, 626; Leake, 664).

Ipomoea. Variety with white margin to coloured flower ('hukurin') is dominant (Miyazawa, 777).

Lathyrus. 'Marbling' (due to lack of F) causes colour to be broken up; also keel and under surface of wings pure white (Punnett, 873).

Mimulus quinquevulnerus. "Here the dark brown spots vary between nearly complete deficiency up to such predominancy as almost to hide the pale yellow ground-color" (de Vries, 601).

Papaver orientale. The dark patches of anthocyanin at the base of the petals are absent in some varieties (de Vries, 669).

P. Rhoeas. In some varieties there is a white margin to the petals due to the presence of a dominant inhibiting factor (Shull, 693).

P. somniferum. Some varieties have dark basal patches of pigment (anthocyanin) on the petals. From others it is absent. On crossing dark "'Mephisto' with the white-hearted 'Danebrog' the hybrid shows the dark pattern" (de Vries, 601).

Phaseolus. In some strains the mottling of anthocyanin in the testa is due to a factor M, but the pattern only shows when the plant is heterozygous for M. In other strains the mottling occurs in both heterozygous and homozygous forms (Emerson, 631, 632; Shull, 616, 624; Tschermak, 582, 695).

Phlox. Dark eye factor dominant (Gilbert, 729; Kelly, 808).

Pisum. The mottling of anthocyanin in the testa is due to a definite Mendelian factor (Lock, 621).

Primula sinensis. The patches of anthocyanin outside the yellow eye are a definite Mendelian character. They do not develop to their fullest intensity unless the stigma is red, or the flower is deep-coloured; the spots are pale in pale-coloured flowers and in flowers with a green stigma.

The 'Duchess' distribution of pigment, i.e. a ring of colour round the eye, spreading to a flush, may be regarded as pattern. So also the 'Sirdar' variety, where the colour is present in minute dots giving the flower a dusty appearance, the margins of the petals being white (Gregory, 660).

Solanum tuberosum. The flowers of several white varieties have 'tongues' of colour radiating out from the throat to the tips of the corolla segments. The 'tongues' show different degrees of intensity of coloration, governed, probably, by different factors (Salaman, unpublished work).

Tropaeolum majus. There is at present very little published work on the flower-colour of this genus, but, as far as observations go, there is a definite inheritance of the blotch at the base of the petals. The original type has an orange-red (anthocyanin and yellow plastids) flower, rather deeper in colour at the base of the petals, and with dark 'honey-guides' running into the 'claw.' There are varieties in which the anthocyanin has disappeared from the main part of the flower so that the flower is yellow with orange-red blotches at the base of the petals. If the flower-colour varies to pale yellow (see p. 152) the orange-red blotch then appears as carmine. The red anthocyanin may give rise to a purple variation and there are correspondingly coloured blotches, maroon on a deep yellow and purple on a pale yellow ground.

Reviewing the cases of pattern set out above we are able to distinguish several different types. One type, for instance, includes all those spots, lines and streaks which form a normal part of flower-coloration and which, in the opinion of biologists, have a significance as signals or guides to insect visitors. This group would include the markings in *Cypripedium*, *Digitalis*, *Erodium*, *Mimulus*, *Papaver*, *Primula* and *Tropaeolum*. As far as can be judged from the evidence at hand, some of these patterns (*Papaver*, *Tropaeolum*) are inherited independently of the ground colour of the flower; the factors for others are intimately associated in various ways with the ground-colour factors (*Digitalis*, *Primula*). A second type of pattern is that which is not a feature of the normal flower-coloration, but is only revealed in varieties from which some of the colour factors are absent, for example the 'delila' variety of *Antirrhinum* and the 'Duchess' and 'Sirdar' varieties of *Primula*. A third type of pattern, for instance the spots on *Arum* leaves, the leaf-spot of *Gossypium* and the mottling of seeds of certain strains of *Pisum* and *Phaseolus*, forms a normal attribute of the plant to which we can assign no special significance.

Striped Varieties and Bud-variation.

A common form of variation in many flowers is striping, i.e. the arrangement of colour in bands or stripes; these markings may vary in thickness from the narrowest hair-like streaks to broad bands, or elongated patches, which may then occupy almost the whole of the flower. It is difficult to define the limits of striping for, on the one hand, among striped varieties we frequently find sectorial variations in which colour is definitely and symmetrically confined to a half, a third, or some other fraction of the flower. On the other hand, striping may pass into spotting or blotching, and it is questionable whether spotted and blotched flowers should be placed in the same category, though their genetical behaviour may be similar. Some of the genera and species in which striping occurs are the following: *Antirrhinum majus*, *Cheiranthus Cheiri*, *Dahlia variabilis*, *Dianthus Caryophyllus*, *D. barbatus*, *Mirabilis Jalapa*, *Papaver*, *Petunia*, *Primula sinensis*, *Tagetes*, *Tulipa*, *Zinnia elegans*.

As further examples of striping and sectorial variation, we may add those mentioned by de Vries (669), i.e. *Celosia variegata cristata* (inflorescence), *Centaurea Cyanus*, *Clarkia pulchella*, *Convolvulus tricolor*, *Cyclamen persicum*, *Delphinium Ajacis*, *D. Consolida*, *Geranium pratense*, *Helichrysum bracteatum*, *Hesperis matronalis*, *Impatiens Balsamina*, *Nemophila insignis*, *Papaver nudicaule*, *Portulaca grandiflora* and *Verbena*.

Striping may be regarded as a variation under normal conditions[1]; there is, however, a kindred phenomenon called flaking which appears in flowers, otherwise self-coloured, towards the end of the vegetative season, or when the plant is in an unhealthy condition. Ex. *Dahlia* (Hildebrand, 561, 569), *Matthiola*.

Striping is apparently not a phenomenon which occurs in plants in the wild state, and, according to the theory of Louis Vilmorin, as expounded by de Vries (669), it only appears in coloured plants which have already produced a white or yellow variety, that is a variety free from anthocyanin pigment. It is certainly true that striping is most usual in connection with anthocyanin; as a rule it is not shown by soluble yellow pigments (*Antirrhinum*, *Althaea*), though in *Papaver nudicaule*, according to de Vries (669), one finds a yellow variety with dark orange stripes. In *Mirabilis*, also, yellow pigment may be found striped upon a white ground. Flowers with yellow plastids rarely, if ever, occur in the striped condition; such yellow flowers may, however, occasionally show light segments containing paler derivative plastid pigments. The common form of striping is that of anthocyanin on an albino ground, either white (*Primula sinensis*) or yellow; the yellow may be a soluble pigment (*Antirrhinum majus* var. yellow striped crimson), or a plastid pigment (*Cheiranthus Cheiri*, *Tagetes*, *Tulipa* spp.). Less commonly one finds deep anthocyanin stripes on a pale anthocyanin ground of the same colour (*Antirrhinum*, see p. 162), or stripes of one coloured anthocyanin on the ground of another colour, as in the case mentioned by de Vries (669) of *Delphinium Consolida striatum plenum*

[1] De Vries (669) notes, however, that striping in *Camellia japonica* may depend on the time of flowering.

in which the flowers may be red striped with blue[1]. *Mirabilis Jalapa* presents a remarkable series of striped varieties: there may be stripes of magenta (anthocyanin) on a white ground, or of soluble yellow pigment on a white ground, or both magenta and yellow stripes on white (tricolour). Moreover *Mirabilis* is characterised by a number of heterozygous forms, and we may find a homozygous stripe colour on a heterozygous ground colour, for instance, deep yellow stripes on a pale yellow, or orange-red stripes on a magenta-rose ground (Marryat, 636).

Another curious phenomenon is the limitation of striping in most species to one colour, for instance in *Antirrhinum* the anthocyanin of rose doré and bronze varieties does not occur in stripes, and the same is true of the pale 'tingeing' of anthocyanin. It is only the factor for full colour that is affected by the striped condition.

With regard to sectorially coloured flowers de Vries (669) notes that they appear to manifest a tendency towards a simple proportion between the two parts. "Frequently exactly half of the flower is atavistic, sometimes a quarter or three quarters. I observed the proportion $\frac{3}{8}$ in white and red striped tulips and in partially dark blue and partially pale blue flowers of *Iris xiphioides*, etc."

Cockerell (754) notes irregular sectorial variegation in red *Helianthus*.

Variegation with anthocyanin occurs in many 'beautiful-leaved' plants, for example *Coleus* (Küster, 735, 775, 776).

The inheritance of striping has received a certain amount of attention. De Vries (669) has investigated the problem in *Antirrhinum* and other striped flowers. He noted that the striped variety of the larkspur (*Delphinium Ajacis* and *D. Consolida*) produces self-coloured flowers as well as striped ones; the self-coloured, moreover, may appear on the same racemes as the striped ones, or on different branches, or some seedlings from the same parent may be self-coloured while others are striped. Seeds, on the other hand, of these blue sports give rise again to the striped variety. Such a variety as this Larkspur, de Vries called 'eversporting,' for as he says: "Here the variability is a thing of absolute constancy, while the constancy consists in eternal changes." His more definite cultures were made with *Antirrhinum majus luteum rubrostriatum.* In *Antirrhinum* (see p. 162) the flower may be ivory striped with magenta anthocyanin, or yellow striped with crimson, this latter colour being merely magenta superposed upon yellow; the yellow pigment (luteolin) does not show striping. In the offspring of de Vries's plant, which was yellow striped with crimson, there were no pure yellows,

[1] There is also a tricolour of red, white and blue.

though all grades of striping were found, ranging from yellow, with a few of the finest stripes, through intermediates, to those with the broadest stripes, or even whole sections of red; and with these occurred a certain number of pure red flowers. The red flowers will appear suddenly as a sport (though without graduated links with striped forms) among the striped individuals, and conversely, the striped flowers will appear among the offspring of the self-form. Thus red can be obtained from striped, and striped from red, and both variations can be obtained by seed; but de Vries noted bud-variations and variations within the spike, only in striped individuals, though he is of the opinion that they might be found on red plants if sufficiently large numbers of individuals were dealt with.

The following table[1] represents the successive families obtained by de Vries from one striped plant:

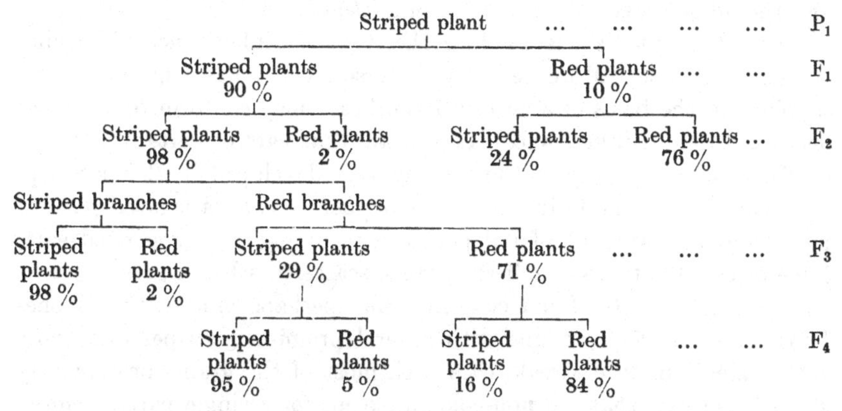

So far then de Vries expresses the facts of inheritance of striping thus:

1. The striped race is an inconstant one and consists of striped and red-flowered plants.

2. Striping itself varies continuously, but there is discontinuity between the striped and red forms.

3. The intensity of inheritance of finely striped plants is about 95–98 %, and they give rise to the red type either by seed or bud-variation.

4. The broadly-striped individuals produce more reds than the narrow-striped, the average being about 39 %.

5. The red individuals thrown by the striped resemble the red type, but differ in again throwing striped. The intensity of inheritance of the red character is about 70–85 %.

[1] See also Emerson (712).

6. The yellow variety (without anthocyanin) does not arise from the striped race.

De Vries draws attention to the fact that striped varieties do not occur in nature. Other examples he gives of striping are in Stocks, Liver-leaf (*Hepatica*), Dame's Violet (*Hesperis*), Sweet William (*Dianthus barbatus*) and Periwinkle (*Vinca minor*); also in *Hyacinthus*, *Cyclamen*, *Azalea*, *Camellia*, and in the Meadow Crane's-bill (*Geranium pratense*) when cultivated. Also these, he says, are known to come true to striping when seed is taken from striped individuals, and from time to time to throw self-coloured individuals. He made pedigree cultures of the Dame's Violet (*Hesperis matronalis*) for five years, and of *Clarkia pulchella* for four years, and they both behaved in exactly the same way as *Antirrhinum*. He further points out that other parts of plants may be striped, as, for instance, the red and white striped roots of Radishes, and the inflorescence of the Cockscomb (*Celosia cristata*).

From these results one would gather that the inheritance of striping is non-Mendelian. More recently Emerson (712) has suggested that striping, on the basis of a certain hypothesis, may conform to the Mendelian scheme of inheritance. It appears that ears of certain varieties of Maize show striping in the red pigment developed in the pericarp. Emerson says: "Plants in which this pigment has a variegated pattern may show any amount of red pericarp, including wholly self-red ears, large or small patches of self-red grains, scattered self-red grains, grains with a single stripe of red covering from perhaps nine-tenths to one-tenth of the surface, grains with several prominent stripes and those with a single minute streak, ears with most of the grains prominently striped and ears that are non-colored except for a single partly colored grain, and probably also plants with wholly self-red and others with wholly colorless ears." A number of selfings were made for several generations of both homozygous and heterozygous variegated cobs; homozygous and heterozygous variegated cobs were also crossed with true-breeding white male plants. From his experiments Emerson deduces the following results:

1. That the amount of pigment in the pericarp of a variegated grain bears a definite relationship to the amount of pigment in the grains of the plant which grows from it. The relationship is such that the more red pigment there is in the grains planted, the more likely are the plants which come from them to produce self-coloured red ears, and the less likely to produce variegated ears.

2. That when F_1 red ears produced by selfing a homozygous varie-

gated-eared plant are selfed and sown they give rise to red-eared and variegated-eared plants in Mendelian proportions; in the same way when crossed with white-eared races they behave as if they had been produced by a cross between red-eared and variegated-eared races.

3. That the F_1 red ears arising from selfed heterozygous variegated-eared plants behave in some cases as if they were hybrids between red-eared and variegated-eared races, and in other cases as if they were hybrids between red-eared and white-eared races.

4. That the F_1 reds arising from crosses between both homozygous and heterozygous variegated-eared plants and white-eared races behave as if they were hybrids between red-eared and white-eared races.

Thus, says Emerson, any interpretation of the above results must take into account these facts: (1) that the more red there is in the pericarp, the more frequently do red ears occur in the progeny; (2) that such red ears behave just as if they were F_1 hybrids between red and variegated, or red and white races.

To explain these phenomena Emerson suggests the following hypothesis. The zygotic formula for a plant homozygous for variegated pericarp may be regarded as VV; heterozygous for variegated pericarp as V –. If in any somatic cell VV, from some unknown cause, a V factor were transformed into a factor for self-colour, S, that cell would then be represented as VS. Any pericarp cells descended from such a cell would be red, and if all the pericarp cells of a grain were thus descended the grain would be self-red, just as if the plant bearing it were a hybrid between pure red and variegated races. Of the gametes, moreover, arising from such somatic cells, one-half would carry V, and one-half S, again just as if the plant were a hybrid between red and variegated races. If both the V factors were changed to S, the grain would be red as before, but all, instead of half, of the gametes would carry S. If the modification from VV to VS should occur very early in the life of the plant, or even in the embryo, then all the ears of the plant might be self-red, and one-half of all the gametes, both male and female, might carry S, and the other half V as in an ordinary hybrid. If the modification should occur much later, for instance when the ear is beginning to form, there might be then only a patch of red grains on a variegated ear, and only those gametes arising from these red masses of tissues would carry half S and half V. Finally if the modification should occur after the grains begin to form, the latter would have broad or narrow stripes according to the amount of pericarp directly descended from the modified cell, and the larger the amount of modified tissue the greater the chance that the gametes concerned would carry S. Similarly in any cell of a heterozygous, variegated-eared plant, V –, he assumes that the V factor may be changed to S. The effect on the pericarp colour would be the same as before, and of the gametes arising from the modified tissue one-half, and never more, would carry S, the other half would carry no factor, and would be represented by –.

Emerson then applies his hypothesis to results obtained in the F_2 and F_3 generations. As stated above F_1 red-eared plants which had arisen from selfed homozygous variegated ears gave in F_2 only red-eared and variegated-eared offspring. On the hypothesis quoted, the constitution of the F_1 red-eared plants would be either VS or SS, the former being more frequent than the latter on account of the rarity of S in the male gametes. The F_1 red ears tested were evidently of the composition VS. Of two F_2 reds from selfed F_1's, one gave reds and variegated, the other bred true to red. Hence the hypothesis is in accordance with the results. Again F_1 red-eared plants which had arisen from heterozygous variegated selfed plants, as we have seen, behaved in some cases like hybrids of red and variegated races, in other cases like hybrids of red and white races. On the hypothesis that variegated-eared plants were V – and their red grains S –, the F_1 plants would be SV, SS, S –, V –, or – –. Of the F_1 reds tested some were evidently SV, and others S –. Of the F_2 reds, one bred true in F_3, and others segregated into reds and variegated.

Finally when F_1 red-eared plants arose from either homo- or heterozygous variegated ears that had been crossed with whites they gave only red-eared and white-eared offspring, never variegated. By hypothesis the parent variegated-eared plants were V – and VV, and the red grains S – and SV (or SS possibly) and the male parents were – –. The F_1 plants therefore would be S –, V – and – –, and only S – would be red-eared. The red-eared F_1 plants tested gave red- and white-eared in Mendelian ratios. Of the F_2 red-eared, one bred true in F_3, and the others segregated into reds and whites.

Emerson further maintains that de Vries's results for striping in *Antirrhinum* indicate that this case is of a similar nature to *Zea*. Also the results obtained by Correns (640) in striped-flowered plants of *Mirabilis*; these results show that plants with self-coloured flowers behave as if they had occurred in an F_2 from a cross of striped by self-coloured plants. But flowers from self-coloured branches on striped plants produce few, if any, more self-coloured plants than flowers on branches with striped flowers. As an explanation of this anomaly Emerson suggests that in the case of seed sports the factors for variegation are affected, whereas in somatic variations there is no corresponding change in the Mendelian factors.

A curious phenomenon in connection with striping in *Antirrhinum* is one noted by the author and previously mentioned (see p. 162). If the factor for tingeing with anthocyanin be denoted by L, and the factor for full colour by D, then LLDD(d) is magenta and LlDD(d) pale magenta. If in the striped variety we represent D by S, then LLSS is ivory striped with magenta, LlSS, ivory striped with pale magenta, LLSs is tinged ivory striped with magenta and LlSs tinged ivory striped with pale magenta. No explanation of this interesting result can be offered at present.

In *Primula sinensis* (Gregory, 660) striping appears to belong to a different category from that in *Antirrhinum*, *Mirabilis*, etc., for it behaves as a simple recessive to a self-coloured form.

Bearing in mind the facts just recorded one cannot fail to realise that the common occurrence of reversion to self-colour among striped varieties is but one expression of the much more widely distributed phenomenon of bud-variation. Of the latter there are a number of cases known, many of them involving pigmentation. Some are concerned with the distribution of chlorophyll, and these it is not necessary to consider here. Of those connected with anthocyanin there are several recorded by de Vries (601, 669), for instance in *Ribes sanguineum*, *Veronica longifolia* and others. *Ribes sanguineum* has red racemes of flowers and a certain amount of anthocyanin in its twigs and petioles, and there is a variety which has white flowers tinged slightly with red, while the vegetative parts lack pigmentation. Of the variation de Vries says: "Occasionally this white-flowered currant reverts back to the original red type and the reversion takes place in the bud. One or two buds on a shrub bearing perhaps a thousand bunches of white flowers produce twigs and leaves in which the red pigment is noticeable and the flowers of which become brightly colored. If such a twig is left on the shrub, it may grow further, ramify and evolve into a larger group of branches. All of them keep true to the old type." Another case is that of the hybrid from *Veronica longifolia* and its variety *alba*; the blue hybrid occasionally produces white flowers. We may also include the production of green-leaved branches by the purple-leaved Beech, Hazel, etc. Other cases are an individual of purple-winged 'Purple Invincible' (*Lathyrus*) which developed a flower of the variety 'Miss Hunt,' lacking in the blue factor (Bateson, 627); a white-flowered branch of red *Phaseolus* (Reinke, 738); a deep- from pale-flowered Lilac (Cobb & Bartlett, 771); white varieties of

Azalea, streaked with red which give branches with red flowers, and the magenta variety 'Don Juan' (*Pelargonium*) which may bear branches of red flowers (Bateson, 696). Considerable advance in elucidation of certain forms of bud-sports is due to the conception of periclinal chimaeras as applied to cultivated forms of *Bouvardia* and *Pelargonium* (Bateson, 740, 822). These forms are made up of the 'core' of one variety and the 'skin' of another. In such cases adventitious buds from roots produce the 'core' variety.

With regard to bud-variation there are several fundamental points to be borne in mind, and these have been well expressed by Bateson (627, 696) from whom the following quotations are taken. First: "when a bud-sport occurs on a plant, the difference between the sport and the plant which produced it may be exactly that which in the case of a seminal variety is proved to depend on allelomorphism." This is exemplified in the cases given above, i.e. in the presence or absence of factors controlling pigment formation (*Antirrhinum*, *Ribes*, *Veronica*, *Fagus*, etc.). Secondly, on consideration of these cases, we may arrive at the conclusion that the segregation of the allelomorphs which control the production of colour must have taken place at some somatic division, and "we are thus obliged to admit that it is not solely the reduction-divisions which have the power of effecting segregation." The third point is that: "The distribution of colour in this case (bud-variation) lies outside the scheme of symmetry of the plant." For "though the parts included in the sports show all the geometrical peculiarities proper to the sport-variety, yet the sporting-buds themselves are not related to each other according to any geometrical plan," just as striping itself in the Carnation or *Antirrhinum* is not under geometrical control. And it is precisely the plants with this disorderly arrangement of striping which so often give rise to bud-sports[1].

At the present moment it is difficult to form any general conception of the mechanism of the change which underlies bud-variation. In a recent paper Emerson (725) describes the occurrence of anomalous seeds of Maize (*Zea*), two of which are concerned with pigmentation. In one, the seed was half colourless and half purple: in the other, half purple and half red. He admits that the occurrence of this phenomenon could be explained on the hypothesis that, subsequent to normal endosperm

[1] Bateson (696) notes a most interesting case of bud-variation in the *Azalea Vervaeana*. The flowers have symmetrical markings of one shade of red, and red streaks of another shade. When self-red sports arise they are of the shade of the red streaks, not of the symmetrical markings.

fertilisation, there occurs a vegetative segregation of genetic factors. But such a segregation, he maintains, cannot be typically Mendelian because in neither of the cases quoted are all the genetic factors (in a heterozygous condition) involved. He prefers to interpret the phenomenon as a somatic mutation, that is as a change in genetic constitution rather than a segregation of genetic factors. He proceeds to apply this conception to the case of bud-sports in general. His hypothesis in fact has already been mentioned in connection with striping in *Zea* and *Antirrhinum*. In his opinion the somatic mutation may be a gain of at least one new factor, the loss of a factor, or the permanent modification of a factor. Against the segregation hypothesis he brings the following considerations. In material homozygous with respect to the Mendelian factors concerned it is not possible for bud-sports to arise by segregation. Nor is it possible if a new character, previously unknown to the species, should arise; nor if a dominant character appears as a bud-sport in material known to be homozygous in a recessive character which is allelomorphic to the dominant character in question (ex. striping in *Zea*). If however the bud-sport be due to the loss of a character, and the material be also heterozygous for the character, then the case can be interpreted equally well as a segregation or a mutation.

Bateson, on the other hand, inclines to the view that bud-sports are the result of somatic segregation, the sporting branches being the outgrowth from cells containing one only of the segregating factors. He also suggests that the phenomenon of striping may be due to the fact that there is an insufficient amount of the colour factor in the gametes to make the zygote self-coloured. This may also apply to some cases of pattern (see p. 198). In these cases the more intimate the mixing, the more likely are the offspring to be striped, and this, as we have seen, is borne out by observations upon *Zea* and *Antirrhinum*.

A very interesting case of patching in *Lathyrus* is described by Punnett (854). The 'Duke of Westminster' (hooded Purple Invincible) may have areas where the factor for purple (as contrasted with red-purple) is missing. There are all grades between this condition and what is really the red-purple Duke patched with normal purple. In this case of flecking, the patched give all three kinds, normals, patched and reds. The similarity to *Mirabilis* and the differences from *Antirrhinum*, *Zea* and *Primula* are pointed out in a subsequent discussion. Additional aspects on 'bud-sports' have been put forward by Emerson (845) and Stout (799).

BIBLIOGRAPHY[1]

GENERAL WORK ON ANTHOCYANINS

Page of text on which reference is made

1. **1682. Grew, N.,** *The Anatomy of Plants, with an Idea of a Philosophical History of Plants, and several other Lectures, read before the Royal Society*, London, 1682.
General consideration of the colours of plants and views as to their origin. Investigations upon yellow, green, red and blue pigments, as regards solubilities and reactions to acids and alkalies, are described. pp. 4, 49, 105

2. **1799. Senebier, J.,** *Physiologie végétale*, Genève, 1799, T. v, pp. 53–77.
A chapter is devoted to the colours of flowers and other parts of plants. Mention is made of the solubilities of pigments, their reactions towards acids and alkalies, and the influence of light and air on their formation. p. 88

3. **1828. Macaire-Princep,** 'Mémoire sur la coloration automnale des feuilles,' *Mém. Soc. Phys.*, Genève, 1828, iv, pp. 43–53.
A general account of autumnal coloration and its connection with light and gaseous exchange in leaves. A special hypothesis is formulated, viz., that red pigments are oxidation products of chlorophyll. pp. 90, **105**

4. **1833. *Candolle, A. P. de,** *Pflanzenphysiologie oder Darstellung der Lebenskräfte und Lebensverrichtungen der Gewächse*, Uebersetzt von I. Röper, Stuttgart und Tübingen, 1833–1835.
Localisation of red pigment and its appearance in different parts of plants. Vegetative pigments classified into two series, xanthic and cyanic. — —

5. **1835. *Marquart, L. Cl.,** *Die Farben der Blüthen, Eine chemisch-physiologische Abhandlung*, Bonn, 1835.
The term anthocyanin is first used for the red, violet and blue pigments of flowers, and the view is brought forward

[1] The abbreviations for journals, etc., are those used in the International Catalogue of Scientific Literature. Publications marked * are those to which direct access has not been possible; the notice is then taken from a summary or other reference if such has been available.

The publications are numbered 1–879. *When reference is made in the text to any publication, the number of the publication is quoted. Under the name also of every author in the General Index will be found the number or numbers referring to his publications.*

Page of text on which reference is made

that it is formed by the dehydration of chlorophyll. To the pigment of yellow flowers the name anthoxanthin is given. pp. 3, 46, 88, **106**

6. **1837. Meyen, F. J. F.,** *Neues System der Pflanzen-Physiologie*, Berlin, 1837, Bd I, pp. 181–189, Bd II, pp. 428–464.
   One section is devoted to coloured cell-sap, which is noted as the colouring matter of flowers. Another section is concerned with the formation of colour. — —

7. **Mohl, H. von,** 'Untersuchungen über die winterliche Färbung der Blätter,' Dissertation, 1837. *Vermischte Schriften botanischen Inhalts*, Tübingen, 1845, pp. 375–392.
   A summary of the views of previous authors on the origin of anthocyanin from chlorophyll; this hypothesis refuted. General account of the winter reddening of leaves. pp. 105, 106

8. **1838. Mohl, H. von,** 'Recherches sur la coloration hibernale des feuilles,' *Ann. sci. nat.* (*Bot.*), Paris, 1838, sér. 2, IX, pp. 212–235.
   A translation of the preceding paper. pp. 105, 106

9. **1853. Martens,** 'Recherches sur les couleurs des végétaux,' *Bul. Acad. roy.*, Bruxelles, 1853, XX (1), pp. 197–235.
   An erroneous view is held that there are only two funda mental pigments, anthocyanin and anthoxanthin, and that these combine together to give chlorophyll. — —

10. **1879. Hildebrand, F.,** *Die Farben der Blüthen*, Leipzig, 1879, 83 pages.
   A very full account is given of the range of colour and colour varieties in various genera and natural orders. Histological distribution and effect of various factors such as light, temperature, soil, etc., on the formation of pigment. — —

11. **1884. Hansen, A.,** *Die Farbstoffe der Blüthen und Früchte*, Würzburg, 1884, 19 pages.
   A general account, though mainly chemical, of various plastid and soluble pigments in plants. pp. 46, 49

12. **1887. Wigand, A.,** 'Die rothe und blaue Färbung von Laub und Frucht,' *Botanische Hefte, Forschungen aus dem botanischen Garten zu Marburg*, Marburg, 1887, Heft 2, pp. 218–243.
   Account of the morphological and histological distribution of anthocyanin; the chromogen of anthocyanin and the factors connected with reddening. p. 33

13. **1888. Sewell, P.,** 'The Colouring Matters of Leaves and Flowers,' *Trans. Bot. Soc.*, Edinburgh, 1888, XVII (Pt 2), pp. 276–308.
   A short general account of plant pigments. — —

14. **1889. Dennert, E.,** 'Anatomie und Chemie des Blumenblatts,' *Bot. Centralbl.*, Cassel, 1889, XXXVIII, pp. 425–431, 465–471, 513–518, 545–553.

Page of text on which reference is made

An important histological account of various pigments found in the petals of a number of species. Also reactions (chemical) of pigments and their connection with tannin. pp. 33, 35, 40, 42

15. **1896. Newbigin, M. I.**, 'An Attempt to classify Common Plant Pigments with some Observations on the Meaning of Colour in Plants,' *Trans. Bot. Soc.*, Edinburgh, 1896, xx, pp. 534–549.

A short account is included of the physiology and functions of anthocyanin, and some original suggestions are made as to the conditions which bring about its formation. As regards the correlation between the presence of anthocyanin and the absence of chlorophyll, it is suggested that the protoplasm of chlorophyll-containing cells has an alkaline reaction, and this is unfavourable to the formation of anthocyanin. — —

16. **1898. Newbigin, M. I.**, *Colour in Nature, A Study in Biology*, London, 1898.

One chapter (chap. III) is concerned with the colours and pigments of plants. — —

17. **1903. Buscalioni, L.**, e **Pollacci, G.**, *Le antocianine e il loro significato biologico nelle piante*, Milano, 1903, 379 pages.

A very full account of anthocyanin from every point of view. First part descriptive: second part experimental, giving results of experiments and observations on histological distribution, reactions, etc., of anthocyanin. pp. 5, 35, 37, 39, 91

18. **1905. Bidgood, J.**, 'Floral Colours and Pigments,' *J. R. Hort. Soc.*, London, 1905, XXIX, pp. 463–480.

Many interesting observations on flower pigments with suggestions as to their chemical nature. General account of anthocyanin. pp. 12, 40, 42

19. **1906. Gertz, O.**, *Studier öfver Anthocyan*, Akademisk Afhandling, Lund, 1906, lxxxvii + 410 pages.

A very important work on anthocyanin dealing with the pigment from every point of view. The first part is a descriptive account of anthocyanin. The second part gives the results of histological examination of species of every natural order, as regards distribution of anthocyanin. pp. 5, 20, 29, 31, 32, 33, 42, 53, 85, 89, 90

20. **1913. Haas, P.**, and **Hill, T. G.**, *An Introduction to the Chemistry of Plant Products*, 3rd ed., London, 1921.

Short account of anthocyanin from the point of view of chemistry, physiology and genetics. — —

DISTRIBUTION OF ANTHOCYANINS

Page of text on which reference is made

21. **1675. Malpighi, M.,** *Anatome Plantarum*, London, 1675.
First account of the histology of coloured petals. — —

22. **1827. Guibourt** et **Robinet,** 'Rapport sur la coloration des feuilles à diverses époques de la végétation,' *Journal de Pharmacie*, Paris, 1827, XIII, pp. 26–27.
A short note on winter coloration of leaves. It is suggested that there is a correlation between the colour of autumn leaves and the colour of ripe fruits in the same plant. — —

23. **1830. Meyen, F. J. F.,** *Phytotomie*, Berlin, 1830, pp. 141–143.
One of the first accounts of the histological distribution of anthocyanin. The red, purple and blue colours of stems, leaves, hairs and petals are stated to be due to coloured cell-sap, and some cases of localisation of pigment are mentioned. — —

24. **1854. Chevreul,** 'Note sur la couleur d'un assez grand nombre de fleurs,' *C. R. Acad. sci.*, Paris, 1854, XXXIX, pp. 213–214.
Remarks on composite colour in flowers and leaves. — —

25. **1857. Chevreul,** 'Explication de la zone brune des feuilles du *Geranium zonale*,' *C. R. Acad. sci.*, Paris, 1857, XLV, pp. 397–398.
Statement that brown colour in the leaf zone is due to the mixture of the complementary colours of green and red (anthocyanin) pigments. The two pigments can be extracted separately and also detected microscopically. The phenomenon can also be reproduced by an artificial blending of the same colours. — —

26. **1858. *Morren, E.,** *Dissertation sur les feuilles vertes et colorées, envisagées spécialement au point de vue des rapports de la chlorophylle et de l'érythrophylle*, Gand, 1858.
Distribution of anthocyanin in variegated leaves. p. 37

27. **1860. Irmisch, Th.,** 'Ueber einige Crassulaceen,' *Bot. Ztg.*, Leipzig, 1860, XVIII, pp. 85–90 (p. 87).
Violet and blue pigments in root tips of species of *Sedum*, *Umbilicus*, *Sempervivum* and *Saxifraga*. p. 30

28. **1861. Chatin, Ad.,** 'Sur la structure anatomique des pétales comparée à celle des feuilles; une conséquence physiologique des faits observés,' *Bull. soc. bot.*, Paris, 1861, VIII, pp. 22–23.
Pigment of flowers usually present in epidermal cells which are prolonged into papillae. In thicker petals pigment more deep-seated. p. 39

Page of text on which reference is made

29. **Lowe, E. J.**, and **Howard, W.**, *Beautiful Leaved Plants*, London, 1861, 144 pages, 60 figs.
Good illustrations of pigmented (anthocyanin) leaves of tropical species used as ornamental plants. p. 22

30. **1863. Hildebrand, F.**, 'Anatomische Untersuchungen über die Farben der Blüthen,' *Jahrb. wiss. Bot.*, Berlin, 1863, III, pp. 59–75.
An important account of the pigments in flowers; the form and distribution of pigments in blue, purple, red, orange, yellow and green parts of flowers. pp. 33, 34, 35, 39, 40

31. **1865. Jaennicke, F.**, 'Ueber gefleckte Blätter,' *Bot. Ztg.*, Leipzig, 1865, XXIII, pp. 269–271.
Reference to classes of plants with red, green and white leaves. — —

32. **1866. Kny, L.**, 'Ueber Bau und Entwickelung der Riccien,' *Jahrb. wiss. Bot.*, Leipzig, 1866–1867, V, pp. 364–386.
Record of the presence of violet pigment in *Riccia*. — —

33. **Tieghem, Ph. van**, 'Recherches sur la structure des Aroidées,' *Ann. sci. nat.* (*Bot.*), Paris, 1866, sér. 5, VI, pp. 72–200.
Anthocyanin in roots of Aroids. p. 30

34. **1870. *Chatin, A.**, *De l'anthère, Recherches sur le développement, la structure et les fonctions de ses tissus*, Paris, 1870, pp. 17, 22, 34, 35.
Occurrence of anthocyanin in anthers. p. 39

35. **1874. *Chatin, A.**, *Der Naturforscher*, 1874, No. 11 (*Biedermanns Centralbl. Agrik. Chem.*, Leipzig, 1874, VI, p. 111).
Varying localisation of anthocyanin in the leaf at different ages. — —

36. ***Sempolowski, A.**, *Beiträge zur Kenntniss des Baues der Samenschale*, Inaugural-Dissertation, Leipzig, 1874, pp. 16, 23, 38.
Record of red pigments in testa of *Lupinus* and *Vicia* seeds. p. 42

37. **1877. Fraustadt, A.**, 'Anatomie der vegetativen Organe von *Dionaea muscipula* Ell.,' *Beitr. Biol. Pflanzen*, Breslau, 1877, II, pp. 27–64 (pp. 57, 60).
Anthocyanin in cells of root-cap of *Dionaea*. — —

38. **1880. *Borbás, V.**, 'Adatok a leveses (húsos) gyümölesök szövettani szerkezetéhez (Beiträge zur histologischen Structur der saftigen (fleischigen) Früchte),' *Földmivelési Érdekeink.*, 1880, No. 37–45.
Localisation of anthocyanin and other pigments in tissues of fruits, such as *Berberis*, *Rivina*, *Phytolacca*, *Hedera*, *Ribes*, etc. p. 42

39. ***Klein, G.**, 'A virágok szinéröl (Ueber die Farbe der

Page of text on which reference is made

Blüten),' *Népszerü természettudományi elö-adások gyüjteménye (Sammlung populär naturwissenschaftlicher Vorträge)*, 1880, Heft 21.

A publication which treats of flower-pigments and their distribution; colour-varieties and the significance of flower-colours. — —

40. **1882. *Koschewnikow, D.**, 'Zur Anatomie der corollinischen Blüthenhüllen,' *Schriften der neurussischen Gesellschaft der Naturforscher*, Odessa, 1882, VIII (1), pp. 1–199.

Some account of the distribution of anthocyanin in the corolla, petaloid calyx and perianth. p. 39

41. **Meehan, T.**, 'Autumn Color of the Bartram Oak,' *Bot. Gaz.*, Indianapolis, 1882, VII, p. 10.

Anthocyanin in *Quercus heterophylla*. — —

42. **Pirotta, R.**, 'Intorno alla produzione di radici avventizie nell' *Echeveria metallica* Lndl.,' *Atti Soc. nat. mat.*, Modena, 1882, ser. 3, I, pp. 73–75.

Adventitious aerial roots, developed from the leaf-bases, are coloured red with anthocyanin. p. 30

43. **1883. Ascherson, P.**, 'Bemerkungen über das Vorkommen gefärbter Wurzeln bei den Pontederiaceen, Haemodoraceen und einigen Cyperaceen,' *Ber. D. bot. Ges.*, Berlin, 1883, I, pp. 498–502.

List and description of genera and species in which pigmented red and blue roots occur, but no evidence to show that pigment is anthocyanin. p. 30

44. **Hildebrand, F.**, 'Ueber einige merkwürdige Färbungen von Pflanzentheilen,' *Ber. D. bot. Ges.*, Berlin, 1883, I, pp. xxvii–xxix.

Pigment in roots of *Pontederia*, *Wachendorfia* and fruits of *Rivina*; also in bracts of *Euphorbia fulgens*. p. 30

45. **1884. Johow, Fr.**, 'Zur Biologie der floralen und extrafloralen Schau-Apparate,' *Jahrbuch des Königlichen botanischen Gartens und des botanischen Museums zu Berlin*, Berlin, 1884, III, pp. 47–68.

Descriptions of perianths, coloured calyces, bracts, spathes, involucres, floral axes, etc. List of ten different classes of 'Schau-Apparate.' — —

46. **Strasburger, E.**, *Das botanische Practicum*, Jena, 1884, pp. 61–66.

Histological distribution of anthocyanin in various flower petals. — —

47. **1885. Schimper, A. F. W.**, 'Untersuchungen über die Chlorophyllkörper und die ihnen homologen Gebilde,' *Jahrb. wiss. Bot.*, Berlin, 1885, XVI, pp. 1–247.

Page of text on which reference is made

Almost entirely an account of plastid pigments. References to soluble pigment in *Strelitzia, Billbergia, Delphinium* and *Glaucium.* — —

48. **1886. Zopf, W.**, 'Ueber die Gerbstoff- und Anthocyan-Behälter der Fumariaceen und einiger anderen Pflanzen,' *Bibl. bot.*, Cassel, 1886, I (2), 40 pages.

An important histological account of anthocyanin-containing sacs in Fumariaceae. In tissues of certain representatives of the family, i.e. spp. of *Corydalis, Dielytra, Fumaria* and *Adlumia*, there are isolated cells which contain tannin, together with either anthocyanin or a yellow pigment. Underground organs contain yellow pigment, the organs exposed to sunlight, red. Light brings about the change from yellow to red. Yellow pigment is also converted into red by acids. It is suggested that the process is not one of oxidation. pp. 30, 36

49. **1887. Fintelmann, H.**, 'Betrachtungen über die Herbstfärbung der Belaubung unsrer Wald- und im freien Lande ausdauernden Schmuck-Gehölze,' *Gartenflora*, Berlin, 1887, XXXVI, pp. 635–637, 651–656.

List of plants showing various autumnal colorations. — —

50. **Heinricher, E.**, 'Vorläufige Mittheilung über die Schlauchzellen der Fumariaceen,' *Ber. D. bot. Ges.*, Berlin, 1887, v, pp. 233–239.

The author fails to detect tannin in the anthocyanin-containing cells of the Fumariaceae (see No. 48). — —

51. ***Lierau, M.**, *Beiträge zur Kenntniss der Wurzeln der Aracéen*, Inaugural-Dissertation der Universität Breslau, Leipzig, 1887.

Anthocyanin formation in roots of Aroids. p. 30

52. **1888. *Levi-Morenos, D.**, 'Contribuzione alla conoscenza dell' antocianina studiata in alcuni peli vegetali,' *Atti Ist. ven.*, Venezia, 1888, VI, ser. 6.

Note on the distribution of anthocyanin in hairs on radical leaves of *Scabiosa arvensis* and *Hieracium Pilosella.* — —

53. **1889. Aitchison, J. E. T.**, 'The Source of Badsha or Royal Salep,' *Trans. Bot. Soc.*, Edinburgh, 1889, XVII, pp. 434–440.

(?) Anthocyanin in roots of *Allium giganteum.* — —

54. **Claudel, L.**, 'Sur les matières colorantes du spermoderme dans les Angiospermes,' *C. R. Acad. sci.*, Paris, 1889, CIX, pp. 238–241.

Colouring matter of the testa occurs either in the cell-wall or in the cell-cavity. Account of pigmentation in several species. — —

55. **Kny, L.**, 'Ueber Laubfärbungen,' Sonder-Abdruck aus der

Page of text on which reference is made

Naturwissenschaftlichen Wochenschrift, Berlin, 1889, 28 pages.
Distribution of anthocyanin in leaves. — —

56. ***Potonié, H.**, 'Die Verbreitung der Samen insbesondere der Paternoster-Erbse,' *Natw. Wochenschr.*, Berlin, 1889, IV, p. 207.
Anthocyanin in testa of seeds. p. 30

57. **1890. Fischer, H.**, 'Beiträge zur vergleichenden Morphologie der Pollenkörner,' *Jahresber. Ges. vaterl. Cultur*, Breslau, 1890, LXVIII, pp. 72–73.
Statement that pollen grains may be coloured with anthocyanin. — —

58. **Levi-Morenos, D.**, 'Sulla distribuzione peristomatica dell' antocianina in alcuni *Sedum*,' *Nuovo Giorn. bot. ital.*, Firenze, 1890, XXII, pp. 79–80.
Occurrence of groups, round stomata, of specially shaped epidermal cells (two, three and four in number), containing anthocyanin. — —

59. **1892. Kronfeld, M.**, 'Ueber Anthokyanblüten von *Daucus Carota*,' *Bot. Centralbl.*, Cassel, 1892, XLIX, pp. 11–12.
The central purple flower of the inflorescence contains anthocyanin in all its parts. — —

60. **1893. Haberlandt, G.**, *Eine botanische Tropenreise*, Leipzig, 1893 pp. 117–119.
Description of the pendent position and red colour (anthocyanin) of young leaves of certain tropical trees. — —

61. **Müller, L.**, 'Grundzüge einer vergleichenden Anatomie der Blumenblätter,' *Nova Acta der Kaiserlichen Leopoldinisch-Carolinischen Deutschen Akademie der Naturforscher*, Halle, 1893, LIX (1), 356 pages.
A very full account of the structure of floral leaves, including the distribution of anthocyanin in the tissues. Also the investigation of the tannin present in petals and the connection between tannins and anthocyanin. pp. 35, 39

62. **Stahl, E.**, 'Regenfall und Blattgestalt,' *Ann. Jard. bot.*, Buitenzorg, 1893, XI, pp. 98–179.
Examples are mentioned of the development of anthocyanin in young pendent shoots—Hängezweige—in the tropics. pp. 25, 37

63. **Veitch, H. J.**, 'Autumnal Tints,' *J. R. Hort. Soc.*, London, 1893, XV, pp. 46–57.
List of trees and shrubs showing autumnal coloration. — —

64. **1898. Berthold, G.**, *Untersuchungen zur Physiologie der pflanzlichen Organisation*, Leipzig, Pt 1, 1898, pp. 199–242.
Morphological and histological distribution of antho-

Page of text on which reference is made

cyanin in leaves and stems in representatives of a number of natural orders; also in leaves reddened by decortication. pp. 37, 39, 83

65. **Diels, L.**, 'Stoffwechsel und Structur der Halophyten,' *Jahrb. wiss. Bot.*, Berlin, 1898, XXXII, pp. 309–322.
Appearance of anthocyanin in halophytes. — —

66. **1899. *Preyer, A.**, *Ueber die Farbenvariationen der Samen einiger Trifolium-Arten*, Doktor-Dissertation, Berlin, 1899.
Anthocyanin in testa of seeds of species of *Trifolium*. pp. 30, 33, 42

67. **Schmidt, F.**, 'Die Farben der Blüthen,' *Die Natur*, Halle a. S., 1899, XLVIII, pp. 223–225.
Short account (popular) of the pigments of flowers. — —

68. **1900. Möbius, M.**, 'Die Farben in der Pflanzenwelt,' Sonder-Abdruck aus der *Naturwissenschaftliche Wochenschrift*, Berlin, 1900, 24 pages.
A general account of pigments, including anthocyanin, found in all classes of the vegetable kingdom. Examples of the occurrence of anthocyanin in leaves, flowers, fruits and seeds. pp. 30, 33

69. ***Rodrigue, A.**, 'Les feuilles panachées et les feuilles colorées,' *Mémoire de l'Herbier Boissier*, No. 17 B, Genève et Bâle, 1900.
Anthocyanin in variegated leaves. — —

70. **1902. *Massart, J.**, 'Comment les jeunes feuilles se protègent contre les intempéries,' *Bulletin du Jardin Botanique de l'État à Bruxelles*, 1902–1905, I, p. 181.
Localisation of anthocyanin in hairs. — —

71. ***Wulff, Th.**, *Ueber das Vorkommen von Anthocyan bei arktischen Gewächsen, Botanische Beobachtungen aus Spitzbergen*, Akademische Abhandlung, Lund, 1902, II.
Arctic vegetation is characterised by abundant production of anthocyanin. Hence periodicity between spring and autumn coloration is less marked. p. 29

72. **1903. Burbidge, F. W.**, 'The Leaf-marking of *Arum maculatum*,' *Irish Nat.*, Dublin, 1903, XII, p. 137.
The author states that the black spotted form is in general distribution in England, whereas in Ireland the unspotted form is more common. — —

73. **Colgan, N.**, 'The Leaf-Marking of *Arum maculatum*,' *Irish Nat.*, Dublin, 1903, XII, pp. 78–81.
The author states that the prevalent plant in Ireland is unspotted, and is inclined to believe that the same is true for Great Britain, though the spotted form may be more abundant in this country than in Ireland. Description given of form of spots. p. 22

Page of text on which reference is made

74. **Garjeanne, A. J. M.**, 'Buntblätterigkeit bei *Polygonum*,' *Beihefte zum Bot. Centralbl.*, Jena, 1903, XIII, pp. 203–210.
A description of varieties of several species of *Polygonum* in which chlorophyll is absent from portions of the leaves. Hence the usual dark leaf-blotch of anthocyanin shows bright red. p. 22

75. **Gunn, W. F.**, '*Arum maculatum* again,' *Irish Nat.*, Dublin, 1903, XII, p. 219.
Suggestion that black spotted form is a variety. — —

76. **Lindinger, L.**, 'Anatomische und biologische Untersuchungen der Podalyrieensamen,' *Beihefte zum Bot. Centralbl.*, Jena, 1903, XIV, pp. 20–61.
Mention is made of the histological distribution of anthocyanin in the seed-coats of several genera. pp. 30, 42

77. **Parkin, J.**, 'On the Localisation of Anthocyan (red cell-sap) in Foliage Leaves,' *Rep. Brit. Ass.*, London, 1903, p. 862.
Investigation of the seat of anthocyanin in leaves reddened through various causes. See text. pp. 5, **38**, 90

78. **Pethybridge, G. H.**, 'The Leaf-spots of *Arum maculatum*,' *Irish Nat.*, Dublin, 1903, XII, pp. 145–151.
From data collected, the author finds that the form with spotted (anthocyanin) leaves is less widely distributed in Great Britain than the unspotted. Remarks on the structure and significance of the spot. p. 22

79. **1906. Wittmack, L.**, 'Violette Weizenkörner,' *SitzBer. Ges. natf. Freunde*, Berlin, 1906, pp. 105–108.
Anthocyanin in outer layers of wheat grain. — —

80. **1910. Exner, F.** und **S.**, 'Die physikalischen Grundlagen der Blütenfärbungen,' *Anz. Ak. Wiss.*, Wien, 1910, XLVII, pp. 11–12.
The authors maintain that, apart from differences in pigmentation, the colours of flowers may be modified by physical phenomena arising from structural features of the petal. Examples are the heightening of epidermal colour by a sub-epidermal air-containing layer; the deadening of tints when pigments of complementary colours are mixed; the raising of colour saturation by reflexion caused by the epidermal papillae. — —

81. ***Paasche, E.**, *Beiträge zur Kenntnis der Färbungen und Zeichnungen der Blüten und der Verteilung von Anthocyan und Gerbstoff in ihnen*, Dissertation, Göttingen, 1910, 113 pages.
Observations upon the development of anthocyanin in perianth leaves of *Tulipa*, *Anthericum* and *Asphodelus*. Also upon the histological distribution of anthocyanin and tannin. — —

Page of text on which reference is made

82. **1911. Coupin, H.**, 'Sur la localisation des pigments dans le tégument des graines de Haricots,' *C. R. Acad. sci.*, Paris, 1911, CLIII, pp. 1489–1492.
Account of histological structure of testa of species and varieties of *Phaseolus* and *Dolichos*, and distribution of pigments (including anthocyanin) in the testa. p. 42

83. **1915. Nagai, I.**, 'Ueber roten Pigmentbildung bei einigen *Marchantia*-Arten,' *Bot. Mag.*, Tokyo, 1915, XXIX, pp. 90–98. p. 20

84. **1917. Baumgärtel, O.**, 'Die Farbstoffzellen von *Ricinus communis* L.,' *Ber. D. bot. Ges.*, Berlin, 1917, XXXV, pp. 603–611. — —

85. **1919. Ichimura, T.**, 'On the Localisation of Anthocyanin in the Spring Leaves of some Trees and Shrubs in the Temperate Regions of Japan,' *Bot. Mag.*, Tokyo, 1919, XXXIII, pp. 12–15. — —

86. **1923. Möbius, M.**, 'Ueber die Färbung der Antheren und des Pollens,' *Ber. D. bot. Ges.*, Berlin, 1923, XLI, pp. 12–16. p. 29

NATURAL OCCURRENCE OF SOLID AND CRYSTALLINE ANTHOCYANINS

87. **1844. Hartig, Th.**, *Das Leben der Pflanzenzelle*, Berlin, 1844.
Mention of solid anthocyanin in *Solanum nigrum*. p. 35

88. **1851. Mohl, H. von**, *Grundzüge der Anatomie und Physiologie der vegetabilischen Zelle*, Braunschweig, 1851, p. 47.
Cases of occurrence of solid anthocyanin in *Strelitzia Reginae* and *Salvia splendens*. pp. 33, 35

89. **1855. Nägeli, C.**, und **Cramer, C.**, *Pflanzenphysiologische Untersuchungen*, Zürich, 1855, Heft 1, pp. 5, 6–7, 31, 41–42, Taf. 2, figs. 1–8.
Plasmolysis of anthocyanin-containing cells and formation of anthocyanin bodies. pp. 33, 34

90. **Unger, F.**, *Anatomie und Physiologie der Pflanzen*, Pest, Wien und Leipzig, 1855, pp. 110–111.
Anthocyanin bodies in berries of *Passiflora* species. pp. 34, 35

91. **1856. Hartig, Th.**, 'Weitere Mittheilungen, das Klebermehl (Aleuron) betreffend,' *Bot. Ztg.*, Leipzig, 1856, XIV, pp. 257–268, 273–281, 297–305, 313–319, 329–335 (pp. 266, 267).
Pigmented aleurone grains. p. 36

92. **1857. Böhm, J. A.**, 'Physiologische Untersuchungen über blaue Passiflorabeeren,' *SitzBer. Ak. Wiss.*, Wien, 1857, XXIII, pp. 19–37.
Crystalline anthocyanin in berries of *Passiflora*. pp. 34, 35

93. **1858. Trécul, A.**, 'Des formations vésiculaires dans les cellules végétales,' *Ann. sci. nat. (Bot.)*, Paris, 1858, sér. 4, X, pp. 20–74, 127–163, 205–376 (pp. 375, 376).
Anthocyanin bodies in *Rubus* and other species. pp. 34, 35

Page of text on which reference is made

94. **1862. Nägeli, C.**, 'Farbcrystalloide bei den Pflanzen,' *SitzBer. Ak. Wiss.*, München, 1862 (2), pp. 147–154.
References to crystalline anthocyanin in *Solanum*. pp. 35, 37

95. **1864. Weiss, A.**, 'Untersuchungen über die Entwickelungsgeschichte des Farbstoffes in Pflanzenzellen,' *SitzBer. Ak. Wiss.*, Wien, 1864, L (1), pp. 6–35, 1866, LIV (1), pp. 157–217.
Appearance of solid blue colouring matter in fruits of *Solanum*. pp. 33, 34, 35

96. **1870. Rosanoff, S.**, 'Zur Morphologie der Pflanzenfarbstoffe,' *Bot. Ztg.*, Leipzig, 1870, XXVIII, pp. 720–723.
Record of anthocyanin bodies in *Neptunia oleracea*. p. 34

97. **1883. Schimper, A. F. W.**, 'Ueber die Entwickelung der Chlorophyllkörner und Farbkörper,' *Bot. Ztg.*, Leipzig, 1883, XLI, pp. 105–112, 121–131, 137–146, 153–160.
Mention of blue colour-bodies in *Glaucium*. p. 34

98. **1884. Fritsch, P.**, 'Ueber farbige körnige Stoffe des Zellinhalts,' *Jahrb. wiss. Bot.*, Berlin, 1884, XIV, pp. 185–231.
Chiefly concerned with plastid pigment. Solid anthocyanin quoted in *Delphinium*. pp. 34, 35

99. **Pim, G.**, 'Cell-sap Crystals,' *J. Bot.*, London, 1884, XXII, p. 124.
Formation of crystals in cells of *Justicia speciosa* by placing sections of stamen in glycerine gelatine. p. 35

100. **1887. *Hanausek, T. F.**, und **Bernowitz, V.**, 'Ueber die Farbstoffkörper des Pimentsamens,' *Zs. Allg. Oest. ApothVer.*, Wien, 1887, No. 16, pp. 253–256.
Colour of seeds of *Pimenta* is due to the presence, in the cells, of large dark red masses which give the reactions of anthocyanin. — —

101. **1888. Courchet, L.**, 'Recherches sur les chromoleucites,' *Ann. sci. nat. (Bot.)*, Paris, 1888, sér. 7, VII, pp. 263–370.
Summary of cases of occurrence of anthocyanin in the form of globules and deposits. pp. 33, 35

102. ***Hansen, A.**, 'Ueber Sphärokrystalle,' *Arbeiten des botanischen Instituts in Würzburg*, Leipzig, 1888, III, p. 92.
Formation of anthocyanin sphaerites similar to those of inulin. — —

103. **1889. Bokorny, Th.**, 'Ueber Aggregation,' *Jahrb. wiss. Bot.*, Berlin, 1889, XX, pp. 427–473.
Anthocyanin bodies in *Primula sinensis*. p. 35

104. **1898. *Tschirch, A.**, 'Violette Chromatophoren in der Fruchtschale des Kaffees,' *Schweiz. Wochenschr. Chem.*, Zürich, 1898, XXXVI.
Solid anthocyanin in fruit of *Coffea arabica*. Deep violet,

Page of text on which reference is made

or blue-black, globules and crystals in the epidermal and sub-epidermal layers. p. 35

105. **1899. *Wallin, G.,** 'Om egendomliga innehållskroppar hos Bromeliaceerna,' *Kongl. Fysiografiska Sällskapets i Lund Handlingar*, 1899, x (2).
Occurrence of deposits, coloured with anthocyanin, in cells. p. 36

106. **1900. Kroemer, K.,** 'Ueber das angebliche Vorkommen von violetten Chromatophoren,' *Bot. Centralbl.*, Cassel, 1900, LXXXIV, pp. 33–35.
Pigment of pericarp of *Coffea* fruits shown to consist of violet crystals of anthocyanin. p. 35

107. **1904. Spiess, K. von,** 'Ueber die Farbstoffe des Aleuron,' *Oest. Bot. Zs.*, Wien, 1904, LIV, pp. 440–446.
Blue colour, in aleurone grains of Maize, due to anthocyanin. p. 37

108. **1905. Molisch, H.,** 'Ueber amorphes und kristallisiertes Anthokyan,' *Bot. Ztg.*, Leipzig, 1905, LXIII, pp. 145–162.
Important account of a number of cases of crystalline and amorphous anthocyanin in tissues of living plants. Also of preparation of crystalline anthocyanin outside the cell. Summary of chief researches and other work on anthocyanin. pp. 34, 35, 47

109. **1906. Gertz, O.,** 'Ett nytt fall af kristalliseradt anthocyan,' *Bot. Not.*, Lund, 1906, pp. 295–301.
Perianth leaves of *Laportea moroides* (Urticaceae) become red and fleshy after flowering. The author observed crystals of anthocyanin in the tissues; crystals dimorphous. — —

110. **1913. Guilliermond, A.,** 'Sur la formation de l'anthocyane au sein des mitochondries,' *C. R. Acad. sci.*, Paris, 1913, CLVI, pp. 1924–1926. — —

111. **Guilliermond, A.,** 'Nouvelles recherches cytologiques sur la formation des pigments anthocyaniques,' *C. R. Acad. sci.*, Paris, 1913, CLVII, pp. 1000–1002. — —

112. **1914. Gertz, O.,** 'Nya iakttagelser öfver anthocyankroppar,' *Sv. Bot. Tidskr.*, Stockholm, 1914, VIII, pp. 405–435.
Further observations on anthocyanin bodies and crystals. — —

113. **Löwschin, A. M.,** 'Zur Frage über die Bildung des Anthocyans in Blättern der Rose,' *Ber. D. bot. Ges.*, Berlin, 1914, XXXII, pp. 386–393. — —

114. **Moreau, F.,** 'Sur des phénomènes d'autochromatisme dans des cellules à anthocyane, *Bull. soc. bot.*, Paris, 1914, LXI, pp. 386–388.
Staining of cell-contents by pigment in cell. — —

115. **Moreau, F.,** 'L'origine et les transformations des produits

Page of text on which reference is made

anthocyaniques,' *Bull. soc. bot.*, Paris, 1914, LXI, pp. 390–405. — —

116. **1915. Guilliermond, A.**, 'Quelques observations cytologiques sur le mode de formations des pigments anthocyaniques dans les fleurs,' *C. R. Acad. sci.*, 1915, CLXI, pp. 494–497. — —

117. **Guilliermond, A.**, 'Sur l'origine des pigments anthocyaniques,' *C. R. Acad. sci.*, Paris, 1915, CLXI, pp. 567–570. — —

118. **1916. *Gertz, O.**, 'Anthocyanin as a microchemical reagent,' *Univ. Årsskr.*, Lund, n. ser., Sect. 2, 1916, XII, No. 5, p. 57.
Pigment employed, in alkaline medium, for nuclear stain, etc. — —

119. **1918. Guilliermond, A.**, 'Sur la métachromatine et les composés phénoliques de la cellule végétale,' *C. R. Acad. sci.*, Paris, 1918, CLXVI, pp. 958–960. — —

120. **1921. Politis, J.**, 'Sur l'origine mitochondriale des pigments anthocyaniques dans les fruits,' *C. R. Acad. sci.*, Paris, 1921, CLXXII, pp. 1061–1063. — —

CHEMISTRY OF ANTHOCYANINS

121. **1664. Boyle, R.**, *Experiments and Considerations touching Colours*, London, 1664.
An account is included of the colour changes which take place on adding acids and alkalies to extracts from flowers and other parts of plants; also the effect of sulphur dioxide on flower pigments. pp. 3, 10, 50, 56, 58

122. **1670. Wray, J.**, 'Extract of a Letter written to the Publisher,' *Phil. Trans. R. Soc.*, London, 1670, v, pp. 2063–2066.
The following observation on Chicory flowers is recorded: 'Bare an Ant-hill with a stick, and then cast the flowers upon it, and you shall see the Ants creep very thick over them. Now as they creep, they let fall a drop of liquor from them, and where that chanceth to light, there you shall have in a moment a large red stain.' The author was able to show eventually that the same coloration could be obtained by treating the petals with various acids. — —

123. **1671. Lister, M.**, 'Some Observations, touching Colours, in order to the Increase of Dyes, and the Fixation of Colours,' *Phil. Trans. R. Soc.*, London, 1671, VI, pp. 2132–2136.
Among other reactions, it is noted that the green parts of plants give a yellow colour with alkalies. — —

124. **1807. Braconnot, H.**, 'Observations sur le *phytolacca*, vulg. raisin d'amérique,' *Annales de Chimie*, Paris, 1807, LXII, pp. 71–90.
An extract is made of the berries, and the effect of various

Page of text on which reference is made

reagents is tried, such as acids, alkalies, salts of metals. It is noted that the behaviour of the pigment towards reagents is different from that of other fruits. p. 10

125. **1808. Melandri, G.,** 'Extrait d'une lettre de M. G. Melandri, docteur en médecine, à Milan, à M. Bouillon-Lagrange,' *Annales de Chimie*, Paris, 1808, LXV, p. 223.
Use of pigment of fruits of *Atropa Belladonna* as an indicator. — —

126. **1818. Smithson, J.,** 'A few facts relative to the colouring matters of some vegetables,' *Phil. Trans. R. Soc.*, London, 1818 (1), pp. 110–117.
Reactions of pigments of Violet, Black Mulberry and Corn Poppy towards acids, alkalies, etc. The view is held that probably all these pigments are the same substance. p. 46

127. **1820. Chevallier, A.,** 'Sur la manière dont se comporte avec les acides et les alcalis la matière colorante des baies de sureau (*sambucus nigra*) appliquée sur le papier,' *Journal de Pharmacie*, Paris, 1820, VI, pp. 177–178.
Action of acids and alkalies on pigment. Different acids are said to give characteristic reactions with paper soaked with the pigment. p. 10

128. **1822. Payen, A.,** et **Chevallier, A.,** 'Sur la fleur de la mauve sauvage (*malva sylvestris*), et particulièrement sur la matière colorante de ses pétales, employée comme réactif, soit en teinture, soit étendue sur le papier pour démontrer la présence des alcalis,' *Journal de Pharmacie*, Paris, 1822, VIII, pp. 483–488.
The pigment is said to be very sensitive to alkalies, and hence is of value as an indicator. p. 10

129. **Payen, A.,** et **Chevallier, A.,** 'Sur la matière colorante des fruits du bois de Sainte-Lucie, *cerasus mahaleb*,' *Journal de Pharmacie*, Paris, 1822, VIII, pp. 489–490.
Action of acids and alkalies on the pigment. p. 10

130. **1825. Roux, J.,** 'D'analyse chimique de la fleur de tilleul (*Tilia europaea*, L.), et de celle de la belle-de-nuit,' *Journal de Pharmacie*, Paris, 1825, XI, pp. 507–512.
Method for extracting pigment from flowers is described. Also solubilities and reactions with acids and alkalies. p. 10

131. **Schübler, G.,** und **Franck, C. A.,** *Untersuchungen über die Farben der Blüthen*, Inaugural-Dissertation, Tübingen, 1825, 38 pages.
A number of observations on the action of acids, alkalies and other reagents on flower-pigment extracts. Hypothesis of two series of pigments—oxidised and deoxidised. p. 105

132. **1834. Kuhlmann, F.,** 'Betrachtungen über den Einfluss des

Page of text on which reference is made

Sauerstoffs auf die Färbung der organischen Produkte und über die Wirkung der schweflichten Säure als Entfärbungsmittel,' *Annalen der Pharmacie*, Heidelberg, 1834, IX, pp. 275–291.
Remarks on the bleaching of flower-pigments with sulphurous acid. p. 56

133. **1836. Hünefeld**, 'Beiträge zur Chemie der Metamorphose der Pflanzenfarben,' *J. prakt. Chem.*, Leipzig, 1836, IX, pp. 217–238.
Experiments to test the effect on flower-pigments of a number of reagents, such as oxygen, carbon dioxide, water, acids, alkalies, alcohol, metallic salts, etc. — —

134. **Hünefeld**, 'Das mit schwefeliger Säure gesäuerte Wasser als Mittel zur Erleichterung der mikroskopischen Untersuchung von Pflanzentheilen,' *J. prakt. Chem.*, Leipzig, 1836, IX, pp. 238–241.
Bleaching of pigments with sulphurous acid. p. 56

135. **1837. Berzelius, J.**, 'Ueber den rothen Farbstoff der Beeren und Blätter im Herbst,' *Annalen der Pharmacie*, Heidelberg, 1837, XXI, pp. 262–267.
Description and properties of red pigments of berries and autumnal leaves. Conclusion that they are similar substances. — —

136. **1849. Morot, F. S.**, 'Recherches sur la coloration des végétaux,' *Ann. sci. nat.* (*Bot.*), Paris, 1849, sér. 3, XIII, pp. 160–235.
Mainly an account of the properties and chemistry of chlorophyll. Criticism of previous work on soluble pigments. Solid blue anthocyanin is prepared from *Centaurea Cyanus* by extracting flowers, and analyses are made of product. Pigment also prepared from *Scilla nutans*, but more unstable. pp. 10, 49, 59

137. **1851. Schönbein, C. F.**, 'Ueber das Verhalten organischer Farbstoffe zur schweflichten Säure,' *J. prakt. Chem.*, Leipzig, 1851, LIII, pp. 321–331.
Pigment of both blue and red flowers bleached by sulphurous acid. p. 56

138. **Schönbein, C. F.**, 'Ueber die Einwirkung der schweflichten Säure auf Blumenfarbstoffe,' *J. prakt. Chem.*, Leipzig, 1851, LIV, pp. 76–78.
Further observations on the bleaching of pigments of flowers and autumnal leaves. — —

139. **1854. Filhol, E.**, 'Observations sur les matières colorantes des fleurs,' *C. R. Acad. sci.*, Paris, 1854, XXXIX, pp. 194–198.
Observation that white flowers turn yellow with alkalies. Yellow colour destroyed again by acids. Substance which

Page of text on which reference is made

is thus affected is in solution in the cell-sap, is soluble in water and alcohol, and less so in ether. Red flowers turn blue or green with alkalies, the green being due to a mixture of blue and yellow. pp. 10, 12, 49, **51**

140. **Fremy, E.**, und **Cloëz,** 'Ueber die Farbstoffe der Blumen,' *J. prakt. Chem.*, Leipzig, 1854, LXII, pp. 269–275.
An account of the properties of soluble blue (cyanin), red and yellow (xantheïn) plant pigments. pp. 10, 46, 49

141. **1855. Martens,** 'Nouvelles recherches sur la coloration des plantes,' *Bul. Acad. roy.*, Bruxelles, 1855, XXII (1), pp. 157–179.
Observations upon substances in plants which turn yellow with alkalies (now known to be flavones). Erroneous suggestion that they form pigment of yellow plastids. — —

142. **1856. *Mulder, G. J.,** *Chemie des Weines*, Leipzig, 1856. — —

143. **1858. Glénard, A.,** 'Recherches sur la matière colorante du vin,' *Ann. chim. phys.*, Paris, 1858, sér. 3, LIV, pp. 366–376.
Preparation and analysis of pigment of wine. pp. 10, 59

144. **Glénard, A.,** 'Ueber den Farbstoff des Weines,' *J. prakt. Chem.*, Leipzig, 1858, LXXV, pp. 317–318.
Same as previous paper. pp. 10, 59

145. **1859. Nicklès, J.,** 'Sur la matière colorante du troëne et son application à la recherche des eaux potables,' *J. pharm. chim.*, Paris, 1859, XXXV, pp. 328–334.
An account of the properties and extraction of the pigment of Privet (*Ligustrum*) berries. Preparation by precipitation with lead acetate, decomposition with sulphuretted hydrogen and subsequent purification. Found to contain carbon, hydrogen and oxygen only, but no constant analysis results. Reactions with acids, alkalies, salts of metals and various natural waters. Pigment termed liguline. — —

146. **1860. Filhol, E.,** 'Note sur quelques matières colorantes végétales,' *C. R. Acad. sci.*, Paris, 1860, L, pp. 545–547.
Further account of the properties of the substance which gives a yellow colour in white flowers on treatment with alkali. It was found to occur also in green parts of plants and to be analogous to luteolin; absent from some flowers, *Pelargonium* and *Papaver*, which become blue or violet with alkalies. pp. 10, **51**

147. **Filhol, E.,** 'Nouvelles recherches sur les matières colorantes végétales,' *C. R. Acad. sci.*, Paris, 1860, L, pp. 1182–1185.
Author is of the opinion that there is only one kind of red or blue soluble pigment (cyanine). Differences in colour due to other substances in the cell-sap. — —

Page of text on which reference is made

148. **1862. Chevreul,** 'Observations sur la propriété décolorante de l'eau oxygénée mêlée avec plusieurs matières colorées d'origine organique,' *C. R. Acad. sci.*, Paris, 1862, LV, pp. 737–738.

Action of hydrogen peroxide on soluble flower-pigments. Bleaching takes place. — —

149. **Wiesner, J.,** 'Einige Beobachtungen über Gerb- und Farbstoffe der Blumenblätter,' *Bot. Ztg.*, Leipzig, 1862, XX, pp. 389–392.

Researches on reactions given by tissues with iron salts and alkalies. If a colourless cell-sap contains a tannin which gives a green colour reaction with iron salts, it also gives a yellow reaction with alkalies: if a blue reaction with iron salts, no colour with alkalies. Anthocyanin itself gives a blue colour with alkalies, never green. Hence, when the former kind of tannin is present with anthocyanin, the cell-sap gives a green, i.e. blue plus yellow, reaction with alkalies; when the latter, a blue reaction only. pp. 11, 46, **51**

150. **Wigand, A.,** 'Einige Sätze über die physiologische Bedeutung des Gerbstoffes und der Pflanzenfarbe,' *Bot. Ztg.*, Leipzig, 1862, XX, pp. 121–125.

Distribution of tannin and pigments in leaves, stems and flowers. It is stated that anthocyanin has no relation to chlorophyll, but arises from a colourless chromogen, which is a tannin, and gives a yellow colour with alkalies. The chromogen, on oxidation, gives rise to anthocyanin. With alkalies, blue anthocyanin turns green, then yellow; red anthocyanin turns first blue, then green, and finally yellow. pp. 10, 46, 51, 59, 91, **107**

151. **1866. Gmelin, L.,** *Fortsetzung des Handbuchs der organischen Chemie*, Heidelberg, 1866, Bd IV (2), pp. 1421–1430.

Important summary of chemical work on anthocyanin up to this period. p. 106

152. **1867. Nägeli, C.,** und **Schwendener, S.,** *Das Mikroskop*, Leipzig, 1867, II, pp. 500–509.

An account of the reactions of anthocyanin including a discussion on the significance of its reaction with alkalies. p. 51

153. **Sorby, H. C.,** 'On a definite Method of Qualitative Analysis of Animal and Vegetable Colouring-Matters by means of the Spectrum Microscope,' *Proc. R. Soc.*, London, 1867, XV, pp. 433–455.

The spectrum microscope can be applied to the examination of pigments of flowers and fruits. Methods described for use and preparation of pigments to be examined.

Page of text on which reference is made

Numerous reagents are used for developing spectra, and a classification of the latter is outlined. p. 57

154. **1870. Schönn,** 'Ueber Blattgrün und Blumenblau,' *Zs. anal. Chem.*, Wiesbaden, 1870, IX, pp. 327–328.
Note on spectroscopic examination of pigments. — —

155. **1871. Vries, H. de,** 'Sur la perméabilité du protoplasma des betteraves rouges,' *Arch. Néerl. Sci. Soc. Holl.*, Haarlem, 1871, VI, pp. 117–126.
Relative permeability of living and dead protoplasm to coloured cell-sap. p. 32

156. **1872. Wiesner, J.,** 'Untersuchungen über die Farbstoffe einiger für chlorophyllfrei gehaltenen Phanerogamen,' *Jahrb. wiss. Bot.*, Leipzig, 1872, VIII, pp. 575–594.
Pigment bodies in *Neottia Nidus-avis* and *Orobanche* spp. Red pigment of *Orobanche*. Discussion as to the nature of the anthocyanin reaction with alkalies. p. 51

157. **1873. Kraus, C.,** 'Ueber die Ursache der Färbung der Epidermis vegetativer Organe der Pflanzen,' *Flora*, Regensburg, 1873, XXXI, pp. 316–317.
Suggestions as to chemical substances from which anthocyanin is derived. — —

158. **Sorby, H. C.,** 'On Comparative Vegetable Chromatology,' *Proc. R. Soc.*, London, 1873, XXI, pp. 442–483.
Method of separation of plastid pigments from others by means of carbon bisulphide as a solvent. Spectroscopical examination of all kinds of pigments. Suggestion that red pigment in leaves is produced by the action of light on chlorophyll under circumstances not yet reproduced. Presence or absence of pigment due to equilibrium, or absence of the same, between constructive and destructive agencies. Colour of flowers due to erythrophyll, and frequently exactly the same as in the leaves. pp. 57, 89

159. **1875. Bořščow, El.,** 'Notiz über den Polychroïsmus einer alkoholischen Cyaninlösung,' *Bot. Ztg.*, Leipzig, 1875, XXXIII, pp. 351–352.
A hot alcoholic extract was made from flowers of *Ajuga reptans* and *A. pyramidalis*, and it was found that the pigment showed continuous changes in fluorescence as the solutions cooled. — —

160. **1876. Pellagri, G.,** 'Sull' uso della fillocianina come reattivo,' *Gazz. chim. ital.*, Palermo, 1876, VI, pp. 35–38.
Anthocyanin in dilute solution can be used as a very sensitive reagent for alkalies. p. 56

161. **1877. Church, A. H.,** 'Coleïn,' *Ber. D. chem. Ges.*, Berlin, 1877, X (1), p. 296.

Page of text on which reference is made

Preparation and analysis of red pigment from leaves and stem of *Coleus Verschaffelii.* p. 59

162. **Senier, H.**, 'The Colouring Matter of the Petals of *Rosa gallica*,' *Pharm. J.*, London, 1877, ser. 3, VII, pp. 650–652.
Extraction and analysis of red pigment. Chief point of interest is the preparation of crystalline salts of the pigment with alkali metals. p. 59

163. **1878. Erdmann, J.**, 'Ueber die Veränderlichkeit des Rothweinfarbstoffes,' *Ber. D. chem. Ges.*, Berlin, 1878, XI (2), pp. 1870–1876.
Chemical test for wine pigment. p. 53

164. **Gautier, A.**, 'Sur les matières colorantes des vins,' *C. R. Acad. sci.*, Paris, 1878, LXXXVI, pp. 1507–1510.
View that plant pigments are formed from colourless tannic acids which redden on oxidation. Preparation and analyses of pigments from wines. It has been shown by later investigators that products were mixtures. pp. 11, 59, 107

165. **1879. Hilger, A.**, 'Ueber den Farbstoff der Familie der Caryophyllinen,' *Landw. Versuchstat.*, Berlin, 1879, XXIII, pp. 456–461.
Investigation of pigments of representatives of orders included under Centrospermae, i.e. *Beta, Phytolacca, Chenopodium* and *Amaranthus.* Pigments found to be similar, both as regards spectra and chemical reactions, but to differ from other red and violet pigments. Preparation of pigment from *Phytolacca*, but no analyses. — —

166. **1880. Lepel, F. von**, 'Pflanzenfarbstoffe als Reagentien auf Magnesiumsalze,' *Ber. D. chem. Ges.*, Berlin, 1880, XIII (1), pp. 766–768.
Spectroscopic examination of a number of red and blue anthocyanin pigments. Development of spectra after addition of magnesium salts. p. 57

167. **Schnetzler, J. B.**, 'Ueber Veränderungen des rothen Farbstoffes von *Paeonia officinalis* unter dem Einfluss chemischer Reagentien,' *Bot. Centralbl.*, Cassel, 1880, I, p. 682.
Short note on the action of acids, alkalies, iron salts, etc., on anthocyanin. — —

168. **Schnetzler, J. B.**, 'Observations sur les matières colorantes des fleurs,' *Bul. Soc. Sci. Nat.*, Lausanne, 1880, XVII, pp. 96–98.
Note on reactions of colouring matter of *Paeonia* flowers. — —

169. **Schnetzler, J. B.**, 'Ueber den rothen Farbstoff der Blätter von *Ampelopsis hederacea*,' *Bot. Centralbl.*, Cassel, 1880, I, pp. 247–248.

Page of text on which reference is made

Slight note on anthocyanin of *Ampelopsis*. Reactions with alkalies, etc. — —

170. ***Schnetzler, J. B.**, 'De la couleur des fleurs,' *Les Mondes*, 1880, LIII, p. 158.

Author contradicts the supposition that various flower-colours are due to different chemical substances, on the ground that *Paeonia* anthocyanin gives different colours with various reagents. Statement is made that reddening of leaves in autumn is due to the action of tannin on chlorophyll, and that chlorophyll is at the basis of all flower-pigments. — —

171. **1882. Husemann, A., Hilger, A.,** und **Husemann, Th.,** *Die Pflanzenstoffe in chemischer, physiologischer, pharmakologischer und toxikologischer Hinsicht*, Berlin, 1882, Bd 1, pp. 259–260.

Short account of anthocyanin from the chemical point of view. — —

172. **Maumené, E. J.,** 'Sur l'œnocyanine,' *C. R. Acad. sci.*, Paris, 1882, xcv, p. 924.

Green grapes dried *in vacuo* over sulphuric acid do not redden, but, on exposure to air, pigment is rapidly formed. Hence conclusion that colouring matter (œnocyanin) is formed from a colourless chromogen by oxidation and possibly hydration. — —

173. **1883. Gardiner, W.,** 'On the general occurrence of Tannins in the vegetable cell and a possible view of their physiological significance,' *Proc. Phil. Soc.*, Cambridge, 1883, IV, pp. 388–395.

Slight reference to connection of anthocyanin with tannins. — —

174. **1884. Flesch, M.,** 'Notiz über die Anwendung des Farbstoffes des Rothkohls in der Histologie,' *Zs. wiss. Mikrosk.*, Braunschweig, 1884, I, pp. 253–254.

Anthocyanin as a stain. p. 32

175. **Lavdowsky, M.,** 'Myrtillus, ein neues Tinctionsmittel für thierische und pflanzliche Gewebe,' *Arch. mikr. Anat.*, Bonn, 1884, XXIII, pp. 506–508.

Anthocyanin from berries of *Vaccinium Myrtillus* used as a stain for plant and animal tissues. p. 32

176. ***Marquis, E.,** 'Ueber den Farbstoff des kaukasischen Rothweines, seine Isolirung, quantitative Bestimmung und chemische Reaktion,' *Pharmaceutische Zeitschrift für Russland*, 1884, XXIII, pp. 7, 20. — —

177. ***Marquis, E.,** 'Ueber die Zersetzung des isolirten Pigments des kaukasischen Rothweines durch Wärme,'

Page of text on which reference is made

Pharmaceutische Zeitschrift für Russland, 1884, XXIII, p. 186. — —

178. **1885. Lindt, O.**, 'Ueber den Nachweis von Phloroglucin,' *Zs. wiss. Mikrosk.*, Braunschweig, 1885, II, pp. 495–499.
Connection between phloroglucin and anthocyanin formation. — —

179. **Terreil**, 'Faits pour servir à l'histoire de la matière colorante du vin et des matières colorantes rouges des végétaux,' *Bul. soc. chim.*, Paris, 1885, XLIV, pp. 2–6.
Exposition of Terreil's reaction for anthocyanins. — —

180. **1886. Vries, H. de**, 'Ueber die Aggregation im Protoplasma von *Drosera rotundifolia*,' *Bot. Ztg.*, Leipzig, 1886, XLIV, pp. 1–11, 17–26, 33–43, 57–62.
Some microchemical reactions in anthocyanin-containing cells. — —

181. **1887. *Jonas, V.**, *Photometrische Bestimmung der Absorptionsspektra roter und blauer Blütenfarbstoffe*, Inaugural-Dissertation, Kiel, 1887. — —

182. **1889. Heise, R.**, *Zur Kenntniss des Rothweinfarbstoffes*, Berlin, 1889.
Account of preparation and properties of two pigments obtained from grapes. pp. 11, 59

183. **Molisch, H.**, 'Ueber den Farbenwechsel anthokyanhältiger Blätter bei rasch eintretendem Tode,' *Bot. Ztg.*, Leipzig, 1889, XLVII, pp. 17–23.
Red leaves of *Coleus* and *Perilla*, when boiled with water, lose red colour and give yellowish or greenish solution, to which red colour returns on acidification. Author maintains that loss of colour is due to alkalinity of dead tissues, and that acidity of cell-sap is essential for preservation of red colour. — —

184. **Müller, N. J. C.**, 'Spectralanalyse der Blüthenfarben,' *Jahrb. wiss. Bot.*, Berlin, 1889, XX, pp. 78–101.
Observations on spectra and fluorescence of various anthocyanins in solution in both sulphuric acid and potash. pp. 46, 57

185. **1890. Lacour, E.**, 'Note sur le suc et la matière colorante du phytolacca,' *J. pharm. chim.*, Paris, 1890, XXI, pp. 243–245.
Note on the pigment of *Phytolacca* which is used for colouring wines. Action of acids, alkalies, salts of metals; solubilities of pigment, etc. — —

186. **Macchiati, L.**, 'Ricerche preliminari sulle sostanze coloranti delle gemme foglifere del castagno indiano,' *Nuovo Giorn. bot. ital.*, Firenze, 1890, XXII, pp. 76–78.
Slight note on extraction of red pigment from bud-scales of Horse Chestnut. — —

Page of text on which reference is made

187. **Waage, Th.**, 'Ueber das Vorkommen und die Rolle des Phloroglucins in der Pflanze,' *Ber. D. bot. Ges.*, Berlin, 1890, VIII, pp. 250–292.
Connection of phloroglucin with anthocyanin. — —

188. ***Zopf, W.**, *Die Pilze in morphologischer, physiologischer, biologischer und systematischer Beziehung*, Breslau, 1890.
Decomposition of anthocyanin by Moulds. — —

189. **1891. *Lidforss, B.**, 'Ueber die Wirkungssphäre der Glycose- und Gerbstoffreagentien,' *Kongl. Fysiografiska Sällskapets i Lund Handlingar*, Ny Följd., Bd 3, Lund, 1891–1892, IX.
Reduction of Fehling's solution by anthocyanin in its capacity of a glucoside. — —

190. **1892. Gautier, A.**, 'Sur l'origine des matières colorantes de la vigne; sur les acides ampélochroïques et la coloration automnale des végétaux,' *C. R. Acad. sci.*, Paris, 1892, CXIV, pp. 623–629.
Account of isolation, purification and properties of three pigments from Vine leaves. pp. 11, 27, 47, 59, 83, 85, 107

191. **Glan, R.**, *Ueber den Farbstoff des schwarzen Malve* (Althaea rosea), Inaugural-Dissertation, Erlangen, 1892.
Full account of preparation, properties and analysis of pigment. See text. p. 59

192. ***Haverland, Fr.**, *Beiträge zur Kenntniss der in den Früchten von* Phytolacca decandra (*Kermesbeeren*) *enthaltenen Bestandtheile*, Inaugural-Dissertation, Erlangen, 1892. — —

193. **1894. Heise, R.**, 'Zur Kenntniss des Heidelbeerfarbstoffes,' *Arb. Gesundhtsamt*, Berlin, 1894, IX, pp. 478–491.
Preparation, purification and analysis of pigments from Bilberry fruits. pp. 11, 48, 59

194. **1895. Weigert, L.**, 'Beiträge zur Chemie der rothen Pflanzenfarbstoffe,' *Jahresbericht und Programm der k. k. önologischen und pomologischen Lehranstalt in Klosterneuburg*. Wien, 1895, pp. i–xxxi.
Scheme for differentiation of anthocyanin pigments by means of qualitative tests into two groups, 'Weinroth' and 'Rübenroth.' pp. 11, 46, **53**

195. **1897. Beilstein, F.**, *Handbuch der organischen Chemie*, 1897, Bd III, pp. 651–652.
Short account of anthocyanin from chemical point of view. — —

196. **Carles, P.**, et **Nivière, G.**, 'Influence des matières colorantes sur la fermentation des vins rouges très colorés,' *C. R. Acad. sci.*, Paris, 1897, CXXV, pp. 452–453.
Explanation offered is that tannin nature of colouring matters acts as an antiseptic and inhibits the fermentation. — —

Page of text on which reference is made

197. **Sostegni, L.**, 'Sulle materie coloranti delle uve rosse,' *Gazz. chim. ital.*, Roma, 1897, XXVII (2), pp. 475–485.
Extraction and analysis of pigments from grapes. — —

198. **1898. Dippel, L.**, *Das Mikroskop und seine Anwendung*, Braunschweig, 1898, Th. 2, pp. 65–66, 105–106, 108–109.
Solubilities and other properties of anthocyanin. — —

199. **1899. Claudius, M.**, 'Ueber die Anwendung einiger gewöhnlicher Pflanzenfarbstoffe in der mikroskopischen Färbungstechnik,' *Centralbl. Bakt.*, Jena, 1899, v, pp. 579–582.
Use of anthocyanin for histological staining. p. 32

200. **Keegan, P. Q.**, 'Experiments on the Floral Colours,' *Nature*, London, 1899, LXI, pp. 105–106.
Reactions of anthocyanin with alkalies, salts, etc. — —

201. **1900. Formánek, J.**, 'Der Farbstoff der rothen Rübe und sein Absorptionsspectrum,' *J. prakt. Chem.*, Leipzig, Neue Folge, 1900, LXII, pp. 310–314.
Account of normal spectrum and of those developed with acid and alkali respectively. p. 57

202. **Miyoshi, M.**, 'Ueber die künstliche Aenderung der Blütenfarben,' *Bot. Centralbl.*, Cassel, 1900, LXXXIII, pp. 345–346.
Reactions of alum salts, acids and alkalies on various anthocyanins. Bearing on Molisch's results with *Hydrangea* (see No. 570). — —

203. **1901. Goppelsroeder, Fr.**, *Capillaranalyse beruhend auf Capillaritäts- und Adsorptionserscheinungen mit dem Schlusskapitel: Das Emporsteigen der Farbstoffe in den Pflanzen*, Basel, 1901.
Method which can be adapted to the separation of mixtures of pigments in solution. Strips of specially prepared filter paper are allowed to dip slightly into the solutions, and the various pigments rise by capillarity to different heights. By cutting off zones of paper, and by repetition of the process, a certain amount of pure pigment can be obtained. — —

204. **1902. Bouffard, A.**, 'Action de l'acide sulfureux sur l'oxydase et sur la matière colorante du vin rouge,' *C. R. Acad. sci.*, Paris, 1902, CXXXIV, pp. 1380–1383.
Action of acid considered to be twofold: first, preservation of the colour of the wine, and secondly, a destructive action on the oxidising enzyme. — —

205. **Sostegni, L.**, 'Sulle materie coloranti delle uve rosse,' *Gazz. chim. ital.*, Roma, 1902, XXXII (2), pp. 17–19.
Continuation of analysis of pigment from grapes. — —

206. **1903. Griffiths, A. B.**, 'Die Pigmente des Geraniums und anderer

Page of text on which reference is made

Pflanzen,' *Ber. D. chem. Ges.*, Berlin, 1903, XXXVI (4), pp. 3959–3961.

Account of analyses and properties of pigment. pp. 47, 59

207. **Ichimura, T.**, 'On the Formation of Anthocyan in the Petaloid Calyx of the Red Japanese Hortense,' *J. Coll. Sci.*, Tokyo, 1903–1904, XVIII, art. 3, pp. 1–18.

Stages of development of anthocyanin in the calyx. View that pigment arises from a tannin-like substance, protanthocyan, which gives a yellow colour with alkalies, and is probably identical with Wigand's cyanogen (see No. 150). In final stages, violet crystals of pigment were observed. p. 34

208. **1904. Kraemer, H.**, 'The Origin and Nature of Color in Plants,' *Proc. Amer. Phil. Soc.*, Philadelphia, Pa., 1904, XLIII, pp. 257–277.

Method for extraction and separation of plastid and soluble pigments. Reactions with acids, alkalies, and other reagents are represented in tables. — —

209. **Naylor, W. A. H.**, and **Chappel, E. J.**, 'Note on the Colouring Matters of *Rosa Gallica*,' *Pharm. J.*, London, 1904, ser. 4, XIX, pp. 231–233.

Preparation of the pigment. — —

210. **1905. Czapek, Fr.**, *Biochemie der Pflanzen*, Jena, 1905, Bd I, pp. 471–477.

Good general account of anthocyanin, chiefly from the chemical point of view. — —

211. **Kastle, J. H.**, 'A Method for the Determination of the Affinities of Acids Colorimetrically, by Means of Certain Vegetable Coloring Matters,' *Amer. Chem. J.*, Baltimore, Md., 1905, XXXIII, pp. 46–59.

Solutions of anthocyanin from purple grapes, red *Pelargonium* and purple *Petunia* flowers were bleached with sulphur dioxide, and then used colorimetrically to test the 'strength' of a number of acids by amount of return of colour. The resulting order corresponds closely to that given by Ostwald's methods. pp. 56, 57

212. **1906. Grafe, V.**, 'Studien über das Anthokyan (I. Mitteilung),' *SitzBer. Ak. Wiss.*, Wien, 1906, CXV (Abt. 1), pp. 975–993.

Important paper on preparation and analyses of pigments from flowers of *Althaea rosea*. pp. 11, 47, 48, 52, 56, 59

213. **1907. Toni, J.-B. de**, 'Observations sur l'anthocyane d'*Ajuga* et de *Strobilanthes*,' *C. R. ass. franç. avanc. sci.*, Paris, 1907 (2e partie), pp. 415–418.

Repetition of Bořšcow's experiment on fluorescence (see No. 159). Phenomenon explained by Molisch's view of

Page of text on which reference is made

alkalinity of dying tissues (see No. 183). Solution of *Strobilanthes* pigment heated to 90° loses properties of anthocyanin. — —

214. **1908. Laborde, J.**, 'Sur l'origine de la matière colorante des raisins rouges et autres organes végétaux,' *C. R. Acad. sci.*, Paris, 1908, CXLVI, pp. 1411–1413.

Author obtains tannin-like chromogens from unripe grapes of both green and red varieties. Chromogens become red on heating with dilute hydrochloric acid at 120°. Investigation of these substances should explain reddening of leaves and fruits, and also why white grapes do not develop pigment, although containing chromogen. p. 107

215. **Laborde, J.**, 'Sur les transformations de la matière chromogène des raisins pendant la maturation,' *C. R. Acad. sci.*, Paris, 1908, CXLVII, pp. 753–755.

Chromogen of pigment, which is in insoluble form in green fruit, becomes more and more soluble as fruits mature. Change probably brought about by a diastase. p. 107

216. **Laborde, J.**, 'Sur le mécanisme physiologique de la coloration des raisins rouges et de la coloration automnale des feuilles,' *C. R. Acad. sci.*, Paris, 1908, CXLVII, pp. 993–995.

Conclusion that anthocyanin is formed from tannin-like substances. Red products are obtained by the action of various reagents on tannins. p. 107

217. **Malvezin, Ph.**, 'Sur l'origine de la couleur des raisins rouges,' *C. R. Acad. sci.*, Paris, 1908, CXLVII, pp. 384–386.

Water extracts of unripe red (or white) grapes, heated for 24 hours at 85°, give red pigment, but not in absence of air. Absence of pigment in white grapes in nature is due to lack of diastase, but high temperature will cause oxidation (artificially). — —

218. **Palladin, W.**, 'Ueber die Bildung der Atmungschromogene in den Pflanzen,' *Ber. D. bot. Ges.*, Berlin, 1908, XXVI *a*, pp. 389–394.

Suggestion that anthocyanin is a respiration pigment.
pp. 13, 59, 91, 100, 102, **108**

219. **Portheim, L. von**, und **Scholl, E.**, 'Untersuchungen über die Bildung und den Chemismus von Anthokyanen,' *Ber. D. bot. Ges.*, Berlin, 1908, XXVI *a*, pp. 480–483.

Method described for purification of anthocyanin by dialysis. Anthocyanin is also prepared from testas of seeds of *Phaseolus multiflorus* by extraction with alcohol. By fractional crystallisation, yellow and red crystalline products are obtained. p. 47

220. **Russo, Ph.**, 'Des pigments floraux,' *C. R. soc. biol.*, Paris, 1908, LXV, pp. 579–581.

Page of text on which reference is made

Examination of pigments of many flowers with tournesol paper leads to the conclusion that, in the cyanic series, flower-colour varies in tint according to the greater or less acidity of the cell-sap. — —

221. **Sorby, H. C.**, 'On the Colouring Matters of Flowers,' *Nature*, London, 1908, LXXVII, pp. 260–261.

Note on the solubilities of various flower-pigments; also observations as regards spectra. — —

222. **1909. Combes, R.**, 'Rapports entre les composés hydrocarbonés et la formation de l'anthocyane,' *Ann. sci. nat.* (*Bot.*), Paris, 1909, sér. 9, IX, pp. 275–303.

View held that an increase in amount of sugars and glucosides in tissues leads to increase in formation of anthocyanin. Estimation is made of glucosides and sugars in various green and red leaves. Found that red leaves contain considerably greater quantities of these substances than green control leaves. pp. 6, 84

223 **Combes, R.**, 'Recherches biochimiques sur le développement de l'anthocyane chez les végétaux,' *C. R. Acad. sci.*, Paris, 1909, CXLVIII, pp. 790–792.

Same results as in preceding paper. — —

224. **Grafe, V.**, 'Studien über das Anthokyan (II. Mitteilung),' *SitzBer. Ak. Wiss.*, Wien, 1909, CXVIII (Abt. 1), pp. 1033–1044.

Further researches on anthocyanin from flowers of *Althaea*. pp. 47, 59

225. **Palladin, W.**, 'Ueber das Wesen der Pflanzenatmung,' *Biochem. Zs.*, Berlin, 1909, XVIII, pp. 151–206.

Full account of the significance of anthocyanin as a respiration pigment. pp. 13, **108**

226. **Wheldale, M.**, 'The Colours and Pigments of Flowers, with Special Reference to Genetics,' *Proc. R. Soc.*, London, 1909, LXXXI B, pp. 44–60.

Attempt to show that there is some correlation between the chemical reactions of pigments and their behaviour in genetics. Recognition of a certain number of types which give rise to a definite series of colour varieties. pp. 40, 53, 58, 91

227. **Wheldale, M.**, 'On the nature of anthocyanin.' *Proc. Phil. Soc.*, Cambridge, 1909, XV, pp. 137–161.

Suggestion, made for the first time, that anthocyanins are formed from chromogens, which are glucosides of flavones, or possibly, xanthones, by the action of oxidase. Successive oxidation stages may give rise to the series, red, purplish-red and purple pigments. pp. 13, 58, 91, **110**

228. **Wheldale, M.**, 'Note on the Physiological Interpretation

Page of text on which reference is made

of the Mendelian Factors for Colour in Plants,' *Rep. Evol. Com. Roy. Soc.*, London, 1909, Rpt. 5, pp. 26–31.
Expression of colour factors in terms of oxidases and reductases. — —

229. **1910. *Dezani, S.**, 'Le sostanze cromogene dell' uva bianca,' *Staz. sper. agr. ital.*, Modena, 1910, XLIII, pp. 428–436.
Substance found in white grapes which gives oenocyanin-like products on treatment with hydrochloric acid. — —

230. ***Sacher, J. F.**, 'Der Farbstoff der roten Radieschen,' *Chem. Ztg.*, Cöthen, 1910, XXXIV, p. 1333.
Properties of colouring matter of radish as indicator. p. 56

231. ***Schwertschlager, J.**, 'Der Farbstoff der roten Radieschen,' *Chem. Ztg.*, Cöthen, 1910, XXXIV, p. 1257.
Properties of colouring matter of the radish. — —

232. **Wheldale, M.**, 'Plant Oxidases and the Chemical Interrelationships of Colour-Varieties,' *Progr. rei bot.*, Jena, 1910, III, pp. 457–473.
Chiefly a résumé of papers published by author in 1909. p. 13

233. **1911. Abderhalden, E.**, *Biochemisches Handlexikon*, Berlin, 1911, Bd VI, pp. 182–183.
Short account, chiefly chemical, of anthocyanin. — —

234. **Combes, R.**, 'Recherches sur la formation des pigments anthocyaniques,' *C. R. Acad. sci.*, Paris, 1911, CLIII, pp. 886–889.
Microchemical tests for anthocyanin and its chromogen in tissues of *Ampelopsis*. — —

235. **Combes, R.**, 'Recherches microchimiques sur les pigments anthocyaniques,' *C. R. ass. franç. avanc. sci.*, Paris, 1911, pp. 464–471.
Same as preceding paper. — —

236. **Gautier, A.**, 'Sur les mécanismes de la variation des races et les transformations moléculaires qui accompagnent ces variations,' *C. R. Acad. sci.*, Paris, 1911, CLIII, pp. 531–539.
Connection between variation and chemical constitution. Analyses of pigments from different varieties of grapes. — —

237. **Grafe, V.**, 'Studien über das Anthokyan (III. Mitteilung),' *SitzBer. Ak. Wiss.*, Wien, 1911, CXX (Abt. 1), pp. 765–807.
Paper on the red pigment of *Pelargonium* flowers. pp. 11, 48, 56, 57, 59

238. **Kastle, J. H.**, and **Haden, R. L.**, 'On the Color Changes occurring in the Blue Flowers of the Wild Chicory, *Cichorium Intybus*,' *Amer. Chem. J.*, Baltimore, Md., 1911, XLVI, pp. 315–325.

Page of text on which reference is made

Author shows that changes in colour exhibited by flowers on withering are due, in part, to variations in amount of acid in the pigmented cells, and, in part, to the action of an oxidase which oxidises and destroys the plant pigment. — —

239. **Nierenstein, M.**, und **Wheldale, M.**, 'Beitrag zur Kenntnis der Anthocyanine. 1. Ueber ein anthocyaninartiges Oxydationsprodukt des Quercetins,' *Ber. D. chem. Ges.*, Berlin, 1911, XLIV (3), pp. 3487–3491.

Description of a red pigment formed on oxidation of quercetin with chromic acid. Properties of oxidised product resemble anthocyanin in some respects. — —

240. **Politis, I.**, 'Sopra speciali corpi cellulari che formano Antocianine,' *Rend. Acc. Lincei*, Roma, 1911, XX, pp. 828–834.

Hypothesis, supported by microscopical and microchemical observations, that anthocyanin is produced by special bodies termed cyanoplasts, which contain tannin substances capable of forming anthocyanin. On disintegration of cyanoplast, anthocyanin colours the cell-sap. — —

241. **Wheldale, M.**, 'On the Formation of Anthocyanin,' *J. Genetics*, Cambridge, 1911, I, pp. 133–157.

It is suggested that anthocyanin is formed from glucosides of xanthones and flavones, and that the reaction takes place in two stages. First, the hydrolysis of glucosides by an enzyme, and then the oxidation of the free flavone by oxidase. Summary of evidence in favour of the hypothesis. pp. 13, 112

242. ***Wissemann, E.**, *Beiträge zur Kenntnis des Auftretens und der topographischen Verteilung von Anthocyan und Gerbstoff in vegetativen Organen*, Dissertation, Göttingen, 1911, 110 pages.

Respective distributions of anthocyanin and tannin and connection between the same. — —

243. **1912.** ***Chodat, R.**, 'Les pigments végétaux,' *Verhandl. schweiz. Naturforsch. Ges.*, 95 Jahresversamml., Altdorf, 1912, pp. 79–95.

Action of tyrosinase on gallic acid produces a series of pigments resembling anthocyanin in their reactions. Suggestion made that anthocyanin is derived from tannins. — —

244. **Keeble, F.**, and **Armstrong, E. F.**, 'The Oxydases of *Cytisus Adami*,' *Proc. R. Soc.*, London, 1912, LXXXV B, pp. 460–465.

Oxidase reactions of flowers of *C. purpureus*, *C. Adami* and *C. Laburnum* support view of Baur that *Adami* is a

Page of text on which reference is made

periclinal chimera, composed externally of *purpureus* and internally of *Laburnum*. — —

245. **Keeble, F.**, and **Armstrong, E. F.**, 'The Distribution of Oxydases in Plants and their Rôle in the Formation of Pigments,' *Proc. R. Soc.*, London, 1912, LXXXV B, pp. 214–218.
Oxidase tests upon tissues of plants of all varieties of *P. sinensis* support view that anthocyanin is the product of activity of an oxidase, and that albinism in *Primula* is due to loss of chromogen. pp. 13, **117**

246. **Keeble, F.**, and **Armstrong, E. F.**, 'The Rôle of Oxydases in the Formation of the Anthocyan Pigments of Plants,' *J. Genetics*, Cambridge, 1912, II, pp. 277–309.
Previous paper in greater detail. pp. 13, 18, **117**

247. **Nierenstein, M.**, 'Beitrag zur Kenntnis der Anthocyanine. II. Ueber ein anthocyanin-artiges Oxydationsprodukt des Chrysins,' *Ber. D. chem. Ges.*, Berlin, 1912, XLV (1), pp. 499–501.
Description of a red pigment, formed on oxidation of chrysin by chromic acid, and similar to that prepared by the same method from quercetin (see No. 239). Product resembles anthocyanin in some of its properties, since it gives a blue colour with alkalies, and red with concentrated sulphuric acid. — —

248. **1913. Atkins, W. R. G.**, 'Oxidases and their Inhibitors in Plant Tissues,' *Sci. Proc. R. Soc.*, Dublin, 1913, XIV (N. S.), pp. 144–156.
Account of oxidase tests with tissues of various plants. Special attention given to comparison of distribution of oxidases and pigment in the genus *Iris*. p. 119

249. **Combes, R.**, 'Production expérimentale d'une anthocyane identique à celle qui se forme dans les feuilles rouges en automne, en partant d'un composé extrait des feuilles vertes,' *C. R. Acad. sci.*, Paris, 1913, CLVII, pp. 1002–1005.
By treating yellow pigment (flavone) occurring in *Ampelopsis* with nascent hydrogen, the author obtains a purple substance which he maintains to be identical with natural anthocyanin. Hence, contrary to previous views, anthocyanin formation from a flavone is brought about by reduction, not oxidation. pp. 16, **120**, 125

250. **Combes, R.**, 'Passage d'un pigment anthocyanique extrait des feuilles rouges d'automne au pigment jaune contenu dans les feuilles vertes de la même plante,' *C. R. Acad. sci.*, Paris, 1913, CLVII, pp. 1454–1457.
Crystalline anthocyanin from *Ampelopsis* is converted into a flavone by oxidation with hydrogen peroxide. pp. 16, **120**

Page of text on which reference is made

251. **Combes, R.**, 'Untersuchungen über den chemischen Prozess der Bildung der Anthokyanpigmente,' *Ber. D. bot. Ges.*, Berlin, 1913, XXXI, pp. 570–578.
Repetition of last two papers. — —

252. **Jones, W. N.**, 'The Formation of Anthocyan Pigments of Plants. Part 5. The Chromogens of White Flowers,' *Proc. R. Soc.*, London, 1913, LXXXVI B, pp. 318–323.
Investigations on presence of chromogens and oxidases in white flowers. p. 120

253. **Jones, W. N.**, 'Some Investigations in Anthocyan Formation,' *Rep. Brit. Ass.*, London, 1913, p. 713.
Anthocyanin formation considered to be the result of an oxidase system. Summary of previous papers. — —

254. **Keeble, F., Armstrong, E. F.**, and **Jones, W. N.**, 'The Formation of the Anthocyan Pigments of Plants. Part 4. The Chromogens,' *Proc. R. Soc.*, London, 1913, LXXXVI B, pp. 308–317.
Hypothesis, that anthocyanin is formed by oxidation of chromogen by an oxidase, is further supported by phenomena connected with the behaviour of pigment in alcohol and water solutions. pp. 49, **119**

255. **Keeble, F., Armstrong, E. F.**, and **Jones, W. N.**, 'The Formation of the Anthocyan Pigments of Plants. Part 6.' *Proc. R. Soc.*, London, 1913, LXXXVII B, pp. 113–131.
Purple pigments are obtained by treatment of flower extracts with nascent hydrogen, and subsequent oxidation. Suggestion that preliminary reduction plays a part in pigment formation. pp. 15, **122**, 123

256. **Molisch, H.**, *Mikrochemie der Pflanze*, Jena, 1913, pp. 236–241.
Microchemical reactions of anthocyanin; occurrence as crystals, etc. — —

257. **Peche, K.**, 'Ueber eine neue Gerbstoffreaktion und ihre Beziehung zu den Anthokyanen,' *Ber. D. bot. Ges.*, Berlin, 1913, XXXI, pp. 462–471.
A new microchemical reaction for tannins which gives a colour resembling anthocyanin. Cells containing iron-greening tannins give, with potash solution and formol, a blue-green pigment which becomes red with acids, and is similar to anthocyanin except in its solubilities. — —

258. **Tswett, M.**, 'Beiträge zur Kenntnis der Anthocyane. Ueber künstliches Anthocyan,' *Biochem. Zs.*, Berlin, 1913, LVIII, pp. 225–235.
Artificial anthocyanin-like substances can be obtained from alcoholic extracts of many colourless parts of

Page of text on which reference is made

plants by treatment with aldehyde and hydrochloric acid. p. 122

259. **Wheldale, M.**, 'The Flower Pigments of *Antirrhinum majus*. 1. Method of Preparation,' *Biochemical Journal*, Cambridge, 1913, VII, pp. 87–91.
Preparation and purification of anthocyanin pigments from varieties of *Antirrhinum majus*. pp. 14, 59, **60**, 114

260. **Willstätter, R.**, und **Everest, A. E.**, 'Ueber den Farbstoff der Kornblume,' *Liebigs Ann. Chem.*, Leipzig, 1913, CCCCI, pp. 189–232.
Method of extraction and preparation of blue pigment from Cornflower. Suggestions as to constitutional formulae for red, purple and blue anthocyanins.
pp. 15, 47, 48, 49, 52, 54, 57, 59, **63**, 113

261. **1914. Atkins, W. R. G.**, 'Oxidases and their Inhibitors in Plant Tissues. Part 2. The Flowers and Leaves of *Iris*,' *Sci. Proc. R. Soc.*, Dublin, 1914, XIV (N. S.), pp. 157–168.
Correlation between distribution of anthocyanin and that of oxidases and inhibitors. p. 119

262. **Atkins, W. R. G.**, 'Some Recent Work on Plant Oxidases,' *Sci. Progr.*, London, 1914, IX, pp. 112–126.
Summary of situation as regards connection between pigment formation and oxidase reactions. — —

263. **Combes, R.**, 'Sur la présence, dans des feuilles et dans des fleurs ne formant pas d'anthocyane, de pigments jaunes pouvant être transformés en anthocyane,' *C. R. Acad. sci.*, Paris, 1914, CLVIII, pp. 272–274. pp. 120, 125

264. **Everest, A. E.**, 'The Production of Anthocyanins and Anthocyanidins,' *Proc. R. Soc.*, London, 1914, LXXXVII B, pp. 444–452.
Account of the production of pigments, which are claimed to be true anthocyanins, by reduction of flavones with nascent hydrogen. pp. 113, **123**

265. **Everest, A. E.**, 'The Production of Anthocyanins and Anthocyanidins. Part 2.' *Proc. R. Soc.*, London, 1914, LXXXVIII B, pp. 326–332.
Further evidence to prove identity of artificial and natural anthocyanins which has been questioned by Willstätter and also by Wheldale and Bassett. pp. 18, 123

266. **Everest, A. E.**, 'A Note on Wheldale and Bassett's Paper "On a supposed Synthesis of Anthocyanin,"' *J. Genetics*, Cambridge, 1914, IV, pp. 191–192.
Further remarks on the identity of artificial and natural anthocyanins. — —

267. **Hall, A. D., Armstrong, E. F.** and **H. E., Keeble, F.**,

Page of text on which reference is made

and **Russell, E. J.**, 'The Study of Plant Enzymes, particularly with relation to Oxidation,' *Rep. Brit. Ass.*, London, 1914, pp. 108–109.

It is pointed out that every oxidation involves also a reduction. Hence the concurrence in distribution between oxidases and pigments need not necessarily prove the pigments to be formed by oxidation. Some such view becomes necessary if the reduction hypothesis of anthocyanin formation be retained. — —

268. **Tswett, M.**, 'Zur Kenntnis des "vegetabilischen Chamaeleons,"' *Ber. D. bot. Ges.*, Berlin, 1914, XXXII, pp. 61–68.

Discussion of the acid and alkali reactions of anthocyanin, and Willstätter's hypothesis as regards them. Also Keeble and Armstrong's explanation of loss of colour of anthocyanin in alcohol which is disbelieved by author (Tswett). Further remarks on author's artificial anthocyanin. p. 119

269. **Wheldale, M.**, 'Our Present Knowledge of the Chemistry of the Mendelian Factors for Flower-Colour,' *J. Genetics*, Cambridge, 1914, IV, pp. 109–129.

Summary of work on subject. — —

270. **Wheldale, M.**, and **Bassett, H. Ll.**, 'The Flower Pigments of *Antirrhinum majus*. 3. The Red and Magenta Pigments,' *Biochemical Journal*, Cambridge, 1914, VIII, pp. 204–208.

Purification and analyses of anthocyanins. pp. 14, 48, 52, 59, **60**, 121

271. **Wheldale, M., and Bassett, H. Ll.**, 'The Chemical Interpretation of some Mendelian Factors for Flower-Colour,' *Proc. R. Soc.*, London, 1914, LXXXVII B, pp. 300–311. pp. 57, 119

272. **Wheldale, M.**, and **Bassett, H. Ll.**, 'On a supposed Synthesis of Anthocyanin,' *J. Genetics*, Cambridge, 1914, IV, pp. 103–107.

Criticism of a paper by Everest who maintains that natural anthocyanins can be obtained by reduction of the flavones. pp. 17, **123**

273. **Willstätter, R.**, 'Ueber die Farbstoffe der Blüten und Früchte,' *SitzBer. Ak. Wiss.*, Berlin, 1914, pp. 402–411.

Preparation and analyses of anthocyanins from *Delphinium* and other flowers, and fruits of Grape Vine, Bilberry and Cranberry. Suggestions for the constitution of anthocyanin. pp. 15, 47, 59, **63**

274. **Willstätter, R.**, und **Mallison, H.**, 'Ueber die Verwandtschaft der Anthocyane und Flavone,' *SitzBer. Ak. Wiss.*, Berlin, 1914, pp. 769–777.

Constitution for anthocyanins suggested. Anthocyanins

Page of text on which reference is made

regarded as forming a reduced series from flavones. Production of anthocyanin experimentally by reduction of a flavone. pp. 15, 17, 47, 59, **123**

275. **Willstätter, R.**, und **Zechmeister, L.**, 'Synthese des Pelargonidins,' *SitzBer. Ak. Wiss.*, Berlin, 1914, XXXIV, pp. 886–893. pp. 15, 67

276. **1915. Atkins, W. R. G.**, 'Oxidases and Inhibitors in Plant Tissues. Part 4. The Flowers of *Iris*,' *Sci. Proc. R. Soc.*, Dublin, 1915, XIV (N. S.), pp. 317–327.
Further observations on the distribution of oxidases in *Iris* flowers; it is found that the distribution does not coincide with that of anthocyanin. Difference possibly due to inhibitors of oxidases. p. 119

277. **Burdick, C. L.**, 'Ueber die Anthocyane der Petunie und Aster,' Basel, 1915, Dissertation, 44 pages.
See references 302 and 303. — —

278. **Cockerell, T. D. A.**, 'Characters of *Helianthus*,' *Torreya*, Lancaster, Pa., 1915, XV, pp. 11–16.
Some reactions of *Helianthus* anthocyanin are described. — —

279. **Everest, A. E.**, 'The Anthocyan Pigments,' *Sci. Progr.*, London, 1915, IX, pp. 597–612.
General review of work on the subject. — —

280. **Everest, A. E.**, 'Recent Chemical Investigations of the Anthocyan Pigments and their Bearing upon the Production of these Pigments in Plants,' *J. Genetics*, Cambridge, 1915, IV, pp. 361–367.
Chiefly a summary of Willstätter's work. — —

281. **Horowitz, B.**, 'Plant Pigments. Their color and interrelationships,' *Biochem. Bull.*, 1915, IV, 161–172.
Short account only of anthocyanin pigments. — —

282. **Schiemann, E.**, 'Neuere Arbeiten über Bildung der Blütenfarbstoffe. Sammelreferat vom Standpunkte der Mendelspaltung,' *Zs. indukt. Abstammungslehre*, Leipzig, 1915, XIV, pp. 80–96.
Résumé of work on the subject. — —

283. **Shibata, K.**, 'Untersuchungen über das Vorkommen und die physiologische Bedeutung der Flavonderivate in den Pflanzen. (I. Mitteilung),' *Bot. Mag.*, Tokyo, 1915, XXIX, pp. 118–132. p. 124

284. **Shibata, K.**, und **Kishida, M.**, 'Untersuchungen über das Vorkommen und die physiologische Bedeutung der Flavonderivate in den Pflanzen (II. Mitteilung), Ein Beitrag zur chemischen Biologie der alpinen Gewächse,' *Bot. Mag.*, Tokyo, 1915, XXIX, pp. 301–308, 316–332. p. 124

285. **Wheldale, M.**, 'Our Present Knowledge of the Chemistry

Page of text on which reference is made

of the Mendelian Factors for Flower-Colour. Part 2,' *J. Genetics*, Cambridge, 1915, IV, pp. 369–376.
Summary of work on anthocyanin continued. — —

286. **Willstätter, R.**, und **Bolton, E. K.**, 'Ueber den Farbstoff der Scharlachpelargonie,' *Liebigs Ann. Chem.*, Leipzig, 1915, CCCCVIII, pp. 42–61. p. **71**

287. **Willstätter, R.**, und **Mallison, H.**, 'Ueber den Farbstoff der Preiselbeere,' *Liebigs Ann. Chem.*, Leipzig, 1915, CCCCVIII, pp. 15–41. p. **72**

288. **Willstätter, R.**, und **Mallison, H.**, 'Ueber Variationen der Blütenfarben,' *Liebigs Ann. Chem.*, Leipzig, 1915, CCCCVIII, pp. 147–162. p. **68**

289. **Willstätter, R.**, und **Martin, K.**, 'Ueber den Farbstoff der *Althaea rosea*,' *Liebigs Ann. Chem.*, Leipzig, 1915, CCCCVIII, pp. 110–121. p. **74**

290. **Willstätter, R.**, und **Mieg, W.**, 'Ueber ein Anthocyan des Rittersporns,' *Liebigs Ann. Chem.*, Leipzig, 1915, CCCCVIII, pp. 61–82. p. **73**

291. **Willstätter, R.**, und **Mieg, W.**, 'Ueber den Farbstoff der wilden Malve,' *Liebigs Ann. Chem.*, Leipzig, 1915, CCCCVIII, pp. 122–135. p. **74**

292. **Willstätter, R.**, und **Nolan, T. J.**, 'Ueber den Farbstoff der Rose,' *Liebigs Ann. Chem.*, Leipzig, 1915, CCCCVIII, pp. 1–14. p. **72**

293. **Willstätter, R.**, und **Nolan, T. J.**, 'Ueber den Farbstoff der Päonie,' *Liebigs Ann. Chem.*, Leipzig, 1915, CCCCVIII, pp. 136–146. p. **73**

294. **Willstätter, R.**, und **Zollinger, E. H.**, 'Ueber die Farbstoffe der Weintraube und der Heidelbeere,' *Liebigs Ann. Chem.*, Leipzig, 1915, CCCCVIII, pp. 83–109. pp. **73, 74**

295. **1916. Haas, A. R.**, 'The Permeability of Living Cells to Acids and Alkalies,' *J. Biol. Chem.*, New York, 1916, XXVII, pp. 225–232.
Anthocyanin used for studying penetration of acids and alkalies. — —

296. **Haas, A. R.**, 'The Acidity of Plant Cells as shown by Natural Indicators,' *J. Biol. Chem.*, New York, 1916, XXVII, pp. 233–241.
Reaction of living cells may be acid. Blue colour does not always indicate alkaline reaction of sap. — —

297. **Hooker, H. D.**, 'Physiological Observations on *Drosera rotundifolia*,' *Bull. Torrey Bot. Cl.*, New York, 1916, XLIII, pp. 1–27.
Reactions given of anthocyanin present. — —

298. **Shibata, K.**, und **Nagai, I.**, 'Untersuchungen über das

Page of text on which reference is made

Vorkommen und die physiologische Bedeutung der Flavonderivate in den Pflanzen (III. Mitteilung), Ueber den Flavongehalt der Tropenpflanzen,' *Bot. Mag.*, Tokyo, 1916, xxx, pp. 149–178. p. 124

299. **Shibata, K., Nagai, I.,** and **Kishida, M.,** 'The Occurrence and physiological Significance of Flavone Derivatives in Plants,' *J. Biol. Chem.*, New York, 1916, xxviii, pp. 93–108. p. 124

300. **Willstätter, R.,** und **Bolton, E. K.,** 'Ueber das Anthocyan der rotblühenden Slaviaarten,' *Liebigs Ann. Chem.*, Leipzig, 1916, ccccxii, pp. 113–136. p. **71**

301. **Willstätter, R.,** und **Bolton, E. K.,** 'Ueber ein Anthocyan der Winteraster (Chrysanthemum),' *Liebigs Ann. Chem.*, Leipzig, 1916, ccccxii, pp. 136–148. p. **71**

302. **Willstätter, R.,** und **Burdick, Ch. L.,** 'Ueber zwei Anthocyane der Sommeraster,' *Liebigs Ann. Chem.*, Leipzig, 1916, ccccxii, pp. 149–164. p. **70**

303. **Willstätter, R.,** und **Burdick, Ch. L.,** 'Ueber den Farbstoff der Petunie,' *Liebigs Ann. Chem.*, Leipzig, 1916, ccccxii, pp. 217–230. p. **74**

304. **Willstätter, R.,** und **Weil, F. J.,** 'Ueber das Anthocyan des violetten Stiefmütterchens,' *Liebigs Ann. Chem.*, 1916, ccccxii, pp. 178–194. p. **73**

305. **Willstätter, R.,** und **Weil, F.,** 'Ueber die Mohnfarbstoffe, I,' *Liebigs Ann. Chem.*, Leipzig, 1916, ccccxii, pp. 231–251. p. **72**

306. **Willstätter, R.,** und **Zollinger, E. H.,** 'Ueber die Farbstoffe der Kirsche und der Schlehe,' *Liebigs Ann. Chem.*, Leipzig, 1916, ccccxii, pp. 164–178. pp. **72, 75**

307. **Willstätter, R.,** und **Zollinger, E. H.,** 'Ueber die Farbstoffe der Weintraube und der Heidelbeere, II,' *Liebigs Ann. Chem.*, Leipzig, 1916, ccccxii, pp. 195–216. pp. **73, 74**

308. **1917. Combes, R.,** 'Recherches biochimiques expérimentales sur le rôle physiologique des glucosides chez les végétaux,' *Rev. gén. bot.*, Paris, 1917, xxix, pp. 321–380 and 1918, xxx, pp. 70–124, etc.
Includes anthocyanin among glucosides reviewed. — —

309. **Gertz, O.,** 'Ueber die vorübergehende Rotfärbung einiger Blätter mit Saltpetersäure bei der Xanthoproteinprobe,' *Biochem. Zs.*, Berlin, 1917, lxxxiii, pp. 129–132. — —

310. **Léger, E.,** 'Les anthocyanes. Matières colorantes des fleurs et des fruits,' *J. pharm. chim.*, Paris, 1917, xv, pp. 312–317.
Very brief account. — —

311. **Nagai, I.,** 'The Action of Oxidase on Anthocyanin,' *Bot. Mag.*, Tokyo, 1917, xxxi, pp. 65–74.

Page of text on which reference is made

The colour disappears from crude extracts of anthocyanins in presence of oxidases. — —

312. **Schroeder, H.**, 'Die Anthocyanine nach den neuen chemischen Untersuchungen,' *Zs. Bot.*, Jena, 1917, IX, pp. 546–558.
Short résumé. — —

313. **1918. Beaverie, J.**, 'L'état actuel de la question de l'anthocyanine,' *Rev. gén. sci.*, Paris, 1918, XXIX, pp. 572–579.
Short review. — —

314. **Everest, A. E.**, 'The Production of Anthocyanins and Anthocyanidins. Part III,' *Proc. R. Soc.*, London, 1918, XC B, pp. 251–265. p. 125

315. **Noack, K.**, 'Untersuchungen über den Anthocyanstoffwechsel auf Grund der chemischen Eigenschaften der Anthocyangruppe,' *Zs. Bot.*, Jena, 1918, X, pp. 561–628. pp. 18, 125

316. **Perkin, A. G.**, and **Everest, A. E.**, *The natural Organic Colouring Matters*, London, 1918.
Detailed account of chemistry of anthocyanin pigments. — —

317. **Rosenheim, O.**, 'Biochemical Changes due to Environment,' *Biochem. J.*, Cambridge, 1918, XII, pp. 283–289. p. 124

318. **Schudel, G.**, 'Ueber die Anthocyane von *Beta vulgaris* L. und *Raphanus sativus*,' Dissertation, Zürich, 1918, 64 pages. p. 75

319. **1919. Costantin, J.**, 'Physiologie de l'anthocyane et chimie de la chlorophylle,' *Ann. sci. nat.* (*Bot.*), Paris, 1919, sér. 10, I, pp. xxxviii–lii.
Short review. — —

320. **Nierenstein, M.**, 'The Colouring Matter of the Red Pea Gall,' *Trans. Chem. Soc.*, London, 1919, CXV, pp. 1328–1332.
Consists of purpurogallin and two molecules of dextrose. — —

321. **Shibata, K., Shibata, Y.**, and **Kasiwagi, I.**, 'Anthocyanins: Color Variation in Anthocyanins,' *J. Amer. Chem. Soc.*, Easton, Pa., 1919, XLI, pp. 208–220.
Different colours of anthocyanins due to introduction of various metals into molecule. Compare reference 326. — —

322. **1920. Klein, G.**, 'Studien über das Anthochlor (I. Mittheilung),' *SitzBer. Ak. Wiss.*, Wien, 1920, CXXIX (1), pp. 341–395.
Pigments to be included among flavones and flavonols. See reference 332. — —

323. **Kryz, F.**, 'Anthocyan Pigments of the Beet-red Group,' *Oester. Chem. Ztg.*, 1920, XXIII, pp. 55–56. p. 53

324. **Rosenheim, O.**, 'Note on the Use of Butyl Alcohol as a Solvent for Anthocyanins,' *Biochem. J.*, Cambridge, 1920, XIV, pp. 73–74.
Many anthocyanins (glucosides) are more soluble in this solvent than in amyl alcohol. — —

Page of text on which reference is made

325. **Rosenheim, O.**, 'Observations on Anthocyanins. I. The Anthocyanins of the young Leaves of the Grape Vine,' *Biochem. J.*, Cambridge, 1920, XIV, pp. 178–188. p. 48

326. **1921. Everest, A. E.**, and **Hall, A. J.**, 'Anthocyanins and Anthocyanidins. Part IV. Observations on: (*a*) Anthocyan Colours in Flowers, and (*b*) the Formation of Anthocyans in Plants,' *Proc. R. Soc.*, London, 1921, XCII B, pp. 150–162.
Errors in work of Shibata and others (see reference 321) are pointed out. Blue colour in flowers due to anthocyan phenolates of the alkali or alkaline-earth metals. — —

327. **Jonesco, St.**, 'Contribution à l'étude du rôle physiologique des anthocyanes,' *C. R. Acad. sci.*, Paris, 1921, CLXXII, pp. 1311–1313.
Loss of red colour of Buckwheat seedlings kept in darkness is accompanied by diminution in amount of flavone and anthocyan glucosides. — —

328. **Jonesco, St.**, 'Sur l'existence d'anthocyanidines à l'état libre dans les fruits de *Ruscus aculeatus* et de *Solanum Dulcamara*,' *C. R. Acad. sci.*, Paris, 1921, CLXXIII, pp. 168–171. p. 48

329. **Jonesco, St.**, 'Les anthocyanidines, à l'état libre, dans les fleurs et les feuilles rouges de quelques plantes,' *C. R. Acad. sci.*, Paris, 1921, CLXXIII, pp. 426–429. p. 48

330. **Jonesco, St.**, 'Formation de l'anthocyane dans les fleurs de *Cobaea scandens* aux dépens des glucosides préexistants,' *C. R. Acad. sci.*, Paris, 1921, CLXXIII, pp. 850–852.
Amount of glucosides the same both before and after coloration of flowers. — —

331. **Jonesco, St.**, 'Transformation, par oxydation, en pigment rouge, des chromogènes de quelques plantes,' *C. R. Acad. sci.*, Paris, 1921, CLXXIII, pp. 1006–1009. pp. 18, 126

332. **Klein, G.**, 'Studien über das Anthochlor (II. Mittheilung),' *SitzBer. Ak. Wiss.*, Wien, 1921, CXXX (1), pp. 247–261. — —

333. **Kohler, D.**, 'Étude de la variation des acides organiques au cours de la pigmentation anthocyanique,' *Rev. gén. bot.*, Paris, 1921, XXXIII, pp. 295–315, 337–356.
See next reference. — —

334. **Kohler, D.**, 'Variation des acides organiques au cours de la pigmentation anthocyane,' *C. R. Acad. sci.*, Paris, 1921, CLXXII, pp. 709–711.
Abstract of previous paper. p. 102

335. **Kozlowski, A.**, 'Formation du pigment rouge de *Beta vulgaris* par oxydation des chromogènes,' *C. R. Acad. sci.*, Paris, 1921, CLXXIII, pp. 855–857. pp. 18, 127

Page of text on which reference is made

336. **1922. Combes, R.**, 'La recherche des pseudo-bases d'anthocyanidines dans les tissus végétaux,' *C. R. Acad. sci.*, Paris, 1922, CLXXIV, pp. 58–61. pp. 19, 126, 127
337. **Combes, R.**, 'La formation des pigments anthocyaniques,' *C. R. Acad. sci.*, Paris, 1922, CLXXIV, pp. 240–242. pp. 19, 127
338. **Currey, G.**, 'The Colouring Matter of Red Roses,' *Proc. R. Soc.*, London, 1922, XCIII B, pp. 194–197. p. 75
339. **Currey, G. S.**, 'The Colouring Matter of the Scarlet Pelargonium,' *J. Chem. Soc.*, London, 1922, CXXI, pp. 319–323. p. 75
340. **Jonesco, St.**, 'Sur la répartition des anthocyanidines dans les organes colorés des plantes,' *C. R. Acad. sci.*, Paris, 1922, CLXXIV, pp. 1635–1637. p. 48
341. **Jonesco, St.**, 'Transformation d'un chromogène des fleurs jaunes de *Medicago falcata* sous l'action d'une oxydase,' *C. R. Acad. sci.*, Paris, 1922, CLXXV, pp. 592–595. pp. 18, 127
342. **Jonesco, St.**, 'Les pigments anthocyaniques et les phlobatanins chez les végétaux,' *C. R. Acad. sci.*, Paris, 1922, CLXXV, pp. 904–907. p. 127
343. **Jonesco, St.**, 'Recherches sur le rôle physiologique des anthocyanes,' *Ann. sci. nat.* (*Bot.*), Paris, 1922, sér. 10, IV, pp. 301–403. — —
344. **Mirande, M.**, 'Sur la relation existant entre l'anthocyanine et les oxydases,' *C. R. Acad. sci.*, Paris, 1922, CLXXV, pp. 595–597. p. 128
345. **Mirande, M.**, 'Sur la relation existant entre l'acidité relative des tissus et la présence de l'anthocyanine dans les écailles de bulbes de Lis exposées à la lumière,' *C. R. Acad. sci.*, Paris, 1922, CLXXV, pp. 711–713. p. 102
346. **Noack, K.**, 'Physiologische Untersuchungen an Flavonolen und Anthocyanen,' *Zs. Bot.*, Jena, 1922, XIV, pp. 1–74. — —
347. **Sando, C. E.**, and **Bartlett, H. H.**, 'Pigments of the Mendelian Color Types in Maize: Isoquercitrin from Brown-husked Maize,' *J. Biol. Chem.*, New York, 1922, LIV, pp. 629–645.
References to chemistry of anthocyanin pigments. — —
348. **1923. Chodat, R.**, et **Rouge, E.**, 'Sur l'analogie des anthocyanines et des flavones,' *C. R. des séances de la Société de physique et d'histoire naturelle*, Genève, 1923, XL, pp. 16–18. — —
349. **Combes, R.**, 'À propos de publications récentes sur la formation des pigments anthocyaniques,' *Bul. soc. bot.*, Paris, 1923, XXIII, pp. 222–276. pp. 19, 127
350. **Anderson, R. J.**, 'Concerning the Anthocyans in Norton

Page of text on which reference is made

and Concord Grapes. A Contribution to the Chemistry of Grape Pigments,' *J. Biol. Chem.*, New York, 1923, LVII, pp. 795–813. p. 75

FACTORS AND CONDITIONS INFLUENCING THE FORMATION OF ANTHOCYANINS

351. **1816. *Voigt, F. S.,** *Die Farben der organischen Körper*, Jena, 1816, p. 19.
Reference to the effect of light on the formation of soluble pigments. p. 90

352. **1848. Middendorff, A. Th. von,** *Reise in den äussersten Norden und Osten Sibiriens*, St Petersburg, 1848–1875, Bd IV (1), p. 674.
Appearance of anthocyanin in plants in Arctic regions. — —

353. **1860. Duchartre, P.,** 'Note sur le Lilas blanchi par la culture forcée,' *Bul. soc. bot.*, Paris, 1860, VII, pp. 152–155.
Purple lilacs, cultivated at 35° and in darkness, produce white flowers, but any lowering of temperature, or exposure to light, causes development of colour. — —

354. **Treviranus, L. C.,** 'Ueber den Wechsel des Grünen und Rothen in den Lebenssäften belebter Körper,' *Bot. Ztg.*, Leipzig, 1860, XVIII, pp. 281–288.
Discussion as to factors and causes bringing about reddening of leaves. — —

355. **1863. Corenwinder, B.,** 'Expiration nocturne et diurne des feuilles. Feuilles colorées,' *C. R. Acad. sci.*, Paris, 1863, LVII, pp. 266–268.
Red leaves behave in exactly the same way as green leaves with respect to assimilation of carbon dioxide. — —

356. **Sachs, J.,** 'Ueber den Einfluss des Tageslichts auf Neubildung und Entfaltung verschiedener Pflanzenorgane,' *Beilage zu Bot. Ztg.*, Leipzig, 1863, XXI, 30 pages.
Author, by experiment, detects two classes of flowers, of which one develops normal coloration without the buds being previously exposed to light; the other only develops normally coloured flowers when the buds have been exposed to light up to the time of unfolding. Experiments conducted in total darkness. pp. 7, **88**, 89

357. **Wied, Prinz Maximilian zu,** 'Eine Frage an die Herren Botaniker über die Ursachen der schönen Herbstfärbung der Baumvegetation im nördlichen Amerika,' *Arch. Natg.*, Berlin, 1863, XXIX, pp. 261–266.
Relation between geographical position and intensity of autumnal coloration. — —

Page of text on which reference is made

358. **1865. Sachs, J.**, 'Wirkung des Lichts auf die Blüthenbildung unter Vermittlung der Laubblätter,' *Bot. Ztg.*, Leipzig, 1865, XXIII, pp. 117–121, 125–131, 133–139.
Experiments on the effect on development and coloration of flowers of darkening certain shoots only, the remainder of the plant being in the light. pp. 7, 85

359. **1868. *Hallier, E.**, *Phytopathologie, Die Krankheiten der Culturgewächse*, Leipzig, 1868, p. 101.
Anthocyanin in roots of *Salix*. pp. 89, 90

360. **1870. *Weretennikow, J.**, *Arbeiten der St. Petersburger Gesellschaft der Naturforscher*, 1870, I (1), p. 57.
Reddening of seedlings of *Polygonum* and other plants is entirely absent in the dark. p. 90

361. **1871. Colladon, D.**, 'Effets de la foudre sur les arbres et les plantes ligneuses,' *Mém. Soc. Phys.*, Genève, 1871–1872, XXI, pp. 501–584.
Injury brought about by lightning may act like artificial ringing, and cause leaves above point of injury to turn red. — —

362. **Wiesner, J.**, 'Untersuchungen über die herbstliche Entlaubung der Holzgewächse,' *SitzBer. Ak. Wiss.*, Wien, 1871, LXIV (1), pp. 465–510.
Paper includes a short account of autumnal coloration and various causes and conditions connected with it. Greater acidity noted in autumnal leaves. — —

363. **1872. Kraus, C.**, 'Weitere Mittheilungen über die winterliche Färbung immergrüner Gewächse,' *SitzBer. physik. Soc.*, Erlangen, 1872, Heft 4, pp. 62–65.
Low temperature is the most important factor in anthocyanin production. — —

364. **1873. *Kraus, C.**, 'Studien über die Herbstfärbung der Blätter und über Bildungsweise der Pflanzensäuren,' *Buchner's Neues Repertorium für Pharmacie*. München, 1873, XXII, p. 273.
View that production of anthocyanin is augmented by acidity of cell-sap. — —

365. **Rafarin**, 'Coloration des feuilles à l'automne,' *Rev. hortic.*, Paris, 1873, pp. 50–51.
Red and yellow autumn pigments exist in a latent state in the green leaves. — —

366. **1874. Chargueraud, A.**, 'Observations sur la coloration des feuilles à l'automne,' *Rev. hortic.*, Paris, 1874, pp. 34–36.
Remarks on causes of autumnal coloration. — —

367. **Chargueraud, A.**, 'Influence du froid sur la coloration des feuilles du *Phalaris picta*,' *Rev. hortic.*, Paris, 1874, pp. 248–249.

Page of text on which reference is made

Formation of anthocyanin in white portions of variegated leaves of *P. arundinacea picta* when exposed to frost. — —

368. **1875. Kraus, C.**, 'Pflanzenphysiologische Untersuchungen. 7. Ueber die Einwirkung von Pflanzensäuren auf Chlorophyll innerhalb der Pflanzen,' *Flora*, Marburg, 1875, xxxiii, pp. 365–368.
Acidity of cell-sap augments the formation of anthocyanin. — —

369. **1876. Askenasy, E.**, 'Ueber den Einfluss des Lichtes auf die Farbe der Blüthen,' *Bot. Ztg.*, Leipzig, 1876, xxxiv, pp. 1–7, 27–31.
Investigations on various plants as to whether they require light for the production of normal colour in flowers. pp. 7, 85, **89**

370. **Haberlandt, G.**, 'Untersuchungen über die Winterfärbung ausdauernder Blätter,' *SitzBer. Ak. Wiss.*, Wien, 1876, lxxiii (1), pp. 267–296.
Account of winter reddening of leaves. Reddening due to low temperature at night and bright light by day. — —

371. **Mer, E.**, 'Des phénomènes végétatifs qui précèdent ou accompagnent le dépérissement et la chute des feuilles,' *Bul. soc. bot.*, Paris, 1876, xxiii, pp. 176–191.
Observations on the reddening of leaves of *Cissus quinquefolia*. pp. 34, 92

372. **Schell, J.**, 'Wirkung einiger Einflüsse auf die Färbung der Pflanzen,' *Beilage zu dem Protocolle der* 75. *Sitzung der Naturforscher-Gesellschaft an der Universität zu Kazan*, Kazan, 1876. Reference in *Justs bot. Jahresber.*, Berlin, 1876, iv, p. 717.
Effect of light and temperature on reddening of seedlings of *Polygonum*, *Rumex*, etc. p. 90

373. **1877. Mer, E.**, 'Recherches sur les causes des colorations diverses qui apparaissent dans les feuilles en automne et en hiver,' *Bul. soc. bot.*, Paris, 1877, xxiv, pp. 105–114.
Anthocyanin no connection with chlorophyll. Formation of pigment as a result of injury caused by animal and vegetable parasites. Oxygen necessary for the appearance of pigment. — —

374. **Schell, J.**, 'Ueber die Pigmentbildung in den Wurzeln einiger *Salix*-Arten,' *Beilage zu dem Protocolle der* 95. *Sitzung der Naturforscher-Gesellschaft an der Universität zu Kazan*, Kazan. Reference in *Justs bot. Jahresber.*, Berlin, 1877, v, p. 562.
Light causes formation of colour in adventitious roots of *Salix* sp. grown in water in glasses. pp. 30, 90

Page of text on which reference is made

375. **1878. Bonnier, G.**, et **Flahault, Ch.**, 'Sur les variations qui se produisent avec la latitude dans une même espèce végétale,' *Bul. soc. bot.*, Paris, 1878, XXV, pp. 300–306.
Colours of flowers more intense in Norway than those of the same species in France. Discussion of the subject. p. 8

376. **Bonnier, G.**, et **Flahault, Ch.**, 'Observations sur les modifications des végétaux,' *Ann. sci. nat.* (*Bot.*), Paris, 1878, sér. 6, VII, pp. 93–125.
Intensity of flower-colour increases in Northern latitudes. p. 8

377. **Flahault, Ch.**, 'Nouvelles observations sur les modifications des végétaux,' *Ann. sci. nat.* (*Bot.*), Paris, 1878, sér. 6, IX, pp. 159–207.
Effect of latitude on formation of pigment. Of a number of species grown both in Upsala and Paris, individuals in the former district produced the more vividly coloured flowers. p. 8

378. **Pellat, A.**, 'Sur quelques variations que présentent les végétaux avec l'altitude,' *Bul. soc. bot.*, Paris, 1878, XXV, pp. 307–308.
Greater intensity of colour at high altitudes. — —

379. **1879. Flahault, Ch.**, 'Sur la formation des matières colorantes dans les végétaux,' *Bul. soc. bot.*, Paris, 1879, XXVI, pp. 268–273.
Connection between pigment formation and both light and reserves of nutritive substances. p. 8

380. **1880. Batalin, A.**, 'Die Einwirkung des Lichtes auf die Bildung des rothen Pigmentes,' *Acta horti Petr.*, St Petersburg, 1880, VI (Fas. 2), pp. 279–286.
Effect of light on pigment production in seedlings. p. 90

381. **Bonnier, G.**, 'De la variation avec l'altitude des matières colorées des fleurs chez une même espèce végétale,' *Bul. soc. bot.*, Paris, 1880, XXVII, pp. 103–105.
Colours of flowers of various species increase in intensity with altitude. pp. 8, 27, **85**

382. ***Pynaert, E.**, 'De l'influence de la lumière sur la coloration des feuilles,' *Congrès de Botanique et d'Horticulture de* 1880 *tenu à Bruxelles*, Partie 2, p. 53. — —

383. **1881. Meehan, Th.**, 'Color in Autumn Leaves,' *Proc. Acad. Nat. Sci.*, Philadelphia, Pa., 1881, pp. 454–456.
Description of plants producing anthocyanin in salt marshes of New Jersey. Discussion as to the greater production of anthocyanin in American as compared with European species of the same genus. No definite conclusion as to causes. — —

384. **1882. Costerus, J. C.**, 'Seasonal Order in Colours of Flowers,' *Nature*, London, 1882, XXV, pp. 481–482.

Page of text on which reference is made

Observations on the development of colour in flowers and fruits in the dark. Conclusion that formation of pigment largely depends upon the supply of chromogen at the time of darkening. For production of chromogen, light is necessary. — —

385. ***Müller-Thurgau, H.**, 'Ueber den Einfluss der Belaubung auf das Reifen der Trauben,' *Weinbaucongress zu Dürkheim a. d. H.*, 1882.

Coloration of fruits of *Vitis* takes place in the dark. p. 89

386. **Schunck, E.**, 'Remarks on the Terms used to denote Colour, and on the Colours of Faded Leaves,' *Chem. News*, London, 1882, XLV, pp. 17–20.

Short, somewhat popular, account of colour in autumnal leaves. — —

387. **1883. Heckel, E.**, 'Sur l'intensité du coloris et les dimensions considérables des fleurs aux hautes altitudes,' *Bul. soc. bot.*, Paris, 1883, XXX, pp. 144–154.

High mountain plants owe intense colour to strong insolation rather than to insect visitation. p. 85

388. **Kraus, C.**, 'Beiträge zur Kenntniss des Verhaltens der leicht oxydablen Substanzen des Pflanzensaftes,' *Ber. D. bot. Ges.*, Berlin, 1883, I, pp. 211–216.

Development of pigment in *Dahlia* tubers as a result of oxidation. — —

389. **1884. Sorby, H.**, 'On the Autumnal Tints of Foliage,' *Nature*, London, 1884, XXXI, pp. 105–106.

Slight note on autumnal coloration. — —

390. **1885. Beyerinck, M. W.**, 'Die Galle von *Cecidomyia Poae* an *Poa nemoralis*. Entstehung normaler Wurzeln in Folge der Wirkung eines Gallenthieres,' *Bot. Ztg.*, Leipzig, 1885, XLIII, pp. 305–315, 321–331.

Coloured roots formed in the region of an injury caused by gall insect. p. 30

391. **1886. Sorauer, P.**, *Handbuch der Pflanzenkrankheiten*, Berlin, 1886, Bd I, p. 324.

Connection of anthocyanin formation with infection by parasites. pp. 34, 84

392. **1887. Dufour, L.**, 'Influence de la lumière sur la forme et la structure des feuilles,' *Ann. sci. nat.* (*Bot.*), Paris, 1887, sér. 7, V, pp. 311–413 (pp. 331, 352, 353, 405).

Observations made on various plants indicate that anthocyanin is only developed in parts of organs exposed to the sun. Anthocyanin noted in roots of Maize exposed to the light. pp. 30, 90

393. **Martin, W. K.**, and **Thomas, S. B.**, 'The Autumnal

Page of text on which reference is made

Changes in Maple leaves,' *Bot. Gaz.*, Crawfordsville, Indiana, 1887, XII, pp. 78–81.

Slight note on autumnal coloration and its effect on cell-contents, etc. — —

394. **1888. Bonnier, G.,** 'Étude expérimentale de l'influence du climat alpin sur la végétation et les fonctions des plantes,' *Bul. soc. bot.*, Paris, 1888, XXXV, pp. 436–439.

Of individuals of the same species grown both in the High Alps and in a lowland station, the former had more highly coloured flowers. pp. 8, 27, **85**

395. **Devaux,** 'De l'action de la lumière sur les racines croissant dans l'eau,' *Bul. soc. bot.*, Paris, 1888, XXXV, pp. 305–308.

Reddening of roots of Maize growing in water and exposed to light. pp. 30, 90

396. **1889. Emery,** 'Sur les variations de l'eau dans les périanthes,' *Bul. soc. bot.*, Paris, 1889, XXXVI, pp. 322–333.

Anthocyanin is not found in flowers kept under water although exposed to sunlight. p. 92

397. **Goiran, A.,** 'Di una singolare esperienza praticata sopra le corolle di *Cyclamen persicum*,' *Nuovo Giorn. bot. ital.*, Firenze, 1889, XXI, p. 415.

Petals, with purplish-red base, of a white variety of *Cyclamen persicum giganteum* were partially cut across. After a time, the cut edge developed anthocyanin of the same colour as that at the base of the petals. p. 101

398. **Kraus, G.,** *Grundlinien zu einer Physiologie des Gerbstoffs*, Leipzig, 1889.

Artificial ringing of the stem and consequent production of anthocyanin in the leaves above. p. 83

399. **1890. *Benecke, Fr.,** 'Over de bordeaux-roode kleur der suikerrietwortels,' *Mededeelingen van het Proefstation 'Midden-Java' te Semarang*, 1890, 77 pages.

Question as to whether 'Bordeaux-red' colour (anthocyanin) of sugar cane is normal or pathological. Author finds development of pigment is dependent on light, and concludes, from experiments, that colour is not caused by disease, but is a normal development and protects root-tip from too intense light. p. 30

400. **Curtel, G.,** 'Recherches physiologiques sur les enveloppes florales,' *C. R. Acad. sci.*, Paris, 1890, CXI, pp. 539–541.

Greater respiratory activity in coloured than in white flowers. This energetic oxidation leads, among other things, to formation of anthocyanin pigments from tannins. — —

401. ***Hieronymus, G.,** 'Beiträge zur Kenntniss der euro-

Page of text on which reference is made

päischen Zoocecidien und der Verbreitung derselben,' *Jahresber. Ges. vaterl. Cultur*, Breslau, 1890, p. 49.

Work on galls and connection of anthocyanin development with these structures. p. 84

402. ***Laurent, E.**, 'Influence de la radiation sur la coloration des raisins,' *Comptes rendus des séances de la Société Royale de Botanique de Belgique*, Bruxelles, 1890, XXIX (2), p. 71.

Colour of grapes develops in the dark. p. 89

403. ***Molisch, H.**, 'Blattgrün und Blumenblau,' *Vorträge des Vereines zur Verbreitung naturwissenschaftlicher Kenntnisse in Wien*, Wien, 1890, XXX (3).

General account of anthocyanin is included, in which author emphasises fact that drought favours development of the pigment. p. 93

404. **1891. Cockerell, T. D. A.**, 'The Alpine Flora: with a Suggestion as to the Origin of Blue in Flowers,' *Nature*, London, 1891, XLIII, p. 207.

Preponderance of blue in high mountain flowers. Suggestion that blue is metabolically the most complex of the colours, and that great concentration of metabolism, correlated with dwarf structure, may lead to the formation of the most complex pigments in the inflorescence. — —

405. ***Heim, F.**, 'Influence de la lumière sur la coloration du périanthe de l'*Himantophyllum variegatum*,' *Bulletin mensuel de la Société Linnéenne de Paris*, 1891, II, p. 932.

Intensity of coloration of flower is directly proportional to the amount of light received. This is contrary to Sachs's view that light does not influence colour so long as leaves can assimilate. — —

406. **Massalongo, C.**, 'Sull' alterazione di colore dei fiori dell' *Amarantus retroflexus* infetti dalle oospore di *Cystopus Bliti*,' *Nuovo Giorn. bot. ital.*, Firenze, 1891, XXIII, pp. 165–167.

Anthocyanin is developed in the inflorescence when infected with *Cystopus*. Suggestion that inflorescence thereby becomes attractive, and possibly spores may be disseminated by insects in consequence. — —

407. **Noll, F.**, 'Einfluss des Lichtes auf die herbstliche Verfärbung des Laubes,' *Verh. nathist. Ver.*, Bonn, 1891, XLVIII, p. 80.

Autumnal coloration dependent on exposure to light. Natural photographs, in green on a red ground, obtained by the shading of one leaf by another. — —

408. **Ráthay, E.**, 'Ueber eine merkwürdige, durch den Blitz an *Vitis vinifera* hervorgerufene Erscheinung,' *Denkschr. Ak. Wiss.*, Wien, LVIII, 1891, pp. 585–610.

Page of text on which reference is made

Action of lightning in bringing about formation of anthocyanin is due to the same effect as 'ringing' of stem. — —

409. **1893. Gain, E.**, 'Sur la matière colorante des tubercules et des organes souterrains,' *Bul. soc. bot.*, Paris, 1893, XL, pp. 95–102.

Author measured the amount of pigment in tubers of *Solanum tuberosum* (red-tubered variety) and *Helianthus tuberosus* grown both in wet and dry ground. More pigment developed in dry ground when conditions otherwise the same. — —

410. **Landel, G.**, 'Influence des radiations solaires sur les végétaux,' *C. R. Acad. sci.*, Paris, 1893, CXVII, pp. 314–316.

Effect of sun and shade on development of anthocyanin. — —

411. ***Nienhaus, C.**, 'Die Bildung der violetten Pflanzenfarbstoffe,' *Schweiz. Wochenschr. Chem.*, Zürich, 1893. No. 39.

Author investigates the formation of anthocyanin in fruits of *Solanum nigrum*, and notes that coloration begins at places where air penetrates, i.e. wounds, stomata. Concludes, however, that oxidation is not essential to formation of all blue and violet pigments. — —

412. **Vöchting, H.**, 'Ueber den Einfluss des Lichtes auf die Gestaltung und Anlage der Blüthen,' *Jahrb. wiss. Bot.*, Berlin, 1893, XXV, pp. 149–208.

When leaves of certain plants were deprived of light, and were thus unable to assimilate, it was found that flowers did not develop anthocyanin. p. 85

413. **1894. Küstenmacher, M.**, 'Beiträge zur Kenntniss der Gallenbildungen mit Berücksichtigung des Gerbstoffes,' *Jahrb. wiss. Bot.*, Berlin, 1894, XXVI, pp. 82–185.

Work on galls, and hence connection of anthocyanin with the same. p. 84

414. ***Warming, E.**, 'Exkursionen til Fanø og Blaavand i Juli 1893,' *Bot. Tids.*, Kjøbenhavn, 1894–1895, XIX, p. 52.

Formation of anthocyanin may vary in the same species according to whether individuals are growing in water or not. p. 93

415. **1895. Bonnier, G.**, 'Recherches expérimentales sur l'adaptation des plantes au climat alpin,' *Ann. sci. nat. (Bot.)*, Paris, 1895, sér. 7, XX, pp. 217–358.

Comparative cultures of plants in Alps and in lowlands. Among other differences, flowers of the former are more intensely coloured. pp. 8, 85, **103**

416. **Tubeuf, K. Freiherr von**, *Diseases of Plants induced by*

Page of text on which reference is made

Cryptogamic Parasites, English edition by W. G. Smith, London, 1897. (*Pflanzenkrankheiten durch kryptogame Parasiten verursacht*, Berlin, 1895.)
Anthocyanin in connection with pathological conditions caused by Fungi. p. 84

417. **1897. Curtel, G.**, 'Recherches physiologiques sur la fleur,' *Ann. sci. nat.* (*Bot.*), Paris, 1897, sér. 8, VI, pp. 221–306.
Connection between flower pigmentation and various physiological processes, i.e. respiration, transpiration, etc. — —

418. **1899. Griffon, E.**, 'L'assimilation chlorophyllienne et la coloration des plantes,' *Ann. sci. nat.* (*Bot.*), Paris, 1899, sér. 8, X, pp. 1–123.
Comparison of assimilation in red (anthocyanin) leaves with that in green. p. 37

419. **Mirande, M.**, *Recherches physiologiques et anatomiques sur les Cuscutacées*, Paris, 1899.
Variation in the amount of anthocyanin developed in these parasites when growing on different hosts. pp. 6, 31, **86**, 90

420. **Overton, E.**, 'Beobachtungen und Versuche über das Auftreten von rothem Zellsaft bei Pflanzen,' *Jahrb. wiss. Bot.*, Leipzig, 1899, XXXIII, pp. 171–231.
Important series of experiments connected with effect of light and temperature on anthocyanin formation. Also effect of many and various nutrient solutions, i.e. sugars, alcohol, glycerine, etc., on its development. pp. 6, 7, 26, 33, 47, 52, 59, 86, **93**, 107

421. **Overton, E.**, 'Experiments on the Autumn Colouring of Plants,' *Nature*, London, 1899, LIX, p. 296.
Same as previous paper. p. 6

422. **Rathbone, M.**, 'Colouring of Plants,' *Nature*, London, 1899, LIX, p. 342.
Red and green leaves of different plants of *Sempervivum* were tested, and red leaves found to contain more starch than green. Contrary to Overton's results (see No. 420). — —

423. **Schenkling-Prévôt**, 'Die herbstliche Färbung des Laubes,' *Die Natur*, Halle a. S., 1899, XLVIII, pp. 584–585.
Slight note on autumnal coloration and its causes. — —

424. **1900. Daniel, L.**, 'L'Incision annulaire du Chou,' *Bul. soc. sci. méd.*, Rennes, 1900, IX, pp. 135–140.
In *Brassica*, anthocyanin is formed in leaves after decortication of the stem. p. 83

425. ***Timpe, H.**, *Beiträge zur Kenntnis der Panachierung*, Inaugural-Dissertation, Göttingen, 1900.
Author shows that red-brown (anthocyanin) zone in leaf

Page of text on which reference is made

of *Pelargonium zonale* is increased by culture of the leaf in sugar solution. — —

426. **1901. Beulaygue, L.**, 'Influence de l'obscurité sur le développement des fleurs,' *C. R. Acad. sci.*, Paris, 1901, CXXXII, pp. 720–722.
Experiments on development of anthocyanin in flowers in the dark. In some cases slightly less colour, in others considerably less: in yet other cases, no colour produced at all. p. 89

427. **Genau, K.**, 'Physiologisches über die Entwicklung von *Sauromatum guttatum*, Schott,' *Oest. Bot. Zeitschr.*, Wien, 1901, LI, pp. 321–325.
Anthocyanin develops in spathe and leaves of *Sauromatum* even when kept in the dark. — —

428. **Linsbauer, L.**, 'Einige Bemerkungen über Anthokyanbildung,' *Oest. Bot. Zeitschr.*, Wien, 1901, LI, pp. 1–10.
Discussion of influence of outside factors on anthocyanin formation. p. 83

429. ***Lüdi, R.**, 'Beiträge zur Kenntniss der Chytridiaceen,' *Hedwigia*, Dresden, 1901, XL, pp. 1–44.
Appearance of anthocyanin as a result of infection. p. 84

430. ***Warming, E.**, und **Johannsen, W.**, *Den almindelige Botanik*, Fjerde Udgave, Kjøbenhavn, 1901, p. 661.
No colour formed in blue Lilacs when cultivated above optimum temperature for anthocyanin formation. — —

431. **1902. *Daniel, L.**, *La théorie des capacités fonctionnelles et ses conséquences en agriculture*, Rennes, 1902.
Views as to the effect of water-supply on anthocyanin formation. — —

432. ***Rostrup, E.**, *Plantepatologi*, Kjøbenhavn, 1902.
Formation of anthocyanin under pathological conditions. p. 84

433. ***Warming, E.**, 'Exkursionen til Fanø og Blaavand i Juli 1899,' *Bot. Tids.*, Kjøbenhavn, 1902, XXV, p. 53.
Reddening varies in the same species according to whether it is growing in or out of water. p. 93

434. **1903. Eberhardt, Ph.**, 'Influence de l'air sec et de l'air humide sur la forme et sur la structure des végétaux,' *Ann. sci. nat.* (*Bot.*), Paris, 1903, sér. 8, XVIII, pp. 61–153.
Dry atmosphere increases formation of anthocyanin in *Coleus Blumei* and *Achyranthes angustifolia*. p. 93

435. **Keegan, P. Q.**, 'Leaf Decay and Autumn Tints,' *Nature*, London, 1903, LXIX, p. 30.
Note on autumnal coloration. — —

436. **Klebs, G.**, *Willkürliche Entwickelungsänderungen bei Pflanzen*, Jena, 1903.

Page of text on which reference is made

Effect of water vapour in atmosphere on development of colour in flowers. — —

437. ***Küster, E.,** *Pathologische Pflanzenanatomie*, Jena, 1903.
References to anthocyanin in connection with these conditions. pp. 83, 84

438. **Sorauer, P.,** 'Ueber Frostbeschädigungen am Getreide und damit in Verbindung stehende Pilzkrankheiten,' *Landw. Jahrb.*, Berlin, 1903, XXXII, pp. 1–66.
In *Secale* seedlings, anthocyanin formed in diffuse light. — —

439. **1904. Burgerstein, A.,** *Die Transpiration der Pflanzen, Eine physiologische Monographie*, Jena, 1904.
Connection between transpiration and anthocyanin formation. — —

440. **1905. *Guttenberg, H. von,** *Beiträge zur physiologischen Anatomie der Pilzgallen*, Leipzig, 1905, p. 6.
Production of anthocyanin in galls. p. 84

441. **Katić, D. L.,** *Beitrag zur Kenntnis der Bildung des roten Farbstoffs (Anthocyan) in vegetativen Organen der Phanerogamen*, Inaugural-Dissertation, Halle a. S., 1905, 83 pages.
Important account of anthocyanin and the conditions governing its formation. Observations on results obtained by growing certain plants in sugar and other nutritive solutions. pp. 6, 59, 92, **96**, 116

442. **Ravaz, L.,** et **Roos, L.,** 'Sur le rougeot de la vigne,' *C. R. Acad. sci.*, Paris, 1905, CXLI, pp. 366–367.
Discussion as to cause of reddening. Analyses of red leaves. — —

443. **Sorauer, P.,** *Handbuch der Pflanzenkrankheiten*, Berlin, 1905.
References to production of anthocyanin in galls. — —

444. ***Zacharewicz,** 'La maladie rouge de la Vigne et son traitement,' *Revue de Viticulture*, 1905, XXIV, pp. 447–448.
Leaves turn red when attacked by *Tetranychus telarius*. — —

445. **1906. Gautier, A.,** 'Sur la coloration rouge éventuelle de certaines feuilles et sur la couleur des feuilles d'automne,' *C. R. Acad. sci.*, Paris, 1906, CXLIII, pp. 490–491.
Note to the effect that the author had previously stated anthocyanin to be formed on injury to tissues, etc. — —

446. **Karzel, R.,** 'Beiträge zur Kenntnis des Anthokyans in Blüten,' *Oest. Bot. Zeitschr.*, Wien, 1906, LVI, pp. 348–354, 377–379.
Histological account of distribution of anthocyanin in various flowers, and effect of absence of light on anthocyanin formation. — —

447. **Klebs, G.,** 'Ueber Variationen der Blüten,' *Jahrb. wiss. Bot.*, Leipzig, 1906, XLII, pp. 155–320.

Page of text on which reference is made

Effect of temperature, nutrition, differently coloured light and other conditions on colour and development of flowers. pp. 8, 85, **87**

448. **Klebs, G.**, 'Ueber künstliche Metamorphosen,' *Abh. natf. Ges.*, Halle, 1906, xxv, 162 pages.
Some effects of artificial conditions on flower-colour are included. — —

449. **Mirande, M.**, 'Sur un cas de formation d'anthocyanine sous l'influence d'une morsure d'Insecte,' *C. R. Acad. sci.*, Paris, 1906, CXLIII, pp. 413–416.
Ravages of a certain caterpillar in leaves of *Galeopsis Tetrahit* are followed by development of anthocyanin. Suggestions as to cause of appearance of pigment. p. 84

450. ***Suzuki, S.**, 'On the Formation of Anthokyan in the Stalks of Barley,' *Bull. Coll. Agric.*, Tokyo, 1906, VII, pp. 29–37.
Results of water and soil cultures show that insufficient supply of phosphorus and nitrogen leads to formation of anthocyanin. — —

451. **1907**. ***Granier, L.**, et **Brun, G.**, *Rev. hortic.*, Paris, 1907, 1908.
Formation of anthocyanin in vegetative parts correlated with lack of flowering. — —

452. **Mirande, M.**, 'Sur l'origine de l'anthocyanine déduite de l'observation de quelques Insectes parasites des feuilles,' *C. R. Acad. sci.*, Paris, 1907, CXLV, pp. 1300–1302.
Production of anthocyanin under such circumstances is due to formation of excess of tannins and glucose, accompanied by presence of oxidases. pp. 84, 91, 102, 107

453. **Molliard, M.**, 'Action morphogénique de quelques subtances organiques sur les végétaux supérieurs,' *Rev. gén. bot.*, Paris, 1907, XIX, pp. 241–291, 329–349, 357–391.
Formation of anthocyanin in plants cultivated in sugar solutions. — —

454. ***Richter, O.**, 'Ueber Antokyanbildung in ihrer Abhängigkeit von äusseren Faktoren,' *Med. Klinik*, Berlin, 1907, XXXIV, 15 pages. — —

455. **1908**. ***Cordemoy, J. de**, 'À propos de la coloration rouge des feuilles,' *Rev. hortic.*, Paris, 1908, p. 31.
Anthocyanin formed in vegetative parts when plants do not flower. Considered to be due to accumulation of sugars which would otherwise have passed into flowers and fruit. — —

456. **Fischer, H.**, 'Belichtung und Blütenfarbe,' *Flora*, Jena, 1908, XCVIII, pp. 380–385.
Author found that, when developing inflorescences were

Page of text on which reference is made

darkened, both blue and red flowers mostly developed little colour. Since plants were not prevented to any great extent from assimilating, it would appear that assimilation is not the only factor upon which the development of colour depends. — —

457. ***Jumelle, H.**, 'À propos de la coloration automnale des feuilles,' *Rev. hortic.*, Paris, 1908, p. 20.

Same observations and conclusions as previous author (Cordemoy, No. 455). — —

458. ***Linsbauer, L.**, 'Ueber photochemische Induktion bei der Anthokyanbildung,' *Wiesner-Festschrift*, Wien, 1908, pp. 421–436.

Anthocyanin formation regarded as a reaction following on a stimulus. Experiments are conducted on the connection between stimulus and response, using *Fagopyrum* seedlings and artificial light. p. 91

459. **1909. Abbott, G.**, 'The Colours of Leaves (*Fagus sylvatica purpurea*),' *Nature*, London, 1909, LXXX, p. 429.

Young tree of Copper Beech, when partially covered with sacking, developed no anthocyanin in covered leaves. After exposure to light for two days, red colour formed. — —

460. **Colin, H.**, 'Sur le rougissement des rameaux de *Salicornia fruticosa*,' *C. R. Acad. sci.*, Paris, 1909, CXLVIII, pp. 1531–1533.

Red branches contain more soluble carbohydrates than green. Accumulation in cells of inorganic chlorides does not prevent formation of anthocyanin. — —

461. **Combes, R.**, 'Production d'anthocyane sous l'influence de la décortication annulaire,' *Bul. soc. bot.*, Paris, 1909, LVI, pp. 227–231.

Production of anthocyanin noted on decortication of branches of *Spiraea prunifolia* and *S. paniculata*. pp. 6, 27, **83**

462. **Miyoshi, M.**, 'Ueber die Herbst- und Trockenröte der Laubblätter,' *J. Coll. Sci.*, Tokyo, 1909, XXVII, 5 pages.

It is noted that, during the dry period in East Indies and Ceylon, a reddening of the leaves of certain trees is produced. Also classification of conditions under which anthocyanin appears. pp. 27, **93**

463. **Molliard, M.**, 'Production expérimentale de tubercules blancs et de tubercules noirs à partir de graines de Radis rose,' *C. R. Acad. sci.*, Paris, 1909, CXLVIII, pp. 573–575.

Oxygen shown to be necessary for formation of pigment by totally submerging radishes in sugar solution, with the result that no pigment was formed. p. 116

464. **1910. Combes, R.**, 'Du rôle de l'oxygène dans la formation et

Page of text on which reference is made

la destruction des pigments rouges anthocyaniques chez les végétaux,' *C. R. Acad. sci.*, Paris, 1910, CL, pp. 1186–1189.

As anthocyanin develops, there is greater absorption of oxygen; the reverse is the case as plants lose the pigment. — —

465. **Combes, R.**, 'Sur le dégagement simultané d'oxygène et d'anhydride carbonique au cours de la disparition des pigments anthocyaniques chez les végétaux,' *C. R. Acad. sci.*, Paris, 1910, CL, pp. 1532–1534.

Experiments made on gaseous exchange in red and green leaves of *Ailanthus*. It is shown that respiration in red leaves is much more active than in green. — —

466. **Combes, R.**, 'Les échanges gazeux des feuilles pendant la formation et la destruction des pigments anthocyaniques,' *Rev. gén. bot.*, Paris, 1910, XXII, pp. 177–212.

An account is given of numerous experiments on autumnal leaves, young red leaves, leaves reddened by decortication of branches, by attacks of insects and by exposure to strong light. Results go to show that the appearance of anthocyanin in the tissues is correlated with accumulation of oxygen in the tissues. The disappearance of pigment from the tissues is accompanied by a loss of oxygen. pp. 91, **92**, 102, 116

467. **Ravaz, L.**, 'Recherches sur l'influence spécifique réciproque du sujet et du greffon chez la Vigne,' *C. R. Acad. sci.*, Paris, 1910, CL, p. 712.

From relationship between colour in leaves and fruits in certain vines, author concludes that pigment is synthesised in fruits, and not in leaves. p. 85

468. **1911. Chartier, H.**, et **Colin, H.**, 'Sur l'anthocyane des plantules de Crassulacées,' *Rev. gén. bot.*, Paris, 1911, XXIII, pp. 264–266.

Properties of anthocyanin developed in roots of species of Crassulaceae, and factors influencing its formation. — —

469. ***Czartkowski, A.**, 'Einfluss des Phloroglucins auf die Entstehung des Anthokyans bei *Tradescantia viridis*,' *Sitzber. Warschau. Ges. Wiss.*, 1911, IV, pp. 23–30.

Author finds that, in addition to sugar, phloroglucin solution favours the production of anthocyanin. — —

470. **Friedel, J.**, 'De l'action exercée sur la végétation par une obscurité plus complète que l'obscurité courante des laboratoires,' *C. R. Acad. sci.*, Paris, 1911, CLIII, pp. 825–826.

It is shown by experiment that anthocyanin can be produced in complete darkness. — —

471. **Gertz, O.**, 'Om anthocyan hos alpina växter. Ett bidrag

Page of text on which reference is made

till Schneebergflorans ökologi,' *Bot. Not.*, Lund, 1911, pp. 101–132, 149–164, 209–229. — —

472. **1912. Combes, R.**, 'Formation de pigments anthocyaniques déterminée dans les feuilles par la décortication annulaire des tiges,' *Ann. sci. nat.* (*Bot.*), Paris, 1912, sér. 9, XVI, pp. 1–53.
Decortication experiments on a number of species, most of which showed reddening of leaves above decortication point, though some did not. Estimation of dry weight, water, ash and hydrocarbon content of red and green leaves respectively; also gaseous exchange. pp. 6, 27, **83**

473. **Gertz, O.**, 'Några iakttagelser öfver anthocyanbildning i blad vid sockerkultur,' *Ark. Bot.*, Stockholm, 1912, XI, No. 6, 45 pages.
Important contribution to the subject of anthocyanin formation and sugar-feeding. pp. 6, **98**

474. **1913. *Czartkowski, A.**, 'Einfluss der Konzentration der Minerallösung auf die Anthocyanbildung aus dem Zucker bei *Tradescantia viridis*,' *Sitzber. Warschau. Ges. Wiss.*, 1913, pp. 959–979. — —

475. **1914. Gertz, O.**, 'Om anthocyan hos alpina växter. II,' *Bot. Not.*, Lund, 1914, pp. 1–16, 49–64, 97–126. — —

476. **Rosé, E.**, 'Étude des échanges gazeux et de la variation des sucres et des glucosides au cours de la formation des pigments anthocyaniques dans les fleurs de *Cobaea scandens*,' *C. R. Acad. sci.*, Paris, 1914, CLVIII, pp. 955–958. Also *Rev. gén. bot.*, Paris, 1914, XXVI, pp. 257–270. p. 92

477. **1915. Portheim, L. von**, 'Ueber den Einfluss von Temperatur und Licht auf die Färbung des Anthokyans,' *Denkschr. Ak. Wiss.*, Wien, 1915, XCI, pp. 499–533. — —

478. **1918. Nicolas, G.**, 'Anthocyane et échanges gazeux respiratoires des feuilles,' *C. R. Acad. sci.*, Paris, 1918, CLXVII, pp. 130–133. — —

479. **1919. Nicolas, G.**, 'Contribution à l'étude des relations qui existent, dans les feuilles, entre la respiration et la présence de l'anthocyane,' *Rev. gén. bot.*, Paris, 1919, XXXI, pp. 161–178. pp. 92, 102

480. **Molliard, M.**, 'Obtention artificielle de pétales panachés chez l'Œillette blanche,' *C. R. soc. biol.*, Paris, 1919, LXXXII, pp. 403–405. — —

481. **1920. Blaringhem, L.**, 'Couleur et sexe des fleurs,' *C. R. soc. biol.*, Paris, 1920, LXXXIII, pp. 892–893.
In *Dianthus barbatus*, as reproductive organs mature, flowers change colour. — —

482. **1922. Bouget, J.**, 'Observations sur l'optimum d'altitude pour

Page of text on which reference is made

la coloration des fleurs,' *C. R. Acad. sci.*, Paris, 1922, CLXXIV, pp. 1723–1724. — —

483. **Bouget, J.**, 'Sur les variations de coloration des fleurs réalisées expérimentalement à haute altitude,' *C. R. Acad. sci.*, Paris, 1922, CLXXV, pp. 900–901. — —

484. **Bouget, J.**, et **de Virville, A. D.**, 'Influence de la météorologie de l'année 1921 sur le rougissement et la chute des feuilles,' *C. R. Acad. sci.*, Paris, 1922, CLXXIV, pp. 768–770. — —

485. **Mirande, M.**, 'Sur la formation d'anthocyanine sous l'influence de la lumière dans les écailles des bulbes de certains Lis,' *C. R. Acad. sci.*, Paris, 1922, CLXXV, pp. 429–430. p. 91

486. **Mirande, M.**, 'Influence de la lumière sur la formation de l'anthocyanine dans les écailles des bulbes de Lis,' *C. R. Acad. sci.*, Paris, 1922, CLXXV, pp. 496–498. p. 91

PHYSIOLOGICAL SIGNIFICANCE OF ANTHOCYANINS

487. **1879. Comes, O.**, 'Ricerche sperimentali intorno all' azione della Luce sulla Traspirazione delle Piante,' *Rend. Acc. sc.*, Napoli, 1879, XVIII, Fasc. 12, pp. 267–282.

The author gives an account of experiments on the effect of light on transpiration. The results are also published later in other papers (see Nos. 488, 489 below). Of special interest in the present paper are his experiments on the transpiration of differently coloured petals, of which he also examined the spectra. He found that transpiration was greatest in those of which the pigment had absorption bands in the greatest number, width and intensity. He found, in addition, that yellow petals of *Hunnemannia fumariaefolia* and *Eschscholtzia* sp. transpire more in blue light than in yellow, other conditions being equal. The contrary was the case with blue petals of species of *Plumbago, Commelina* and *Tradescantia*. p. 135

488. **1880. Comes, O.**, 'La luce e la traspirazione nelle piante,' *Mem. Acc. Lincei*, Roma, 1880, VII, pp. 55–88.

An investigation of the effect of light, of various intensities and of different colours, on transpiration. p. 135

489. **Comes, O.**, 'Influence de la lumière sur la transpiration des plantes,' *C. R. Acad. sci.*, Paris, 1880, XCI, p. 335.

A short summary of work published in previous papers (Nos. 487, 488). p. 135

490. **1883. Pick, H.**, 'Ueber die Bedeutung des rothen Farbstoffes bei den Phanerogamen und die Beziehungen desselben

Page of text on which reference is made

zur Stärkewanderung,' *Bot. Centralbl.*, Cassel, 1883, XVI, pp. 281–284, 314–318, 343–347, 375–382.

The author agrees with Wigand that there is a connection between the occurrence of red pigment and the presence of tannins. A theory is formulated that the protection from white light, afforded by anthocyanin pigment, facilitates the hydrolysis and transportation of starch, and various evidence is given in support. pp. 24, 37, **134**

491. **1884. Johow, Fr.,** 'Ueber die Beziehungen einiger Eigenschaften der Laubblätter zu den Standortsverhältnissen,' *Jahrb. wiss. Bot.*, Berlin, 1884, XV, pp. 282–310.

Anthocyanin is regarded as a protection for leaves, especially the conducting system, against too intense light. p. 25

492. **Wortmann,** 'Criticism of Pick's paper: "Ueber die Bedeutung des rothen Farbstoffes bei den Phanerogamen und die Beziehungen desselben zur Stärkewanderung" (*Bot. Centralbl.*, Cassel, 1883, XVI, p. 281),' *Bot. Ztg.*, Leipzig, 1884, XLII, p. 237.

Criticism of Pick's hypothesis as to the function of anthocyanin. p. 134

493. **1886. Hassack, C.,** 'Untersuchungen über den anatomischen Bau bunter Laubblätter, nebst einigen Bemerkungen, betreffend die physiologische Bedeutung der Buntfärbung derselben,' *Bot. Centralbl.*, Cassel, 1886, XXVIII, pp. 84–85, 116–121, 150–154, 181–186, 211–215, 243–246, 276–279, 308–312, 337–341, 373–375, 385–387.

An important histological account of the distribution of anthocyanin in red leaves. Favourable reference is made to the screen theory of anthocyanin, and Kerner's observations are quoted as evidence. pp. **22**, 23, 33, 37, 132

494. **1887. Engelmann, Th. W.,** 'Die Farben bunter Laubblätter und ihre Bedeutung für die Zerlegung der Kohlensäure im Lichte,' *Bot. Ztg.*, Leipzig, 1887, XLV, pp. 393–398, 409–419, 425–436, 441–450, 457–463.

Investigations on the spectra of various pigments of coloured leaves. That of anthocyanin is found on the whole to be complementary to that of chlorophyll. pp. 9, 37, 57, **132**

495. **1889. Roze, E.,** 'Contribution à l'étude de l'action de la chaleur solaire sur les enveloppes florales,' *Bul. soc. bot.*, Paris, 1889, XXXVI, pp. ccxii–ccxiv.

The author attempts to arrive at a reason for the great variety of coloration of corollas, perianths, etc. He takes temperature of flowers (differently coloured) which have been exposed to the sun, and finds that some colours absorb

Page of text on which reference is made

more heat than others. He believes this to have a great physiological importance as regards fertilisation, dehiscence of anthers, etc. — —

496. **1890. Jumelle, H.**, 'Sur l'assimilation chlorophyllienne des arbres à feuilles rouges,' *C. R. Acad. sci.*, Paris, 1890, CXI, pp. 380–382.
Less assimilation in red leaves than in green. — —

497. **1892. Kny, L.**, 'Zur physiologischen Bedeutung des Anthocyans,' *Estratto dagli Atti del Congresso Botanico Internazionale*, 1892.
Account of experiments performed in support of the function of anthocyanin as a screen and as a medium for transforming light rays into heat. pp. 132, 136

498. **1894. Kerner von Marilaum, A.**, and **Oliver, F. W.**, *The Natural History of Plants*, London, 1894.
Many references to anthocyanin:
Morphological distribution in leaves, etc. p. 21
Development in Alpine plants. pp. 28, 103
Various physiological functions suggested. p. 130
Description of autumnal coloration. p. 25
Development of pigment in rhizomes. p. 31

499. **Wehrli, L.**, 'Ueber die Bedeutung der Färbung bei den Pflanzen,' *Ber. Schweiz. Bot. Ges.*, Bern, 1894, IV, pp. xxiii–xxviii.
Scheme for classification of anthocyanin and other pigments according to uses. Physiological uses of chief importance. These are classified as (1) protection of chlorophyll, (2) conversion of light into warmth. — —

500. **Wiesner, J.**, 'Pflanzenphysiologische Mittheilungen aus Buitenzorg. 2. Beobachtungen über Einrichtungen zum Schutze des Chlorophylls tropischer Gewächse,' *SitzBer. Ak. Wiss.*, Wien, 1894, CIII (1), pp. 8–36.
Screen theory of anthocyanin is supported. p. 133

501. **1895. Ewart, A. J.**, 'On Assimilatory Inhibition in Plants,' *J. Linn. Soc. Bot.*, London, 1895–1897, XXXI, pp. 364–461.
Author states that, in addition to its protective function against too intense light and heat, the main function of anthocyanin is to protect the assimilating cells against those rays of light which tend to induce in the protoplasm, more especially in the chlorophyll grain, a condition of light rigor, and thereby to diminish or inhibit their power of assimilation. p. 134

502. **Filarszky, F.**, 'Ueber Anthocyan und einen interessanten Fall der Nichtausbildung dieses Farbstoffes,' *Bot. Centralbl.*, Cassel, 1895, LXIV, p. 157.

Page of text on which reference is made

Protective function of anthocyanin by changing light into heat. — —

503. **Keeble, F. W.**, 'The Hanging Foliage of certain Tropical Trees,' *Ann. Bot.*, Oxford, 1895, IX, pp. 59–93.

A suggestion is made that, in addition to its value as a screen, anthocyanin in young leaves in the tropics protects the leaf against the too great heating effects of the sun's rays. Experimental evidence is given in favour of this view. pp. 24, 25, **133**, 138

504. **1896. MacDougal, D. T.**, 'The Physiology of Color in Plants,' *Science*, New York, 1896, IV, pp. 350–351.

Short note on Stahl's hypothesis. — —

505. **Stahl, E.**, 'Ueber bunte Laubblätter,' *Ann. Jard. bot.*, Buitenzorg, 1896, XIII, pp. 137–216.

Important paper on the physiological significance of anthocyanin. The author favours the view that the chief function of the pigment is to convert light rays into heat. By virtue of this property, it has the power of accelerating transpiration under difficult circumstances, and hence its distribution in many shade-loving plants in damp tropical regions. pp. 9, 22, 23, 24, 25, 37, 133, 135, **136**

506. **1897. Ewart, A. J.**, 'The Effects of Tropical Insolation,' *Ann. Bot.*, Oxford, 1897, XI, pp. 439–480.

The author maintains that anthocyanin is highly important in its protective action against rigor produced in the assimilating cell by too strong insolation. Stahl's views are adversely criticised. pp. 25, 135, **138**

507. **Keeble, F. W.**, 'The Red Pigment of Flowering Plants,' *Sci. Progr.*, London, 1897, VI (N. S. I), pp. 406–421.

General account of the physiological significance of anthocyanin. — —

508. **1898. Montemartini, L.**, 'Sopra la struttura del sistema assimilatore nel fusto del *Polygonum Sieboldii*,' *Malpighia*, Genova, 1898, XII, pp. 78–80.

Red spots on the stem are found to correspond to an area of epidermal anthocyanin-containing cells round the stomata. Beneath this area is assimilating tissue. The suggestion is made that anthocyanin protects chlorophyll from too intense light. — —

509. **1899. Macchiati, L.**, 'Ufficio dei peli, dell' antocianina e dei nettarii estranuziali dell' *Ailanthus glandulosa*,' *Boll. Soc. bot. ital.*, Firenze, 1899, pp. 103–112.

Suggestion that anthocyanin acts as a screen against too bright light and excessive transpiration. — —

Page of text on which reference is made

510. **1900. Hansgirg, A.**, 'Zur Biologie der Laubblätter,' *SitzBer. Böhm. Ges. Wiss.*, Prag, 1900, No. 20, 142 pages.
Classification of leaves into physiological types. — —

511. **1901. Linsbauer, L.**, 'Untersuchungen über die Durchleuchtung von Laubblättern,' *Beiheft z. Bot. Centralbl.*, Cassel, 1901, x, pp. 53–89.
Slight reference to the significance of anthocyanin as a light screen. — —

512. **Smith, F. G.**, 'On the Distribution of red Color in vegetative Parts in the New England Flora,' *Bot. Gaz.*, Chicago, Ill., 1901, XXXII, pp. 332–342.
Distribution of anthocyanin in plants growing in various situations, wet and dry, shady and sunny. Short discussion on significance of anthocyanin; no conclusion reached. p. 23

513. **Thomas, Fr.**, 'Anpassung der Winterblätter von *Galeobdolon luteum* an die Wärmestrahlung des Erdbodens,' *Ber. D. bot. Ges.*, Berlin, 1901, XIX, pp. 398–403.
A suggestion that anthocyanin in the under surface of the leaf increases the capacity of the leaf for absorbing warmth from the ground. — —

514. **1902. Casares, Gil A.**, 'Algunas observaciones sobre la coloración rojiza de ciertas hepáticas,' *Boletín de la Sociedad Española de Historia Natural*, 1902, II, pp. 207–211.
Pigment (? anthocyanin) in certain species of Liverworts. It is developed on under surfaces, which roll over, and protects thallus from undue exposure to the more refrangible rays of sunlight. — —

515. ***Thomas, Fr.**, 'Ueber die Winterblätter von *Galeobdolon luteum*,' *Mittheilung des Thüringischen botanischen Vereins*, N. F., 1902, XVI, p. 13.
Probably as in paper above (No. 513). — —

516. **1903. Koning, C. J.**, en **Heinsius, H. W.**, 'De beteekenis en het ontstaan van het anthocyaan in bladeren,' *Ned. Kruidk. Arch.*, Nijmegen, 1903 (ser. 3), II, pp. 1011–1018.
Results support view that anthocyanin favours translocation of starch by protection of diastase. p. 135

517. **1905. *Hryniewiecki, B.**, 'Antocyan a wytrzymatość roślin na zimno (Anthocyan und Winterhärte der Pflanzen),' *Wszechświat*, Warszawa, 1905, XXIV, p. 687.
Red-leaved beech more readily acclimatised than green-leaved. — —

518. **Tischler, G.**, 'Ueber die Beziehungen der Anthocyanbildung zur Winterhärte der Pflanzen,' *Beihefte z. Bot. Centralbl.*, Cassel, 1905, XVIII, pp. 452–471.

Page of text on which reference is made

Greater endurance of red-leaved varieties to low temperatures. Reasons suggested. — —

519. **1908. *Schilberszky, K.**, 'Az anthoczián elöfordulásáról és élettani szerepéröl (Ueber das Vorkommen und über die physiologische Rolle des Anthocyans),' *Pótf. Termt. Közl.*, Budapest, 1908, XL, pp. 153–155. — —

520. **1909. Smith, A. M.**, 'On the Internal Temperature of Leaves in Tropical Insolation with special Reference to the Effect of their Colour on the Temperature,' *Ann. R. Bot. Gard.*, Ceylon, 1909, IV, pp. 229–298.
Author has shown experimentally that red leaves have a higher temperature than green. pp. 9, 25, 133, **142**

521. **1910. Weevers, Th.**, 'Kurze Notizen in Bezug auf die Anthocyanbildung in jungen Schösslingen der tropischen Pflanzen,' *Ann. Jard. bot.*, Buitenzorg, 1910, 3ième sup., 1e partie, pp. 313–318.
List of plants inhabiting damp tropical forests and developing anthocyanin in young shoots. p. 25

522. **1923. Walker, J. C.**, 'Disease Resistance to Onion Smudge,' *J. Agric. Res.*, Washington, D.C., 1923, XXIV, pp. 1019–1039.
Bulbs with pigmented skins resist attacks of disease. — —

BIOLOGICAL SIGNIFICANCE OF ANTHOCYANINS

523. **1877. *Kuntze, O.**, 'Die Schutzmittel der Pflanzen gegen Thiere und Wetterungunst,' *Bot. Ztg.*, Leipzig, 1877, Supplementheft.
Suggestion that red colour in plants may resemble blood to animals, and thereby be protective. — —

524. **1888. Focke, W. O.**, 'Die Verbreitung beerentragender Pflanzen durch die Vögel,' *Abh. natw. Ver.*, Bremen, 1888, X, p. 140.
Examples of cases of distribution of fruits. p. 129

525. **1889. *Ludwig, F.**, 'Extranuptiale Saftmale bei Ameisenpflanzen,' *Humboldt*, 1889, VIII, pp. 294–297.
Red pigmented cushions and lines denote the path to extra-floral nectaries (cp. honey-guides) in *Impatiens*, *Viburnum* and *Sambucus*. — —

526. **1891. Ludwig, F.**, 'Die Beziehungen zwischen Pflanzen und Schnecken,' *Beihefte z. Bot. Centralbl.*, Cassel, 1891, I, pp. 35–39.
Red colour as a warning signal to animals. — —

527. **1898. Knuth, P.**, *Handbuch der Blüthenbiologie*, Leipzig, 1898. (*Handbook of Flower Pollination*, Translated by J. R. Ainsworth Davis, 3 vols., Oxford, 1906, vol. I, pp. 212–380.)

Page of text on which reference is made

Full bibliography relating to biological significance of anthocyanin. pp. 9, 129

528. **1899. Macchiati, L.**, 'Osservazioni sui nettarii estranuziali del *Prunus Laurocerasus*,' *Boll. Soc. bot. ital.*, Firenze, 1899, pp. 144–147.
Anthocyanin as guide for insects to extra-floral nectaries. — —

529. **1914. East, E. M.**, and **Glaser, R. W.**, 'Observations on the Relation between Flower Color and Insects,' *Psyche*, Boston, Mass., 1914, XXI, pp. 27–30.
Record of insect visits to differently coloured flowers of *Nicotiana*. — —

530. **1922. Kostka, G.**, 'Farbenwechsel und Insektenbesuch bei *Pulmonaria officinalis* L.,' *Oest. Bot. Zs.*, Wien, 1922, LXXI, 246–254.
Insects do not visit flowers which have changed colour since these have loosened corolla. — —

ANTHOCYANINS AND GENETICS

531. **1817. Hopkirk, T.**, *Flora Anomoia. A General View of the Anomalies in the Vegetable Kingdom*, London, 1817, 187 pages.
An interesting account of variation in plants. A section is devoted to variation in flower-colour, and lists of species are given in which the colour varies from blue, purple, or red, to white: red to blue, red to yellow, etc. Also variation in colour of fruits. The causes of variation are discussed. — —

532. **1834. Schübler, G.**, und **Lachenmeyer, C.**, 'Untersuchungen über die Farbenveränderungen der Blüthen,' *J. prakt. Chem.*, Leipzig, 1834, I, pp. 46–58.
An account of the effect of different soils on the flower-colour of *Hortensia speciosa*. Cases of change in colour with varying age of flowers are mentioned (ex. Boraginaceae). — —

533. **1849. Gärtner, C. F. von**, *Versuche und Beobachtungen über die Bastarderzeugung im Pflanzenreich*, Stuttgart, 1849.
Inheritance of flower-colour in hybrids. — —

534. **1872. Hoffmann, H.**, 'Ueber Variation,' *Bot. Ztg.*, Leipzig, 1872, XXX, pp. 529–539.
The causes of variation in flower-colour are dealt with. The effects of chemical substances, temperature, light, water and cross-fertilisation on variation are discussed. Also the fixing of varieties. — —

Page of text on which reference is made

535. **1874. Hoffmann, H.**, 'Ueber *Papaver Rhoeas*, L.,' *Bot. Ztg.*, Leipzig, 1874, XXXII, pp. 257–269.
An account of the variations noted in cultures of *Papaver*. — —

536. **Hoffmann, H.**, 'Zur Kenntniss der Gartenbohnen,' *Bot. Ztg.*, Leipzig, 1874, XXXII, pp. 273–283, 289–302.
Variation in flower- and seed-colour in cultures of beans. — —

537. **1875. Hoffmann, H.**, 'Culturversuche,' *Bot. Ztg.*, Leipzig, 1875, XXXIII, pp. 601–605, 617–628.
Variation in flower-colour in cultures of species of *Clarkia*, *Collinsia*, *Datura*, *Gilia*, *Godetia*, *Gypsophylla*, *Hydrangea*, *Myosotis* and others. — —

538. **1876. Hoffmann, H.**, 'Culturversuche,' *Bot. Ztg.*, Leipzig, 1876, XXXIV, pp. 545–552, 561–572.
Variation in flower-culture colour in cultures of *Althaea rosea*, *Cheiranthus Cheiri*, and species of *Linum*, *Lychnis* and *Primula*. — —

539. **1877. Hoffmann, H.**, 'Culturversuche,' *Bot. Ztg.*, Leipzig, 1877, XXXV, pp. 265–279, 281–295, 297–305.
Experiments on the effect of darkness, mechanical bending, temperature, chemical substances, etc., on variation in *Papaver*. — —

540. **1878. Hoffmann, H.**, 'Culturversuche,' *Bot. Ztg.*, Leipzig, 1878, XXXVI, pp. 273–286, 289–299.
Variation in flower-colour in cultures of *Atropa Belladonna* and fruits of *Prunus Avium*. — —

541. **1879. Hoffmann, H.**, 'Culturversuche,' *Bot. Ztg.*, Leipzig, 1879, XXXVII, pp. 177–187, 193–207, 569–576, 585–595, 601–604.
Variations in flower-colour in cultures of *Anagallis coerulea*, *Papaver* and others. — —

542. **1881. Focke, W. O.**, *Die Pflanzen-mischlinge*, Berlin, 1881.
Very complete account of hybridisation among Phanerogams. — —

543. **Hoffmann, H.**, 'Culturversuche über Variationen,' *Bot. Ztg.*, Leipzig, 1881, XXXIX, pp. 105–110, 121–125, 137–143.
Variation in flower-colour in cultures of *Anthyllis Vulneraria* and *Glaucium luteum*. — —

544. **Hoffmann, H.**, 'Rückblick auf meine Variations-Versuche,' *Bot. Ztg.*, Leipzig, 1881, XXXIX, pp. 345–351, 361–368, 377–383, 393–399, 409–415, 425–432.
Résumé of previous experiments. — —

545. **1882. Hoffmann, H.**, 'Culturversuche über Variation,' *Bot. Ztg.*, Leipzig, 1882, XL, pp. 483–489, 499–514.
Variation in flower-colour in cultures of *Papaver alpinum*, *P. somniferum*, *Dianthus superbus* and others. — —

546. **1883. Hoffmann, H.**, 'Culturversuche über Variation,' *Bot.*

Page of text on which reference is made

Ztg., Leipzig, 1883, XLI, pp. 276–281, 289–299, 305–314, 321–330, 337–347.

Variation in flower-colour in cultures of *Anemone nemorosa*, *Phyteuma spicatum* and others. — —

547. **1884. Goff, E. S.**, 'The Relation of Color to Flavor in Fruits and Vegetables,' *Amer. Nat.*, Philadelphia, 1884, XVIII, pp. 1203–1210.

Many cases are quoted in which there is a distinct—often strong and unpleasant—flavour when colour is present in fruits and vegetables. p. 194

548. **Hoffmann, H.**, 'Culturversuche über Variation,' *Bot. Ztg.*, Leipzig, 1884, XLII, pp. 209–219, 225–237, 241–250, 257–266, 275–279.

Variation in flower-colour in cultures of species of *Papaver* and *Raphanus*. — —

549. **1887. Focke, W. O.**, 'Die Culturvarietäten der Pflanzen,' *Abh. natw. Ver.*, Bremen, 1887, IX, pp. 447–468.

General discussion on variation. Examples are given of variation in colour of flowers, fruits, roots, etc., under cultivation. — —

550. **Hoffmann, H.**, 'Culturversuche über Variation,' *Bot. Ztg.*, Leipzig, 1887, XLV, pp. 24–28, 40–45, 55–57, 72–76, 86–90, 169–174, 233–239, 255–260, 288–291, 729–746, 753–761, 769–779.

Variation in flower-colour in cultures of *Anagallis arvensis*, *Anthyllis Vulneraria*, *Atropa Belladonna*, *Dictamnus Fraxinella*, *Digitalis purpurea* and others. — —

551. **Jäger, H.**, 'Zur Färbung der Blutbuche,' *Gartenflora*, Berlin, 1887, XXXVI, pp. 40–41.

Suggested correlation between young wood colour and leaf colour. — —

552. **Webster, A. D.**, 'Change of Colour in the Flowers of *Anemone nemorosa*,' *J. Bot.*, London, 1887, XXV, p. 84.

Colour of flowers is said to depend on nature of soil in which plants were growing. — —

553. **1888. Pammel, L. H.**, 'Color Variation in Flowers of *Delphinium*,' *Bot. Gaz.*, Crawfordsville, Indiana, 1888, XIII, p. 216.

Record of colour varieties of *D. tricorne* occurring naturally. — —

554. **1889. *Focke, W. O.**, 'Der Farbenwechsel der Rosskastanienblumen,' *Verh. bot. Ver.*, Berlin, 1889, XXXI, pp. 108–112.

Changes in colour in flowers after fertilisation. — —

555. **Kerner von Marilaum, A.**, 'Ueber das Wechseln der Blüthenfarbe an einer und derselben Art in verschiedenen Gegenden,' *Oest. Bot. Zs.*, Wien, 1889, XXXIX, pp. 77–78.

Page of text on which reference is made

The distribution of different colour varieties of any species may be determined by colour contrast (and consequent attraction for insects) with flowers of other species amongst which it grows. — —

556. **1890. Jacobasch, E.,** 'Verschiedene Blütezeit der rot-, blau- und weissblütigen Form von *Hepatica triloba* Gil. und Umwandlung der Normalform in die Rote,' *Verh. bot. Ver.*, Berlin, 1890, XXXI, pp. 253–254.

Observations on the time of flowering of different colour-varieties. Author suggests that cold may cause variation in colour. — —

557. ***Körnicke,** 'Varietätenbildung im Pflanzenreiche,' *Sitzungsberichte der niederrheinischen Gesellschaft für Naturkunde*, Bonn, 1890, pp. 14–20.

Mention is made of the influence of soil on the colour of plants. A case is quoted where development of red and violet colour in Maize-cobs, due to effect of soil, is said to have taken place. — —

558. **1891. Eckert, J. P.,** 'Some Peculiar Changes in the Colour of the Flower of *Swainsonia procumbens*,' *Nature*, London, 1891–1892, XLV, p. 185.

The corolla is at first lilac, but changes later to crimson in places and blue. Petals may assume original colour and go through the gradations once more. — —

559. **1892. Guinier, E.,** 'Sur la coloration accidentelle de la fleur du Fraisier commun,' *Bul. soc. bot.*, Paris, 1892, XXXIX, pp. 64–66.

An account of the occurrence of a variety of strawberry with reddish-purple flowers. p. 157

560. **1893. Gillot, X.,** 'Observations sur la coloration rosée ou *érythrisme* des fleurs normalement blanches,' *Bul. soc. bot.*, Paris, 1893, XL, pp. clxxxix–cxciv.

List of species in which flowers of the types are without anthocyanin, and which give rise to varieties with anthocyanin, or to coloration with anthocyanin under certain conditions. p. 157

561. **Hildebrand, F.,** 'Ueber einige Variationen an Blüthen,' *Ber. D. bot. Ges.*, Berlin, 1893, XI, pp. 476–480.

The author notes the occurrence, among the pale blue flowers of *Iris florentina*, of a sudden 'sport' (in some individual flowers) which had a large portion of the perianth dark violet. Also a hybrid *Dahlia* (from red × white) was found to produce both red and white flowers and pure red flowers at different times. p. 199

562. **Newell, J. H.,** 'The Flowers of the Horse Chestnut,'

Page of text on which reference is made

Bot. Gaz., Bloomington, Indiana, 1893, XVIII, pp. 107–109.

Change of colour in flowers with age. — —

563. **Praeger, R. L.**, 'Colour-variation in Wild Flowers,' *Irish Nat.*, Dublin, 1893, II, pp. 174–175.

The author gives a list of colour-variations he has noted in British wild flowers: ex. *Scabiosa succisa* (white, pink), *Orchis mascula* (white, flesh-colour), *Jasione montana* (pink): white of *Scilla verna, Vicia sepium, Trifolium pratense, Epilobium montanum* and many others. — —

564. **1894. Chabert, A.**, 'Les variations à fleurs rouges de certains *Galium*,' *Bul. soc. bot.*, Paris, 1894, XLI, pp. 302–305.

In connection with a red-flowered variety of *Galium silvestre* found by Gillot (see No. 560), the author objects to its being called a new variety; he believes 'erythrism' on the contrary to be merely due to chemical nature of the soil, etc. He quotes cases of occurrence of red-flowered individuals of other species of *Galium*. p. 157

565. **Gillot, X.**, 'Variations parallèles à fleurs rouges des espèces du genre *Galium*,' *Bul. soc. bot.*, Paris, 1894, XLI, pp. 28–30.

Record of red-flowered varieties of *Galium* spp., the types being normally white-flowered. — —

566. **Meehan, T.**, 'Contributions to Life Histories of Plants. On Purple-leaved Plants, No. 11,' *Proc. Acad. Nat. sci.*, Philadelphia, Pa., 1894, pp. 162–171.

Note on grafting of blood-variety of *Betula* on type. — —

567. **Pillsbury, J. H.**, 'On the Color Description of Flowers,' *Bot. Gaz.*, Madison, Wisconsin, 1894, XIX, pp. 15–18.

Attempt to classify and define terms used for colour by means of a mechanical device. — —

568. **1895. Prehn, J.**, 'Ueber das Vorkommen zuweilen weissblühender Pflanzen,' *Schr. natw. Ver.*, Kiel, 1895, X, pp. 259–261.

Author gives a list of white-flowered varieties he has found among wild plants. Includes white varieties of *Carduus crispus, Centaurea Cyanus, Aster Tripolium, Cichorium Intybus, Campanula Trachelium, Echium vulgare* and many others. — —

569. **1896. Hildebrand, F.**, 'Einige biologische Beobachtungen,' *Ber. D. bot. Ges.*, Berlin, 1896, XIV, pp. 324–331.

Further observations on a hybrid *Dahlia* (see No. 561). It was found that the production of completely red flowers, or of red and white flowers, was influenced more or less by nutrition. Same kind of phenomenon observed with purple and white *Petunia*. p. 199

Page of text on which reference is made

570. **1897. Molisch, H.,** 'Der Einfluss des Bodens auf die Blüthenfarbe der Hortensien,' *Bot. Ztg.*, Leipzig, 1897, LV, pp. 49–61.
Effect on flower-colour of adding salts of iron and aluminium, and many other substances, to the soil. Alum, aluminium sulphate and ferrous sulphate turn red flowers blue. — —

571. **Zopf, W.,** 'Ueber den Einfluss des Bodens auf die Farbe der Hortensiablüten,' *Die Natur*, Halle a. S., 1897, XLVI, pp. 318–319.
Blue colour produced in flowers by artificial use of aluminium salts in soil. — —

572. **1898. Hasslinger, J. von,** 'Beobachtungen über Variationen in den Blüten von *Papaver Rhoeas* L.,' *Oest. Bot. Zs.*, Wien, 1898, XLVIII, pp. 139–141.
Record of variation in blotches (purple anthocyanin) at the base of the petals. — —

573. **1899. Buchenau, F.,** 'Zwei interessante Beobachtungen an Topf-Pelargonien,' *Abh. natw. Ver.*, Bremen, 1899, XVI, pp. 274–277.
Author describes a scarlet-flowered *Pelargonium* plant with a branch bearing rose-coloured flowers. — —

574. **Jackson, B. D.,** 'A Review of the Latin Terms used in Botany to denote Colour,' *J. Bot.*, London, 1899, XXXVII, pp. 97–106.
Classification of Latin terms used for various colours. — —

575. **1900. Correns, C.,** 'Ueber Levkojenbastarde,' *Bot. Centralbl.*, Cassel, 1900, LXXXIV, pp. 97–113.
Preliminary experiments on inheritance of flower-colour in Stocks are included (violet, red and white). Relation of flower- to seed-colour. p. 158

576. **Hildebrand, F.,** 'Ueber Bastardirungsexperimente zwischen einigen *Hepatica*-Arten,' *Bot. Centralbl.*, Cassel, 1900, LXXXIV, pp. 65–73.
Inheritance of flower-colour and other characters in crosses between different species of *Hepatica*. p. 158

577. **Vries, H. de,** 'Der Spaltungsgesetz der Bastarde,' *Ber. D. bot. Ges.*, Berlin, 1900, XVIII, pp. 83–90.
Early record of cases of Mendelian dominance and segregation. Certain instances of colour-variation are included, i.e. in *Papaver, Antirrhinum, Polemonium, Lychnis, Datura, Clarkia, Trifolium* and others. pp. 157, 158, 159, 168, 178, 195

578. **1901. Bateson, W.,** and **Saunders, E. R.,** 'Experimental Studies in the Physiology of Heredity,' *Rep. Evol. Com. Roy. Soc.*, London, 1901, Rpt. I, 160 pages.
Experiments on inheritance of colour in flowers of

Page of text on which reference is made

Lychnis: in flowers and fruits of *Atropa* and *Datura*: in flowers of *Matthiola*. pp. 11, 153, 157, 158, **164**, **168**, **174**, 191, 194

579. **Correns, C.**, 'Bastarde zwischen Maisrassen,' *Bibl. bot.*, Stuttgart, 1901, x, pp. xii + 161.
Important account of inheritance of colour (anthocyanin) and other characters. pp. 159, **189**

580. **Coupin, H.**, 'La couleur des fleurs de la flore française,' *C. R. ass. franç. avanc. sci.*, Paris, 1901, 2e partie, pp. 500–520.
Statistical study of the flora of France showing percentages of different flower-colours in various natural orders. Also percentage of colours in various habitats, viz., forests, mountains, dry and wet regions, sea-shore, etc. — —

581. **Taliew, W.**, 'Ueber den Polychroismus der Frühlingspflanzen,' *Beihefte z. Bot. Centralbl.*, Cassel, 1901, x, pp. 562–564.
The author maintains that there is greater variation in flower-colour in species of spring plants than others. He suggests it may be due to the conditions to which they are exposed at this season. Many examples are quoted including *Anemone*, *Tulipa*, *Iris*, *Myosotis* and *Primula* species. — —

582. **Tschermak, E. von**, 'Weitere Beiträge über Verschiedenwerthigkeit der Merkmale bei Kreuzung von Erbsen und Bohnen,' *Ber. D. bot. Ges.*, Berlin, 1901, XIX, pp. 35–51.
A preliminary account of some cross-breeding in *Phaseolus* and *Pisum*. Among the characters observed are flower- and seed-colour. pp. 159, **197**

583. **1902. Bateson, W.**, 'Note on the Resolution of Compound Characters by Cross-breeding,' *Proc. Phil. Soc.*, Cambridge, 1902–1904, XII, pp. 50–54.
Suggestion as to the factors for inheritance of flower-colour in *Antirrhinum*. pp. 157, **160**

584. **Bateson, W.**, *Mendel's Principles of Heredity. A Defence*, Cambridge, 1902.
References to inheritance of colour. — —

585. **Correns, C.**, 'Ueber Bastardirungsversuche mit *Mirabilis*-Sippen,' *Ber. D. bot. Ges.*, Berlin, 1902, XX, pp. 594–608.
Experiments on inheritance of colour in flowers of *Mirabilis Jalapa*. Also crosses between *M. longiflora* and *M. Jalapa*. Scheme of factors suggested. pp. 158, **175**

586. **Emerson, R. A.**, 'Preliminary Account of Variation in Bean Hybrids,' *Annual Report Nebraska Agricultural Experiment Station*, Nebraska, 1902, xv, pp. 30–43.
Pigmentation in testa of seeds of *Phaseolus*. p. 159

Page of text on which reference is made

587. **Tschermak, E. von,** 'Ueber die gesetzmässige Gestaltungsweise der Mischlinge. Fortgesetzte Studien an Erbsen und Bohnen,' *Zs. Landw. VersWes.*, Wien, 1902, 81 pages. — —

588. **1903. Correns, C.,** 'Ueber die dominierenden Merkmale der Bastarde,' *Ber. D. bot. Ges.*, Berlin, 1903, XXI, pp. 133–147.
Definition of dominance. Partial and total dominance. Illustrations among flower-colour cases. p. 158

589. **Correns, C.,** 'Weitere Beiträge zur Kenntnis der dominierenden Merkmale und der Mosaikbildung der Bastarde,' *Ber. D. bot. Ges.*, Berlin, 1903, XXI, pp. 195–201.
Four types of dominance in F_1. Illustrations among flower-colour cases. pp. 159, 182, 183

590. **1904. Bateson, W., Saunders, E. R.,** and **Punnett, R. C.,** 'Experimental Studies in the Physiology of Heredity,' *Rep. Evol. Com. Roy. Soc.*, London, 1904, Rpt. II, 154 pages.
Accounts of inheritance of flower-colour in *Matthiola, Salvia, Pisum* and *Lathyrus*.
pp. 11, 149, 154, 158, 159, **168, 171, 174, 182, 187**, 191, 193

591. **Bidgood, J.,** 'Albinism, with special Reference to Shirley Poppies,' *J. R. Hort. Soc.*, London, 1904, XXVIII, pp. 477–482.
The author discusses albinism in general, and in particular the loss of colour and of the black basal patch in *Papaver*. The black pigment at the base of the petal gives different chemical reactions from that in the rest of the petal. — —

592. ***Bitter, G.,** 'Dichroismus und Pleochroismus als Rassencharaktere,' *Festschrift für P. Ascherson*, Berlin, 1904, pp. 158–167.
From a number of observations the author makes the suggestion that within the species there may be races having either different colouring in all the organs, or in one organ only, viz. in stem, leaf, flower, fruit or seed. — —

593. **Correns, C.,** 'Ein typisch spaltender Bastard zwischen einer einjährigen und einer zweijährigen Sippe des *Hyoscyamus niger*,' *Ber. D. bot. Ges.*, Berlin, 1904, XXII, pp. 517–524.
Crosses are made between *H. niger* and *H. pallidus*, the former having anthocyanin, the later none. F_1 is intermediate in colour. Mendelian segregation in F_2. p. 158

594. **Coutagne, G.,** 'De la polychromie polytaxique florale des végétaux spontanés,' *C. R. Acad. sci.*, Paris, 1904, CXXXIX, pp. 77–79.
Discussion as to existence, side by side in nature, of differently coloured varieties of the same species. — —

Page of text on which reference is made

595. **Emerson, R. A.,** 'Heredity in Bean Hybrids,' *Annual Report Nebraska Agricultural Experiment Station*, Nebraska, 1904, XVII, pp. 33–68.
Appearance of anthocyanin in testa of F_1 from two varieties without pigment. p. 159

596. **Lock, R. H.,** 'Studies in Plant Breeding in the Tropics,' *Ann. R. Bot. Gard.*, Ceylon, 1904, II, pp. 299–356.
Some observations on colour inheritance in peas and maize are included. pp. 159, **189**

597. ***Tschermak, E. von,** 'Weitere Kreuzungsstudien an Erbsen, Leukojen und Bohnen,' *Zs. Landw. VersWes.*, Wien, 1904, VII, pp. 533–638.
Same observation as in No. 595. pp. 158, 159

598. **Wittmack, L.,** '*Daucus Carota*, L. var. *Boissieri Schweinfurth*,' *Gartenflora*, Berlin, 1904, LIII, pp. 281–284.
Description of a variety of Carrot of which the root contains anthocyanin. p. 156

599. **1905. Bateson, W., Saunders, E. R.,** and **Punnett, R. C.,** 'Further Experiments on Inheritance in Sweet Peas and Stocks,' *Proc. R. Soc.*, London, 1905, LXXVII B, pp. 236–238.
Reference to inheritance of colour. Expression of C and R factors. pp. 12, 158, 171, 174

600. **Correns, C.,** 'Zur Kenntnis der scheinbar neuen Merkmale der Bastarde,' *Ber. D. bot. Ges.*, Berlin, 1905, XXIII, pp. 70–85.
Further experiments on inheritance of flower-colour in *Mirabilis Jalapa*. Chiefly second and third generations. Also striped varieties. pp. 158, **175**

601. **Vries, H. de,** *Species and Varieties. Their Origin by Mutation*, Chicago and London, 1905.
Various references to inheritance of colour.
pp. 154, 155, 156, 157, 158, 159, 164, 169, 187, 188, 189, 197, 205

602. **1906. Bateson, W.,** 'Coloured Tendrils of Sweet Peas,' *Gard. Chron.*, London (Ser. 3), 1906, XXXIX, p. 333.
Association with coloured leaf-axils. pp. 158, **192**

603. **Bateson, W., Saunders, E. R.,** and **Punnett, R. C.,** 'Experimental Studies in the Physiology of Heredity,' *Rep. Evol. Com. Roy. Soc.*, London, 1906, Rpt. III, 53 pages.
References to inheritance of flower-colour in Sweet Peas and Stocks. pp. 154, 158, **171, 174**

604. **Biffen, R. H.,** 'Experiments on Hybridisation of Barleys,' *Proc. Phil. Soc.*, Cambridge, 1906, XIII, pp. 304–308.
Purple colour in paleae and grains of Barley. pp. 158, **171**

605. **Hurst, C. C.,** 'Mendelian Characters in Plants and

Page of text on which reference is made

Animals,' *Rep. Conf. Genet. R. Hort. Soc.*, London, 1906, pp. 114–128.

Inheritance of flower-colour in *Antirrhinum*, Sweet Peas, Poppies, and Orchids. pp. 158, 166, 168, 178, 196

606. **Kraemer, H.**, 'Studies on Color in Plants,' *Bull. Torrey Bot. Cl.*, New York, N. Y., 1906, xxxiii, pp. 77–92.

Chiefly an account of the effect on flower-colour of various salts added to the soil. Changes insignificant. — —

607. **Lock, R. H.**, 'Studies in Plant Breeding in the Tropics,' *Ann. R. Bot. Gard.*, Ceylon, 1906, iii, pp. 95–184.

Inheritance of colour in *Zea Mays*. pp. 158, 159, 166, 179, **189**

608. **Ostenfeld, C. H.**, 'Experimental and Cytological Studies in the *Hieracia*. 1. Castration and Hybridisation Experiments with some Species of *Hieracia*,' *Bot. Tids.*, Kjøbenhavn, 1906, xxvii, pp. 225–248.

Inheritance of anthocyanin in some of the hybrids concerned. p. 158

609. **Saunders, E. R.**, 'Certain Complications arising in the Cross-breeding of Stocks,' *Rep. Conf. Genet. R. Hort. Soc.*, London, 1906, pp. 143–147.

Relationship between factors for hoariness and factors for colour. pp. 158, **174**

610. **1907. Bateson, W.**, 'The progress of Genetics since the rediscovery of Mendel's Papers,' *Progr. rei bot.*, Jena, 1907, i, pp. 368–418.

References to the inheritance of colour in plants. — —

611. **Fletcher, F.**, 'Mendelian Heredity in Cotton,' *J. Agric. Sci.*, Cambridge, 1907–1908, ii, pp. 281–282.

Inheritance of anthocyanin in flower is mentioned, and is noted as not being a simple Mendelian case. p. 158

612. **Focke, W. O.**, 'Betrachtungen und Erfahrungen über Variation und Artenbildung,' *Abh. natw. Ver.*, Bremen, 1907, xix, pp. 68–87.

Some observations on colour-variation are included. — —

613. **Gregory, R. P.**, 'On the Inheritance of certain Characters in *Primula sinensis*,' *Rep. Brit. Ass.*, London, 1907, pp. 691–693.

Remarks on inheritance of colour in flower and stem. — —

614. ***Lindman, C. A. M.**, 'Amphichromie bei *Calluna vulgaris*,' *Bot. Not.*, Lund, 1907, pp. 201–207.

Variation in colour in flowers of an individual of *Calluna vulgaris*. — —

615. **Lock, R. H.**, 'On the Inheritance of Certain Invisible Characters in Peas,' *Proc. R. Soc.*, London, 1907, lxxix B, pp. 28–34.

Page of text on which reference is made

Mention is made of the purple spots (due to anthocyanin) on the testa as a Mendelian character. pp. 159, **182**

616. **Shull, G. H.**, 'Some Latent Characters of a White Bean,' *Science*, N. S., New York, N. Y., 1907, xxv, pp. 828–832.
Inheritance of colour (anthocyanin) in the testa. pp. 159, **180**, 197

617. **Wheldale, M.**, 'The Inheritance of Flower-colour in *Antirrhinum majus*,' *Proc. R. Soc.*, London, 1907, LXXIX B, pp. 288–305.
Preliminary account of the colour-varieties and the Mendelian factors for flower-colour. pp. 12, 157, **160**

618. **1908. Balls, W. L.**, 'Mendelian Studies of Egyptian Cotton,' *J. Agric. Sci.*, Cambridge, 1908, II, pp. 346–379.
Inheritance of anthocyanin in spot on leaf and petals. pp. 158, **169**, 194, 197

619. **Bateson, W., Saunders, E. R.**, and **Punnett, R. C.**, 'Experimental Studies in the Physiology of Heredity,' *Rep. Evol. Com. Roy. Soc.*, London, 1908, Rpt. IV, 60 pages.
Gametic coupling in the Sweet Pea. Colour factors involved. pp. 158, 171

620. **Baur, E.**, 'Einige Ergebnisse der experimentellen Vererbungslehre,' *Med. Klinik*, Berlin, 1908, IV, Beiheft 10, pp. 265–292.
Colour inheritance in flowers of *Antirrhinum* and *Mirabilis*. pp. 157, 158, 162, 175

621. **Lock, R. H.**, 'The Present State of Knowledge of Heredity in *Pisum*,' *Ann. R. Bot. Gard.*, Ceylon, 1908, IV, pp. 93–111.
Inheritance of anthocyanin in flowers, leaf-axils, testa, etc. of *Pisum*. pp. 149, 159, **182**, 192, 197

622. **Rawson, H. E.**, 'Colour Changes in Flowers produced by controlling Insolation,' *Rep. Brit. Ass.*, London, 1908, pp. 902–903.
Account of variations of flower-colour produced in *Tropaeolum* by shading plants at certain times of the day. — —

623. **Shull, G. H.**, 'Some new Cases of Mendelian Inheritance,' *Bot. Gaz.*, Chicago, Ill., 1908, XLV, pp. 103–116.
Among other cases an account is given of a cross between white- and purple-flowered *Lychnis dioica*. Inheritance of the purple (anthocyanin) disk in Sunflower. pp. 158, **170**, **174**, 192

624. **Shull, G. H.**, 'A new Mendelian Ratio and several Types of Latency,' *Amer. Nat.*, New York, 1908, XLII, pp. 433–451.
Factors controlling inheritance of mottling (anthocyanin) in Beans. pp. 159, **180**, 195, 197

Page of text on which reference is made

625. **Vouk, V.**, 'Einige Versuche über den Einfluss von Aluminium-salzen auf die Blütenfärbung,' *Oest. Bot. Zs.*, Wien, 1908, LVIII, pp. 236–243.
An account of experiments on treating plants of *Hydrangea hortensis* with solutions of aluminium salts. — —

626. **1909. Balls, W. L.**, 'Studies of Egyptian Cotton,' *Yrbk. Khed. Agric. Soc.*, Cairo, 1909, 158 pages.
Spot on leaf due to anthocyanin. Inheritance of spot. pp. 158, **169**, 194, 197

627. **Bateson, W.**, *Mendel's Principles of Heredity*, Cambridge, 1909. (3rd ed. 1913.)
Full reference to inheritance of colour in plants. pp. 149, 153, 158, 166, 192, 205, 206

628. **Bunyard, E. A.**, 'The Effect of Salts upon Pigments,' *Gard. Chron.*, London, 1909, XLVI, pp. 97–98.
Note on effect of various salts in the soil upon flower-colour in *Hydrangea* and Roses. Lead acetate said to intensify red colour. — —

629. **Chodat, R.**, 'Sur des grappes de raisins panachées,' *Bul. Soc. Bot.*, Genève, 1909, I (sér. 2), pp. 359–363.
Description of cases among vines in which white grapes appear among coloured branches. Also individual berries which are partly purple and partly white. — —

630. **East, E. M.**, 'The Transmission of Variations in the Potato in Asexual Reproduction,' *Report Connecticut Experiment Station*, 1909–1910, pp. 119–160.
A record of the production asexually of white tubers as a variety which breeds true. — —

631. **Emerson, R. A.**, 'Inheritance of Color in the Seeds of the Common Bean, *Phaseolus vulgaris*,' *Annual Report Nebraska Agricultural Experiment Station*, 1909, XXII, pp. 67–101.
Full account of inheritance of pigmentation and mottling. pp. 159, **181**, 197

632. **Emerson, R. A.**, 'Factors for Mottling in Beans,' *Amer. Breed. Ass. Rep.*, Washington, D. C., 1909, V, pp. 368–376.
Suggestion is made that two factors are concerned in the production of mottling in beans. pp. 159, **181**, 197

633. ***Hildebrand, F.**, 'Die Veränderung der Blumenfarben durch die Kultur,' *Umschau*, Frankfurt a. M., 1909, XIII, pp. 612–615.
Author remarks on the appearance of different colour-varieties under cultivation. He contrasts some cases (*Primula acaulis, P. sinensis*) where blue varieties have arisen, with other cases (*Althaea, Antirrhinum, Calceolaria,*

Page of text on which reference is made

among others) where no blue variety has appeared. Sometimes variation only gives rise to different shades of the original colour. — —

634. **Hurst, C. C.,** 'Inheritance of Albinism in Orchids,' *Gard. Chron.*, London, 1909, XLV, pp. 81–82.
Production of colour (anthocyanin) by crossing albinos. pp. 158, 166, 168

635. **Lock, R. H.,** 'A Preliminary Survey of Species Crosses in the Genus *Nicotiana* from the Mendelian Standpoint,' *Ann. R. Bot. Gard.*, Ceylon, 1909, IV, pp. 195–227.
Dominance of blue colour in pollen-grains. pp. 158, 176

636. **Marryat, D. C. E.,** 'Hybridisation Experiments with *Mirabilis Jalapa*,' *Rep. Evol. Com. Roy. Soc.*, London, 1909, Rpt. V, pp. 32–50.
Inheritance of flower-colour in *Mirabilis* and Mendelian factors concerned. pp. 151, 158, **175**, 195, 200

637. **Weiss, F. E.,** 'Note on the Variability in the Colour of the Flowers of a *Tropaeolum* Hybrid,' *Mem. Lit. Phil. Soc.*, Manchester, 1909, LIV, No. 18, 5 pages.
Description of a *Tropaeolum* plant which bore both yellow and red flowers. The offspring, from selfing, consisted of red-flowered and yellow-flowered individuals, and of these, one varied like the original parent. This variation to redness appeared to be affected by weather conditions. p. 159

638. **Wheldale, M.,** 'Further Observations upon the Inheritance of Flower-colour in *Antirrhinum majus*,' *Rep. Evol. Com. Roy. Soc.*, London, 1909, Rpt. V, pp. 1–26.
Full account of the inheritance of flower-colour in *Antirrhinum*. pp. 12, 149, 154, 157, **162**, 195, 196

639. **1910. Baur, E.,** 'Vererbungs- und Bastardierungsversuche mit *Antirrhinum*,' *Zs. indukt. Abstammungslehre*, Berlin, 1910, III, pp. 34–98.
Full account of inheritance of flower-colour in *Antirrhinum*. pp. 12, 157, **162**

640. **Correns, C.,** 'Der Uebergang aus dem homozygotischen in einen heterozygotischen Zustand im selben Individuum bei buntblättrigen und gestreiftblühenden *Mirabilis*-Sippen,' *Ber. D. bot. Ges.*, Berlin, 1910, XXVIII, pp. 418–434.
Connection between striped and whole-coloured flowers in inheritance is included. pp. 158, 175, **204**

641. **East, E. M.,** 'Inheritance in Potatoes,' *Amer. Nat.*, New York, 1910, XLIV, pp. 424–430.
Inheritance of anthocyanin in potatoes. pp. 159, 188

642. **Gates, R. R.,** 'The Material Basis of Mendelian Phenomena,' *Amer. Nat.*, New York, 1910, XLIV, pp. 203–213.

Page of text on which reference is made

Suggestion of a quantitative inheritance of pigment in *Oenothera*. pp. 158, 177

643. **Hurst, C. C.**, 'Mendel's Law of Heredity and its Application to Horticulture,' *J. R. Hort. Soc.*, London, 1910, XXXVI, pp. 22–52.
Mendel's work, cases of Sweet Pea and *Antirrhinum* given in detail. Instances of coloured F_1 produced by mating of two albino Orchids. pp. 158, 166, 168

644. **Keeble, F.**, and **Pellew, C.**, 'White Flowered Varieties of *Primula sinensis*,' *J. Genet.*, Cambridge, 1910, I, pp. 1–5.
A record of the existence of a recessive white-flowered *Primula* with coloured (anthocyanin) stems. pp. 159, 185, 186, 192

645. **Keeble, F.**, **Pellew, C.**, and **Jones, W. N.**, 'The Inheritance of Peloria and Flower-colour in Foxgloves (*Digitalis purpurea*),' *N. Phytol.*, London, 1910, IX, pp. 68–77.
Inheritance of flower-colour (anthocyanin). pp. 158, **168**, 191, 196

646. **Ostenfeld, C. H.**, 'Further Studies on the Apogamy and Hybridization of the *Hieracia*,' *Zs. indukt. Abstammungslehre*, Berlin, 1910, III, pp. 241–285.
Inheritance of anthocyanin in some of the hybrids concerned. p. 158

647. **Salaman, R. N.**, 'The Inheritance of Colour and other Characters in the Potato,' *J. Genet.*, Cambridge, 1910, I, pp. 7–46.
Inheritance of colour (anthocyanin) in the Potato. pp. 30, 156, 159, **188**

648. **Shull, G. H.**, 'Germinal Analysis through Hybridization,' *Proc. Amer. Phil. Soc.*, Philadelphia, Pa., 1910, XLIX, pp. 281–290.
Consideration of nature, chemical and otherwise, of unit characters. — —

649. **Shull, G. H.**, 'Color Inheritance in *Lychnis dioica*,' *Amer. Nat.*, New York, 1910, XLIV, pp. 83–91.
Inheritance of colour in *Lychnis* flowers. pp. 158, **174**

650. **Thoday (Sykes), M. G.**, and **Thoday, D.**, 'On the Inheritance of the Yellow Tinge in Sweet Pea Colouring,' *Proc. Phil. Soc.*, Cambridge, 1910, XVI, pp. 71–84.
Some cases of inheritance of anthocyanin are included. pp. 153, 158, **172**

651. **Wheldale, M.**, 'Die Vererbung der Blütenfarbe bei *Antirrhinum majus*,' *Zs. indukt. Abstammungslehre*, Berlin, 1910, III, pp. 321–333.
Comparison of results obtained by the author with those of Baur (see Nos. 638 and 639). pp. 149, 154, 157, **162**

Page of text on which reference is made

652. **1911. Bateson, W.**, and **Punnett, R. C.**, 'On the Inter-relations of Genetic Factors,' *Proc. R. Soc.*, London, 1911, LXXXIV B, pp. 3–8.
Account of certain 'couplings' which involve some colour factors. — —

653. **Bateson, W.**, and **Punnett, R. C.**, 'On Gametic Series involving Reduplication of Certain Terms,' *J. Genet.*, Cambridge, 1911, I, pp. 293–302.
On the so-called 'coupling' between colour and other factors. — —

654. **Baur, E.**, *Einführung in die experimentelle Vererbungslehre*, Berlin, 1911.
General work on Genetics. Contains details of experiments on inheritance of flower-colour, especially in *Antirrhinum*. — —

655. **Colgan, N.**, 'On the Inheritance of pitted Leaf-blotchings in *Arum maculatum*,' *Irish Nat.*, Dublin, 1911, XX, pp. 210–217.
Segregation among offspring of a blotched plant. pp. 157, **164**, 196

656. **East, E. M.**, and **Hayes, H. K.**, 'Inheritance in Maize,' *Bulletin Connecticut Agricultural Station*, 1911, No. 167, 142 pages.
General account of an inheritance of anthocyanin in Maize. pp. 159, **190**, 193

657. **Emerson, R. A.**, 'Genetic Correlation and Spurious Allelomorphism in Maize,' *Annual Report Nebraska Agricultural Experiment Station*, Nebraska, 1911, XXIV, pp. 59–90.
Inheritance of anthocyanin in separate parts of Maize inflorescence, both independently and otherwise. Discussion of the same. pp. 159, **193**

658. **Gates, R. R.**, 'Studies on the Variability and Heritability of Pigmentation in *Oenothera*,' *Zs. indukt. Abstammungslehre*, Berlin, 1911, IV, pp. 337–371.
Experiments and discussion upon the quantitative inheritance of anthocyanin in *Oenothera rubrinervis* and *Oenothera rubricalyx*. pp. 158, **177**, 192

659. ***Gawalowsky, A.**, 'Künstliche Blatt- und Blütenfärbungen,' *Wiener landw. Zeit.*, 1911, LXI, p. 616.
Author finds that treatment of soil with sodium orthophosphate gives a deep red or blue-violet colour to petals of *Rosa centifolia*; also to *Malva tinctoria* a deep brown-red. Potassium carbonate causes development of anthocyanin in *Lactuca sativa*. — —

Page of text on which reference is made

660. **Gregory, R. P.**, 'Experiments with *Primula sinensis*,' *J. Genet.*, Cambridge, 1911, I, pp. 73–132.
Full account of inheritance of anthocyanin in *Primula*.
pp. 150, 154, 155, 159, **184**, 193, 195, 197, 205

661. **Gregory, R. P.**, 'On Gametic Coupling and Repulsion in *Primula sinensis*,' *Proc. R. Soc.*, London, 1911, LXXXIV B, pp. 12–15.
'Coupling' in *Primula* involving colour factors. — —

662. **Kajanus, B.**, 'Genetische Studien an *Beta*,' *Zs. indukt. Abstammungslehre*, Berlin, 1911, VI, pp. 137–179.
Inheritance of anthocyanin in root and leaves of *Beta*.
pp. 157, **164**, 191

663. ***Kajanus, B.**, 'Ueber die Farben der Blüten und Samen von *Trifolium pratense*,' *Fühlings landw. Ztg.*, Stuttgart, 1911, pp. 763–776.
Inheritance of flower- and seed-colour (anthocyanin). — —

664. **Martin Leake, H.**, 'Studies in Indian Cotton,' *J. Genet.*, Cambridge, 1911, I, pp. 205–271.
Inheritance of colour (anthocyanin) in the leaf and flower. pp. 158, **169**, 191, 194, 195, 197

665. **Saunders, E. R.**, 'Further Experiments on the Inheritance of "Doubleness" and other Characters in Stocks, *J. Genet.*, Cambridge, 1911, I, pp. 303–376.
Notes on inheritance of flower-colour are included. pp. 158, 174

666. **Saunders, E. R.**, 'On Inheritance of a Mutation in the Common Foxglove (*Digitalis purpurea*),' *N. Phytol.*, London, 1911, X, pp. 47–63.
Remarks on inheritance of spots (anthocyanin) on corolla. pp. 158, **168**, 191

667. **Spillman, W. J.**, 'Inheritance of the "Eye" in Vigna,' *Amer. Nat.*, New York, 1911, XLV, pp. 513–523. pp. 159, 188

668. **Tammes, T.**, 'Das Verhalten fluktuierend variierender Merkmale bei der Bastardierung,' *Rec. Trav. Bot. Néerl.*, Nijmegen, 1911, VIII, pp. 201–286.
References to inheritance of colour of flowers in varieties of *Linum*. pp. 158, **173**, 194

669. **Vries, H. de**, *The Mutation Theory*, Translated by J. B. Farmer and A. D. Darbishire, 1911. (*Die Mutationstheorie*, Leipzig, 1901.)
References to inheritance of pigmentation.
pp. 156, 157, 160, 196, 197, 199, 200, 205

670. **Weiss, F. E.**, 'Researches on Heredity in Plants,' *Mem. Lit. Phil. Soc.*, Manchester, 1911–1912, LVI, pp. 1–12.
Inheritance of colour in *Anagallis*. pp. 157, 159

671. **1912. Baur, E.**, 'Vererbungs- und Bastardierungsversuche mit

Page of text on which reference is made

Antirrhinum. 2. Faktorenkoppelung,' *Zs. indukt. Abstammungslehre*, Berlin, 1912, IV, pp. 201–216.
Reduplication in *Antirrhinum* involving factors for flower-colour. — —

672. **Burtt-Davy, J.**, 'Observations on the Inheritance of Characters in *Zea Mays*,' *Trans. R. Soc. S. Africa*, Cape Town, 1912, II, pp. 261–270.
Red pigment in pericarp and in aleurone layer. p. 159

673. **Collins, G. N.**, 'Gametic Coupling as a Cause of Correlations,' *Amer. Nat.*, New York, 1912, XLVI, pp. 569–590. p. 159

674. **Compton, R. H.**, 'Preliminary Note on the Inheritance of Self-sterility in *Reseda odorata*,' *Proc. Phil. Soc.*, Cambridge, 1912–1914, XVII, p. 7.
Anthocyanin in pollen (orange) dominant to its absence (yellow). pp. 159, **187**

675. **East, E. M.**, 'Inheritance of Color in the Aleurone Cells of Maize,' *Amer. Nat.*, New York, 1912, XLVI, pp. 363–365.
Production of purple colour in aleurone layer is due to presence of factor, P, in addition to C and R. p. 159

676. **Emerson, R. A.**, 'The Unexpected Occurrence of Aleurone Colors in F_2 of a Cross between Non-colored Varieties of Maize,' *Amer. Nat.*, New York, 1912, XLVI, pp. 612–615.
F_2, from mating of two whites, contained both red and purple grains. Two factors are necessary for red colour, and a third for purple: colour is also inhibited by a fourth factor. Hence it is possible to show how the present result is obtained. p. 159

677. **Finlow, R. S.**, and **Burkill, I. H.**, 'The Inheritance of Red Colour, and the Regularity of Self-fertilisation, in *Corchorus capsularis*, Linn.,—the Common Jute Plant,' *Mem. Depart. Agric. India, Bot.*, London and Calcutta, 1912, IV, pp. 73–92.
Inheritance of anthocyanin in stem and leaf. pp. 158, **167**, 191, 194

678. **Gates, R. R.**, 'Mutations in Plants,' *The Botanical Journal*, 1912.
Descriptions of the mutant, *Oenothera rubrinervis*. pp. 158, 177

679. **Hedrick, U. P.**, and **Wellington, R.**, 'An Experiment in Breeding Apples,' New York Agricultural Experiment Station, Geneva, N. Y., 1912, *Bull.* 350, pp. 141–186.
Pigmentation in apples. — —

680. **Heribert-Nilsson, N.**, 'Die Variabilität der *Oenothera Lamarckiana* und das Problem der Mutation,' *Zs. indukt. Abstammungslehre*, Berlin, 1912, VIII, pp. 89–231.
References to inheritance of anthocyanin. p. 158

Page of text on which reference is made

681. **Heribert-Nilsson, N.**, 'Ärftlighetsförsök med blomfärgen hos *Anagallis arvensis*,' *Bot. Not.*, Lund, 1912, pp. 229–235.
Inheritance of flower-colour in *Anagallis*. pp. 157, 160

682. **Hill, A. W.**, 'The History of *Primula obconica*, Hance, under Cultivation, with some Remarks on the History of *Primula sinensis*, Sab.' *J. Genet.*, Cambridge, 1912, II, pp. 1–20.
History of variation in flower-colour is involved. pp. 149, 153

683. **Jones, W. N.**, 'Species Hybrids of *Digitalis*,' *J. Genet.*, Cambridge, 1912, II, pp. 71–88.
Inheritance of anthocyanin in hybrids *D. purpurea* and *D. grandiflora*. p. 158

684. **Kajanus, B.**, 'Ueber die Farben der Blüten und Samen von *Trifolium pratense*,' *Fühlings landw. Ztg.*, Stuttgart, 1912, IV, pp. 763–776.
Anthocyanin in flowers and seeds. pp. 153, 159, 188

685. **Kajanus, B.**, 'Mendelistische Studien an *Rüben*,' *Fühlings landw. Ztg.*, Stuttgart, 1912, IV, pp. 142–149.
Inheritance of anthocyanin in root of *Beta* and *Brassica* spp. pp. 157, **164**, 165

686. **Kajanus, B.**, 'Genetische Studien an *Brassica*,' *Zs. indukt. Abstammungslehre*, 1912, VI, pp. 217–237.
Inheritance of anthocyanin in root. pp. 157, **165**

687. **Keeble, F.**, 'Gigantism in *Primula sinensis*,' *J. Genet.*, Cambridge, 1912, II, pp. 163–188.
Various results in colour inheritance are included. p. 159

688. **Lock, R. H.**, 'Notes on Colour Inheritance in Maize,' *Ann. R. Bot. Gard.*, Ceylon, 1912, V, pp. 257–264.
Reply to East and Hayes' criticism of Lock's previous results. p. 159

689. **Punnett, R. C.**, *Mendelism*, London, 1912.
References to inheritance of colour in plants. — —

690. **Salaman, R. N.**, 'A Lecture on the Hereditary Characters in the Potato,' *J. R. Hort. Soc.*, London, 1912, XXXVIII, pp. 34–39.
Inheritance of anthocyanin in flower and tuber. p. 159

691. **Saunders, E. R.**, 'On the Relation of *Linaria alpina* Type to its Varieties *concolor* and *rosea*,' *N. Phytol.*, London, 1912, XI, pp. 167–169.
Inheritance of anthocyanin in the flower. pp. 149, 158, **173**

692. **Saunders, E. R.**, 'Further Contribution to the Study of the Inheritance of Hoariness in Stocks (*Matthiola*),' *Proc. R. Soc.*, London, 1912, LXXXV B, pp. 540–545.
Further observations on the connection between hoariness and sap-colour (anthocyanin). It is shown that two

Page of text on which reference is made

factors are essential for hoariness, and that a certain relation exists between these and the factors for colour. p. 194

693. **Shull, G. H.**, 'The Primary Color-factors of *Lychnis* and Color-inhibitors of *Papaver Rhoeas*,' *Bot. Gaz.*, Chicago, Ill., 1912, LIV, pp. 120–135.
An account of the inheritance of anthocyanin in both genera. pp. 158, **174**, **178**, 197

694. **Trow, A. H.**, 'On the Inheritance of Certain Characters in the Common Groundsel—*Senecio vulgaris*, Linn.—and its Segregates,' *J. Genet.*, Cambridge, 1912, II, pp. 239–276.
Inheritance of anthocyanin in stem. pp. 159, **187**

695. **Tschermak, E. von**, 'Bastardierungsversuche an Levkojen, Erbsen und Bohnen mit Rücksicht auf die Faktorenlehre,' *Zs. indukt. Abstammungslehre*, Berlin, 1912, VII, pp. 81–234.
Inheritance of anthocyanin in flowers and seeds of the above plants. pp. 158, 159, 175, 181, 197

696. **1913. Bateson, W.**, *Problems of Genetics*, New Haven, London and Oxford, 1913.
Includes references to inheritance of pigmentation. p. 206

697. **Davis, B. M.**, 'Genetical Studies on *Oenothera*. 4. The Behaviour of Hybrids between *Oenothera biennis* and *O. grandiflora* in the Second and Third Generations,' *Amer. Nat.*, New York, 1913, XLVII, pp. 449–476, 547–571.
Inheritance of anthocyanin in stem papillae in cross *O. grandiflora* × *biennis* is included. pp. 158, **177**

698. **Emerson, R. A.**, 'The Possible Origin of Mutations in Somatic Cells,' *Amer. Nat.*, New York, 1913, XLVII, pp. 375–377.
Reference to striping of anthocyanin in Maize in connection with origin of mutations in somatic cells. — —

699. **Emerson, R. A.**, 'The simultaneous Modification of distinct Mendelian Factors,' *Amer. Nat.*, New York, 1913, XLVII, pp. 633–636. p. 159

700. **Haig Thomas, R.**, '*Nicotiana* Crosses,' *Comptes Rendus et Rapports, Conférence Internationale de Génétique, Paris*, 1911, Paris, 1913, pp. 450–461.
Inheritance of anthocyanin in *Nicotiana* species. pp. 158, 176

701. ***Holdefleiss, P.**, 'Ueber Züchtungs- und Vererbungsfragen bei Rotklee,' *Kühn Arch.*, Berlin, 1913, pp. 81–115.
Inheritance of colour in flower and seed of *Trifolium pratense*. — —

702. **Hurst, C. C.**, 'The Application of Genetics to Orchid Breeding,' *J. R. Hort. Soc.*, London, 1913, XXXVIII, pp. 412–429.

Page of text on which reference is made

Albinism in Orchids. Production of coloured F_1 by mating of two albinos, and other results connected with colour. pp. 158, 166, 168

703. ***Kajanus, B.**, 'Ueber die kontinuierlich violetten Samen von *Pisum arvense*,' *Fühlings landw. Ztg.*, Stuttgart, 1913, pp. 153–160.
Observations upon varieties with purple pigment in the testa. — —

704. **Kajanus, B.**, 'Ueber die Vererbungsweise gewisser Merkmale der *Beta-* und *Brassica*-Rüben,' *Zs. Pflanzenzüchtg*, Berlin, 1913, I, pp. 125–463.
Inheritance of anthocyanin is included. pp. 157, **165**

705. **Punnett, R. C.**, 'Reduplication Series in Sweet Peas,' *J. Genet.*, Cambridge, 1913, III, pp. 77–103.
Reduplications involving colour factors. — —

706. **Rawson, H. E.**, 'Variation of Structure and Colour of Flowers under Insolation,' *Rep. Brit. Ass.*, London, 1913, pp. 711–713.
Further observations on the variations in flower-colour produced by screening *Tropaeolum* flowers from the sun's rays at different times of the day. — —

707. **Saunders, E. R.**, 'On the mode of Inheritance of Certain Characters in Double-throwing Stocks, A Reply,' *Zs. indukt. Abstammungslehre*, Berlin, 1913, X, pp. 297–310.
Paper includes case of reduplication in factors for flower-colour involving full and pale colour (anthocyanin) and white and cream plastid colour. — —

708. **Spillman, W. J.**, 'Color Correlation in Cowpeas,' *Science*, N. S., New York, N. Y., 1913, XXXVIII, p. 302. pp. 159, 188

709. **1914. Cockerell, T. D. A.**, 'Suppression and Loss of Characters in Sunflowers,' *Science*, N. S., New York, N. Y., 1914, XL, pp. 283–285.
Variation in occurrence and dominance of red pigment over yellow. pp. 151, 155, 158, 170

710. **Cockerell, T. D. A.**, 'Sunflower Problems,' *Science*, N. S., New York, N. Y., 1914, XL, pp. 708–709.
Discussion as to nature of variation which produces red Sunflower. pp. 151, 155, 158, 170

711. **Davis, B. M.**, 'Genetical Studies on *Oenothera*. 5. Some Reciprocal Crosses of *Oenothera*,' *Zs. indukt. Abstammungslehre*, Berlin, 1914, XII, pp. 169–205.
Inheritance of anthocyanin in stem papillae and in sepals in crosses, *O. franciscana* × *biennis* and reciprocal, and *O. biennis* × *grandiflora* and reciprocal, is included. pp. 158, **177**

712. **Emerson, R. A.**, 'The Inheritance of a Recurring Somatic

Page of text on which reference is made

Variation in Variegated Ears of Maize,' *Amer. Nat.*, New York, 1914, XLVIII, pp. 87–115.

Inheritance of variegation of red colour in the pericarp. pp. 159, 201, **202**

713. **Gates, R. R.**, 'Breeding Experiments which show that Hybridisation and Mutation are independent Phenomena,' *Zs. indukt. Abstammungslehre*, Berlin, 1914, XI, pp. 209–279.

Inheritance of anthocyanin in calyx in cross, *O. grandiflora* × *O. rubricalyx* and reciprocal, is included. pp. 158, **177**

714. **Gregory, R. P.**, 'On the Genetics of Tetraploid Plants in *Primula sinensis*,' *Proc. R. Soc.*, London, 1914, LXXXVII B, pp. 484–492.

References to inheritance of anthocyanin are included. p. 159

715. **Honing, J. A.**, 'Experiments on Hybridisation with *Canna indica*,' *Proc. Sci. K. Akad. Wet.*, Amsterdam, 1914, XVI, pp. 835–841.

Inheritance of anthocyanin in stems, leaves, fruits, etc. of *Canna*. pp. 158, 166

716. **Kajanus, B.**, 'Zur Genetik der Samen von *Phaseolus vulgaris*,' *Zs. Pflanzenzüchtg*, Berlin, 1914, II, pp. 377–388. p. 159

717. **Kajanus, B.**, 'Ueber die Vererbung der Blütenfarbe von *Lupinus mutabilis* Swt.,' *Zs. indukt. Abstammungslehre*, Berlin, 1914, XII, pp. 57–58. p. 158

718. **Mann, A.**, 'Coloration of the Seed Coat of Cowpeas,' *J. Agric. Res.*, Washington, 1914, II, pp. 33–56. pp. 159, 188

719. **Richardson, C. W.**, 'A preliminary Note on the Genetics of *Fragaria*,' *J. Genet.*, Cambridge, 1914, III, pp. 171–177. pp. 158, **169**

720. **Shull, G. H.**, 'A Peculiar Negative Correlation in *Oenothera* Hybrids,' *J. Genet.*, Cambridge, 1914, IV, pp. 83–102.

Inheritance of anthocyanin in stems, leaves, calyx, etc. among offspring of selfed plants of *O. rubricalyx*, and from reciprocal crosses between *O. rubricalyx* and *O. Lamarckiana* and *O. rubrinervis*. pp. 158, **177**, 192

721. **Tammes, T.**, 'The Explanation of an apparent Exception to Mendel's Law of Segregation,' *Proc. Sci. K. Akad. Wet.*, Amsterdam, 1914, XVI, pp. 1021–1031.

Explanation of a deficiency of white-flowered plants in F_2 from a cross between blue- and white-flowered varieties of *Linum*. pp. 158, **173**

722. **1915. Cockerell, T. D. A.**, 'An Early Observation on the Red Sunflower,' *Science*, N. S., New York, N. Y., 1915, XLI, pp. 33–34.

Origin of the red Sunflower as a 'sport' in wild state.

Page of text on which reference is made

Connection between pigmentation of inflorescence and stem. pp. 151, 155, 158, 170, 192

723. **Cockerell, T. D. A.**, 'Variation in *Œnothera hewetti*,' *Science*, N.S., New York, N. Y., 1915, XLII, pp. 908–909. pp. 158, 177

724. **Cockerell, T. D. A.**, 'The Marking Factors in Sunflowers,' *J. Heredity*, Washington, D. C., 1915, VI, pp. 542–545. pp. 158, 171

725. **Emerson, R. A.**, 'Anomalous Endosperm Development in Maize and the Problem of Bud Sports,' *Zs. indukt. Abstammungslehre*, Berlin, 1915, XIV, pp. 241–259. pp. 159, **206**

726. **Fruwirth, C.**, 'Versuche zur Wirkung der Auslese,' *Zs. Pflanzenzüchtg*, Berlin, 1915, III, pp. 173–224.
Certain cases of pigmentation are included. — —

727. **Gates, R. R.**, 'On the Origin and Behaviour of *Oenothera rubricalyx*,' *J. Genet.*, Cambridge, 1915, IV, pp. 353–360. pp. 158, 177

728. **Gates, R. R.**, 'On the Modification of Characters by Crossing,' *Amer. Nat.*, New York, 1915, XLIX, pp. 562–569. pp. 158, 177

729. **Gilbert, A. W.**, 'Heredity of Color in *Phlox Drummondii*,' *J. Agric. Res.*, Washington, D. C., 1915, IV, pp. 293–301. pp. 159, **182**, 197

730. **Gmelin, H. M.**, 'Erste Reihe von Untersuchungen über die Bestäubungs- und Befruchtungsverhältnisse beim Rotklee,' *Zs. Pflanzenzüchtg*, Berlin, 1915, III, pp. 67–75. p. 159

731. **Hallqvist, C.**, 'Brassica-Kreuzungen,' *Bot. Not.*, Lund, 1915, pp. 97–112. p. 157

732. **Heribert-Nilsson, N.**, 'Elminierung der positiven Homozygoten bezüglich der Rotnervigkeit bei *Oenothera Lamarckiana*,' *Bot. Not.*, Lund, 1915, pp. 23–25. pp. 158, 177

733. **Heribert-Nilsson, N.**, 'Die Spaltungserscheinungen der *Oenothera Lamarckiana*,' *Univ. Årsskr.*, N. F., Lund, 1915, Avd. 2, Bd 12, 131 pages. pp. 158, 177

734. ***Honing, J. A.**, 'Kreuzungsversuche mit *Canna*-Varietäten,' *Rec. Trav. Bot. Néerl.*, Groningen, 1915, XII, 26 pages. pp. 158, 166

735. **Küster, E.**, 'Ueber Anthocyan-Zeichnung und Zellen-Mutation,' *Ber. D. bot. Ges.*, Berlin, 1915, XXXIII, pp. 536–537. p. 200

736. **Nohara, S.**, 'Genetical Studies on *Oxalis*,' *J. Coll. Agric.*, Tokyo, 1915, VI, pp. 165–181. pp. 158, **178**

737. **Rasmuson, H.**, 'Zur Vererbung der Blütenfarben bei der Balsamine,' *Bot. Not.*, Lund, 1915, pp. 79–83. pp. 158, **171**

738. **Reinke, J.**, 'Eine bemerkenswerte Knospenvariation der Feuerbohne nebst allgemeinen Bemerkungen über Allogonie,' *Ber. D. bot. Ges.*, Berlin, 1915, XXXIII, pp. 324–348. p. 205

Page of text on which reference is made

739. **Tammes, T.**, 'Die genotypische Zusammensetzung einiger Varietäten derselben Art und ihr genetischer Zusammenhang,' *Rec. Trav. Bot. Néerl.*, Groningen, 1915, XII, pp. 217–277. pp. 158, **173**

740. **1916. Bateson, W.**, 'Root-cuttings, Chimaeras and "Sports,"' *J. Genet.*, Cambridge, 1916, VI, pp. 75–80. p. 206

741. **Fujii, K.**, and **Kuwada, Y.**, 'On the Composition of Factorial Formulae for Zygotes in the Study of Inheritance of Seed-Characters of *Zea Mays*, L., with Notes on Seed-pigments,' *Bot. Mag.*, Tokyo, 1916, XXX, pp. 83–88. p. 159

742. **Graham, R. J. D.**, 'Pollination and Cross-fertilization in the Juar Plant (*Andropogon sorghum*, Brot.),' *Mem. Dept. Agric. India, Bot. Ser.*, Calcutta, 1916, VIII, pp. 201–216. pp. 157, **160**, 193

743. **Hector, G. P.**, 'Observations on the Inheritance of Anthocyan Pigment in Paddy Varieties,' *Mem. Dept. Agric. India, Bot. Ser.*, Calcutta, 1916, VIII, pp. 89–101. pp. 158, 177

744. ***Malinowski, E.**, i **Sachsowa, M.**, 'O driedziczenia barw i kształtów kwiatu u Petunii' (Ueber die Veränderung der Farben und der Gestalt bei Petunia Bluten), *Ber. Wiss. Ges.*, Warschau, 1916, IX, pp. 865–894. p. 159

745. **Rasmuson, H.**, 'Zur Vererbung der Blütenfarben bei *Malope trifida*,' *Bot. Not.*, Lund, 1916, V, pp. 227–230. pp. 158, **174**, 193

746. **Rasmuson, H.**, 'Kreuzungsuntersuchungen bei Reben,' *Zs. indukt. Abstammungslehre*, Berlin, 1916, XVII, pp. 1–52. pp. 159, **189**, 193

747. **Schellenberg, H. C.**, 'Die Vererbungsverhältnisse von Rassen mit gestreifte Blüten und Früchten,' *Vierteljahrsch. Natf. Ges.*, Zürich, 1916, LXI, pp. xxix–xxx. — —

748. **Tammes, T.**, 'On the Mutual Effect of Genotypic Factors,' *Proc. Sci. K. Akad. Wet.*, Amsterdam, 1916, XVIII, pp. 1056–1067. pp. 158, **173**

749. **Tammes, T.**, 'Die gegenseitige Wirkung genotypischer Faktoren,' *Rec. Trav. Bot. Néerl.*, Groningen, 1916, XIII, p. 44. pp. 158, **173**

750. **Tschermak, E. von**, 'Ueber den gegenwärtigen Stand der Gemüsezüchtung,' *Zs. Pflanzenzüchtg*, Berlin, 1916, IV, pp. 65–104.
Inheritance of anthocyanin in vegetables. — —

751. **Vargas Eyre, J.**, and **Smith, G.**, 'Some Notes on the Linaceae. The Cross Pollination of Flax,' *J. Genet.*, Cambridge, 1916, V, pp. 189–197. pp. 158, 173

752. **1917. Barker, E. E.**, 'Heredity Studies in the Morning-Glory

Page of text on which reference is made

(*Ipomoea purpurea* [L.] Roth.),' *Cornell Univ. Agric. Expt. Sta. Bull.*, Ithaca, 1917, No. 392, 38 pages. pp. 158, **171**

753. **Cockerell, T. D. A.**, 'Sunflower Seedlings,' *J. Heredity*, Washington, D. C., 1917, VIII, pp. 361–362. pp. 158, **171**

754. **Cockerell, T. D. A.**, 'Somatic Mutations in Sunflowers,' *J. Heredity*, Washington, D. C., 1917, VIII, pp. 467–470. p. 200

755. ***Emerson, R. A.**, 'Genetical Studies of Variegated Pericarp in Maize,' *Genetics*, Princeton, N. J., 1917, II, pp. 1–35. p. 159

756. **Gates, R. R.**, 'Vegetative Segregation in a Hybrid Race,' *J. Genet.*, Cambridge, 1917, VI, pp. 237–253. — —

757. ***Hayes, H. K.**, 'Inheritance of a Mosaic Pericarp Pattern Color in Maize,' *Genetics*, Princeton, N. J., 1917, II, pp. 261–281. p. 159

758. **Kajanus, B.**, 'Ueber Bastardierungen zwischen *Brassica Napus* L. und *Brassica Rapa* L.,' *Zs. Pflanzenzüchtg*, Berlin, 1917, V, pp. 265–322. pp. 157, **165**

759. **Kajanus, B.**, 'Ueber die Farbenvariation der *Beta*-Rüben,' *Zs. Pflanzenzüchtg*, Berlin, 1917, V, pp. 357–372. pp. 157, **165**

760. **Kempton, J. H.**, 'A Correlation between Endosperm Color and Albinism in Maize,' *J. Wash. Acad. Sci.*, Baltimore, Md, 1917, VII, pp. 146–149. p. 159

761. **Lindstrom, E. W.**, 'Linkage in Maize: Aleurone and Chlorophyll Factors,' *Amer. Nat.*, New York, 1917, LI, pp. 225–237. pp. 159, 190

762. ***Lundberg, J. F.**, och **Aakerman, A.**, 'Jakttagelser rörande fröfärgen hos avkommen av en spontan korsning mellan tvenne former av *Phaseolus vulgaris*' (Observations on the Colour of Seeds originating from spontaneous Crossing between two Forms of *Phaseolus vulgaris*), *Sver. utsädes. Tidskr.*, Svälof, 1917, XXVII, pp. 115–121. p. 159

763. **Parnell, F. R.**, **Ayyangar, G. N. R.**, and **Ramiah, K.**, 'The Inheritance of Characters in Rice. I,' *Mem. Dept. Agric. India, Bot. Ser.*, Calcutta, 1917, IX, pp. 75–105. pp. 158, **177**

764. **Pellew, C.**, 'Types of Segregation,' *J. Genet.*, Cambridge, 1917, VI, pp. 317–339. pp. 157, **166**

765. **Punnett, R. C.**, 'Reduplication Series in Sweet Peas. II,' *J. Genet.*, Cambridge, 1917, VI, pp. 185–193. pp. 158, **173**

766. **Roemer, Th.**, 'Ueber Farbenabweichungen bei Zuckerrüben,' *Zs. Pflanzenzüchtg*, Berlin, 1917, V, pp. 381–391. pp. 157, 165

767. **White, O. E.**, 'Inheritance Studies in Pisum. IV. Interrelation of the Genetic Factors of Pisum,' *J. Agric. Res.*, Washington, D. C., 1917, XI, pp. 167–190. pp. 159, **182**

768. **White, O. E.**, 'Studies of Inheritance in Pisum. II. The

Page of text on which reference is made

Present State of Knowledge of Heredity and Variation in Peas,' *Proc. Amer. Phil. Soc.*, Philadelphia, Pa., 1917, LVI, pp. 487–588. pp. 159, **182**

769. **1918. Becker, J.**, 'Vererbung gewisser Blütenmerkmale bei *Papaver Rhoeas* L.,' *Zs. Pflanzenzüchtg*, Berlin, 1918, VI, pp. 215–221. pp. 158, **178**

770. **Bregger, T.**, 'Linkage in Maize: The *C* Aleurone Factor and Waxy Endosperm,' *Amer. Nat.*, New York, 1918, LII, pp. 57–61. pp. 159, 190

771. **Cobb, F.**, and **Bartlett, H. H.**, 'Purple Bud Sport on Pale flowered Lilac (*Syringa persica*),' *Bot. Gaz.*, Chicago, Ill., 1918, LXV, pp. 560–562. p. 205

772. **Cockerell, T. D. A.**, 'The Story of the Red Sunflower,' *Amer. Mus. J.*, New York, 1918, XVIII, pp. 38–47. pp. 158, **171**

773. **Dahlgren, K. V. O.**, 'Ueber einige Kreuzungsversuche mit *Chelidonium majus* L., *Polemonium coeruleum* L. und *Lactuca muralis* L.,' *Sv. Bot. Tidskr.*, Stockholm, 1918, XII, pp. 103–110. pp. 158, 159, **171**, 183

774. **Emerson, R. A.**, 'A fifth Pair of Factors, *Aa*, for Aleurone Color in Maize, and its Relation to the *Cc* and *Rr* Pairs,' *Cornell Univ. Agric. Exp. Sta. Mem.*, Ithaca, N. Y., 1918, No. 16. pp. 159, 190

775. **Küster, E.**, 'Die Verteilung des Anthocyans bei Coleus-spielarten,' *Flora*, Jena, 1918, N. F., X, pp. 1–33. p. 200

776. **Küster, E.**, 'Ueber Mosaikpanaschierung und vergleichbare Erscheinungen,' *Ber. D. bot. Ges.*, Berlin, 1918, XXXVI, pp. 54–61. p. 200

777. **Miyazawa, B.**, 'Studies of Inheritance in the Japanese *Convolvulus*,' *J. Genet.*, Cambridge, 1918, VIII, pp. 59–82. pp. 158, **171**, 197

778. **Rasmuson, H.**, 'Zur Genetik der Blütenfarben von *Tropaeolum majus*,' *Bot. Not.*, Lund, 1918, pp. 253–259. pp. 159, **188**

779. **Rasmuson, H.**, 'Ueber eine Petunia-Kreuzung,' *Bot. Not.*, Lund, 1918, pp. 287–294. pp. 159, **179**

780. **Richardson, C. W.**, 'A further Note on the Genetics of *Fragaria*,' *J. Genet.*, Cambridge, 1918, VII, pp. 167–170. pp. 158, 169

781. **St Clair Caporn, A.**, 'On a Case of permanent Variation in the Glume Lengths of Extracted Parental Types and the Inheritance of Purple Colour in the Cross *Triticum Polonicum* × *T. eloboni*,' *J. Genet.*, Cambridge, 1918, VII, pp. 259–280. pp. 159, **188**, 193

782. ***Shaw, J. K.**, and **Norton, J. B.**, 'The Inheritance of Seed-Coat colors in Garden Beans,' *Agric. Exp. Sta. Mass. Bull.*, 1918, No. 185, pp. 58–194. pp. 159, 181

Page of text on which reference is made

783. **White, O. E.**, 'Inheritance Studies on Castor Beans,' *Mem. Bot. Garden*, Brooklyn, N. Y., 1918, I, pp. 513–520. pp. 159, 187

784. ***White, O. E.**, 'Breeding New Castor Beans,' *J. Heredity*, Washington, D. C., 1918, IX, pp. 195–200. pp. 159, 187

785. **1919. Allard, H. A.**, 'Some Studies in Blossom Color Inheritance in Tobacco, with special Reference to *N. sylvestris* and *N. tabacum*,' *Amer. Nat.*, New York, 1919, LIII, pp. 79–84. pp. 158, **176**

786. **Bach, S.**, 'Zur näheren Kenntnis der Faktoren der Anthozyanbildung bei *Pisum*,' *Zs. Pflanzenzüchtg*, Berlin, 1919, VII, pp. 64–66. pp. 159, **182**

787. ***Collins, J. L.**, 'Chimeras in Corn Hybrids,' *J. Heredity*, Washington, D. C., 1919, X, pp. 2–10. p. 159

788. **Harland, S. C.**, 'Inheritance of Certain Characters in the Cowpea (*Vigna sinensis*),' *J. Genet.*, Cambridge, 1919, VIII, pp. 101–132. pp. 159, **188**, 193

789. **Jones, D. F.**, and **Gallastegui, C. A.**, 'Some Factor Relations in Maize with Reference to Linkage,' *Amer. Nat.*, New York, 1919, LIII, pp. 239–246. p. 159

790. **Kajanus, B.**, 'Genetische Papaver-Notizen,' *Bot. Not.*, Lund, 1919, pp. 99–102. pp. 158, **178**

791. **Kajanus, B.**, 'Genetische Studien über die Blüten von *Papaver somniferum* L.,' *Ark. Bot.*, Stockholm, 1919, XV, pp. 1–87. pp. 158, **179**

792. **Kajanus, B.**, und **Berg, S. O.**, 'Pisum-Kreuzungen,' *Ark. Bot.*, Stockholm, 1919, XV, pp. 1–18. pp. 159, **183**

793. **Kempton, J. H.**, 'Inheritance of spotted Aleurone Color in Hybrids of Chinese Maize,' *Genetics*, Princeton, N. J., 1919, IV, pp. 261–274. pp. 159, 190

794. **Lindhard, E.**, und **Iversen, K.**, 'Vererbung von roten und gelben Farbenmerkmalen bei *Beta*-Rüben,' *Zs. Pflanzenzüchtg*, Berlin, 1919, VII, pp. 1–18. pp. 157, **165**

795. **Miyake, K.**, and **Imai, Y.**, 'On the Inheritance of Flower-Colour and other Characters in *Digitalis purpurea*,' *Bot. Mag.*, Tokyo, 1919, XXXIII, pp. 175–186. (Japanese.) pp. 158, **168**

796. **Rasmuson, H.**, 'Genetische Untersuchungen in der Gattung Godetia,' *Ber. D. bot. Ges.*, Berlin, 1919, XXXVII, pp. 399–403. pp. 158, **169**

797. **Rasmuson, H.**, 'Zur Frage von der Entstehungsweise der roten Zuckerrüben,' *Bot. Not.*, Lund, 1919, pp. 169–180. pp. 157, **165**

798. **Raum, J.**, 'Ein weiterer Versuch über die Vererbung der Samenfarbe bei Rotklee,' *Zs. Pflanzenzüchtg*, Berlin, 1919, VII, pp. 149–155. pp. 159, **188**

Page of text on which reference is made

799. **Stout, A. B.**, 'Bud Variation,' *Proc. Nation. Acad. Sci.*, Baltimore, M. D., 1919, v, pp. 130–134. p. 207

800. ***Stuckey, H. P.**, 'Work with Vitis rotundifolia, a species of Muscadine Grapes,' *Georgia Agric. Exp. Sta. Bull.*, 1919, No. 133, pp. 60–74. p. 159

801. **Tjebbes, K.**, en **Kooiman, H. N.**, 'Erfelijkheidsonderzoekingen bij boonen (Genetical Experiments with Beans),' *Genetica*, 's Gravenhage, 1919, I, pp. 323–346. pp. 159, 181

802. **von Tschermak, E.**, 'Beobachtungen über anscheinende vegetative Spaltungen an Bastarden und über anscheinende Spätspaltungen von Bastardnachkommen, speziell Auftreten von Pigmentierungen an sonst pigmentlosen Deszendenten,' *Zs. indukt. Abstammungslehre*, Berlin, 1919, XXI, pp. 216–232. — —

803. **1920. Coulter, M. C.**, 'Inheritance of Aleurone Color in Maize,' *Bot. Gaz.*, Chicago, Ill., 1920, LXIX, pp. 407–425. p. 159

804. **Frimmel, Fr.**, 'Ueber einen Versuch der Züchtung schwarzer Farbentöne an der Gartenprimel,' *Zs. Pflanzenzüchtg*, Berlin, 1920, VII, pp. 346–356.
Attempt to obtain 'black' Primula by crossing suitable varieties. — —

805. **Harland, S. C.**, 'Inheritance of certain Characters in the Cowpea (*Vigna sinensis*). II,' *J. Genet.*, Cambridge, 1920, X, pp. 193–205. pp. 159, **189**, 193

806. **Harland, S. C.**, 'Inheritance in *Ricinus communis*, L. Part I,' *J. Genet.*, Cambridge, 1920, X, pp. 207–218. pp. 159, **187**

807. **Harland, S. C.**, 'Inheritance in *Dolichos lablab*, L. Part I,' *J. Genet.*, Cambridge, 1920, X, pp. 219–226. pp. 158, **169**, 193

808. **Kelly, J. P.**, 'A genetical Study of Flower Form and Flower Color in *Phlox Drummondii*,' *Genetics*, Princeton, N. J., 1920, V, pp. 189–248. pp. 159, **182**, 197

809. **Martin Leake, H.**, and **Ram Pershad, B.**, 'A preliminary Note on the Flower Colour and associated Characters of the Opium Poppy,' *J. Genet.*, Cambridge, 1920, X, pp. 1–20. pp. 158, **178**, 193, 194

810. **Meunissier, A.**, 'Observations faites à Verrières par Philippe de Vilmorin, sur le caractère "hile noir" chez le pois,' *J. Genet.*, Cambridge, 1920, X, pp. 53–60. pp. 159, 182

811. **Miyake, K.**, and **Imai, Y.**, 'On the Inheritance of Flower Colour and other Characters in *Digitalis purpurea*,' *J. Col. Agric.*, Tokyo, 1920, VI, pp. 391–402. pp. 158, **168**

812. **Rasmuson, H.**, 'Ueber einige genetische Versuche mit *Papaver Rhoeas* und *Papaver laevigatum*,' *Hereditas*, Lund, 1920, I, pp. 107–115. pp. 158, **178**

Page of text on which reference is made

813. **Rasmuson, H.**, 'On some Hybridisation Experiments with Varieties of Collinsia species,' *Hereditas*, Lund, 1920, I, pp. 178–185. pp. 158, **167**, 193, 196

814. **Rasmuson, H.**, 'Die Hauptergebnisse von einigen genetischen Versuchen mit verschiedenen Formen von Tropaeolum, Clarkia und Impatiens,' *Hereditas*, Lund, 1920, I, pp. 270–276. pp. 158, 159, **167**, **171**, **188**

815. **Richardson, C. W.**, 'Some Notes on Fragaria,' *J. Genet.*, Cambridge, 1920, X, pp. 39–46. pp. 158, 169

816. **Riolle, Y. T.**, 'Les hybrides de *Raphanus*,' *Rev. gén. bot.*, Paris, 1920, XXXII, pp. 438–447. pp. 159, **187**

817. **Sirks, M. J.**, 'De Analyse van een spontane boonenhybride,' *Genetica*, 's Gravenhage, 1920, II, pp. 97–114. pp. 159, 181

818. ***Sylvén, N.**, 'Om själv- och korsbefruktning hos rapsen,' *Sver. utsädes. Tidskr.*, Svälof, 1920, pp. 225–244. — —

819. **Tedin, H.**, 'The Inheritance of Flower Colour in Pisum,' *Hereditas*, Lund, 1920, I, pp. 68–97. pp. 159, **182**, 194

820. **Yasui, K.**, 'Genetical Studies in *Portulaca grandiflora*,' *Bot. Mag.*, Tokyo, 1920, XXXIV, pp. 55–65. pp. 159, **183**

821. **1921. Anderson, E. G.**, 'The Inheritance of Salmon Silk Color in Maize,' *Cornell Univ. Agric. Exp. Sta. Mem.*, Ithaca, N. Y., 1921, No. 48, pp. 539–554. pp. 159, 190

822. **Bateson, W.**, 'Root-cuttings and Chimaeras. II,' *J. Genet.*, Cambridge, 1921, XI, pp. 91–97. p. 206

823. **Blakeslee, A. F.**, 'A chemical Method of distinguishing genetic Types of yellow Cones of Rudbeckia,' *Zs. indukt. Abstammungslehre*, Berlin, 1921, XXV, pp. 211–221. pp. 159, **187**

824. **Burlingame, L. L.**, 'Variation and Heredity in Lupinus,' *Amer. Nat.*, New York, LV, pp. 427–448. p. 158

825. **Clausen, R. E.**, and **Goodspeed, T. H.**, 'Inheritance in *Nicotiana tabacum*. II. On the Existence of Genetically Distinct Red-flowering Varieties,' *Amer. Nat.*, New York, 1921, LV, pp. 328–334. pp. 158, **176**

826. **Emerson, R. A.**, 'The Genetic Relations of Plant Colors in Maize,' *Cornell Univ. Agric. Exp. Sta. Mem.*, 1921, No. 39, 156 pages. pp. 159, **190**

827. **Emerson, R. A.**, 'Genetic Evidence of Aberrant Chromosome Behaviour in Maize Endosperm,' *Amer. J. Bot.*, Brooklyn, N. Y., 1921, VIII, pp. 411–424. p. 159

828. **Hallqvist, C.**, 'The Inheritance of the Flower Colour and the Seed Colour in *Lupinus angustifolius*,' *Hereditas*, Lund, 1921, II, pp. 299–363. pp. 158, **174**, 193

829. ***Hutchison, C. B.**, 'Heritable Characters of Maize. VII. Shrunken Endosperm,' *J. Heredity*, Washington, D.C., 1921, XII, pp. 76–83. p 159

Page of text on which reference is made

830. **Ikeno, S.**, 'Studies on the Genetics of Flower-Colours in *Portulaca grandiflora*,' *J. Coll. Agric.*, Tokyo, 1921, VIII, pp. 93–133. pp. 159, **183**, 193

831. **Kempton, J. H.**, 'Linkage between brachytic Culms and Pericarp and Cob Color in Maize,' *J. Wash. Acad. Sci.*, Baltimore, M. D., 1921, XI, pp. 13–20. p. 159

832. ***Kristofferson, K. B.**, 'Undersökning av F_1 och F_2 generationerna av en spontan bastard mellan vitkal och grönkal (mit deutschen Résumé),' *Sver. utsädes. Tidskr.*, Svälof, 1921, pp. 31–52. pp. 157, 166

833. **Miyake, K.**, and **Imai, Y.**, 'On the Inheritance of Flower-Colour in *Sisyrinchium angustifolium*,' *Bot. Mag.*, Tokyo, 1921, XXXV, pp. 261–265. pp. 159, **188**

834. **Miyazawa, B.**, 'Studies of Inheritance in the Japanese Convolvulus. Part II,' *J. Genet.*, Cambridge, 1921, XI, pp. 1–15. pp. 158, **171**, 193, 196

835. **Nagai, I.**, 'A Genetico-Physiological Study on the Formation of Anthocyanin and Brown Pigments in Plants,' *J. Coll. Agric.*, Tokyo, 1921, VIII, pp. 1–92. p. 124

836. **Rasmuson, H.**, 'Beiträge zu einer genetischen Analyse zweier Godetia-arten und ihrer Bastarde,' *Hereditas*, Lund, 1921, II, pp. 143–289. pp. 158, **169**, 195, 196

837. **Raum, J.**, 'Weissblühender Rotklee eine "umschlagende Sippe"?' *Zs. Pflanzenzüchtg*, Berlin, 1921, VIII, pp. 73–77. pp. 159, 188

838. **Renner, O.**, 'Das Rotnervenmerkmal der Önotheren,' *Ber. D. bot. Ges.*, Berlin, 1921, XXXIX, pp. 264–270. p. 158

839. **Sachs-Skalińska, M.**, 'Krzyżowanie ras wielopostaciowych' (Croisement des races polymorphes), *Mém. Inst. Génétique*, Varsovie, 1921, I, pp. 34–46. pp. 159, 179

840. **Sachs-Skalińska, M.**, 'Badania nad mieszańcami Tytuniu' (Recherches sur les hybrides du Nicotiana), *Mém. Inst. Génétique*, Varsovie, 1921, I, pp. 47–122. pp. 158, 176

841. **Tjebbes, K.**, en **Kooiman, H. N.**, 'Erfelijkheidsonderzoekingen bij boonen. IV en V,' *Genetica*, 's Gravenhage, 1921, III, pp. 28–49. pp. 159, 181

842. **Vilmorin, J. de**, 'Sur des croisements de pois à cosses colorées,' *C. R. Acad. sci.*, Paris, 1921, CLXXII, pp. 815–817. pp. 159, **182**

843. **Yamaguchi, Y.**, 'Études d'hérédité sur la couleur des glumes chez le riz,' *Bot. Mag.*, Tokyo, 1921, XXXV, pp. 106–112. p. 158

844. **1922. Becker, J.**, 'Ueber vegetative Bastardspaltung,' *Zs. Pflanzenzüchtg*, Berlin, 1922, VIII, pp. 402–420.

845. **Emerson, R. A.**, 'The Nature of Bud Variations as

Page of text on which reference is made

indicated by their Mode of Inheritance,' *Amer. Nat.*, New York, 1922, LVI, pp. 64–79. p. 207

846. **Fruwirth, C.**, 'Zur Hanfzüchtung,' *Zs. Pflanzenzüchtg*, Part II.,' Berlin, 1922, VIII, pp. 340–401. pp. 158, 173

847. **Harland, S. C.**, 'Inheritance in *Ricinus communis* L. Part II.,' *J. Genet.*, Cambridge, 1922, XII, pp. 251–253. pp. 159, **187**

848. **Harland, S. C.**, 'Inheritance of certain Characters in the Cowpea (*Vigna sinensis*). Part III. The very small-eye Pattern of the Seed-coat,' *J. Genet.*, Cambridge, 1922, XII, p. 254. p. 159

849. ***Hutchison, C. B.**, 'The Linkage of certain Aleurone and Endosperm Factors in Maize and their Relation to other Linkage Groups,' *Cornell Univ. Agr. Exp. Sta. Mem.*, Ithaca, 1922, No. 60, pp. 1421–1473. pp. 159, 190

850. **Kristofferson, K. B.**, 'Studies on Mendelian Factors in *Aquilegia vulgaris*,' *Hereditas*, Lund, 1922, III, pp. 178–190. pp. 157, **164**, 191, 193, 196

851. **Lathouwers, M. V.**, 'Recherches expérimentales sur l'hérédité chez "Campanula medium L.,"' *Mém. Acad. Roy.*, Bruxelles, 1922, 33 pages. pp. 157, **166**

852. **Martin Leake, H.**, and **Ram Pershad, B.**, 'The Coloration of the Testa of the Poppy Seed (*Papaver somniferum* L.),' *J. Genet.*, Cambridge, 1922, XII, pp. 247–249. pp. 158, 179

853. **Przyborowski, J. von**, 'Genetische Studien über *Papaver somniferum* L. I,' *Zs. Pflanzenzüchtg*, Berlin, 1922, VIII, pp. 211–236. pp. 158, 179

854. **Punnett, R. C.**, 'On a Case of Patching in the Flower Colour of the Sweet Pea (*Lathyrus odoratus*),' *J. Genet.*, Cambridge, 1922, XII, pp. 255–281. pp. 158, **172**, 192, 193, 207

855. ***Remy, E.**, 'Vergleichende Untersuchungen über weissen, gelben, roten und violetten Mais,' *Zs. Unters. Nahrgsmittel*, Berlin, 1922, XLIV, pp. 209–213.

Chemical analyses show no difference in corns with presence or absence of anthocyanin. p. 159

856. **Setchell, W.**, **Goodspeed, Th.**, and **Clausen, R.**, 'Inheritance in *Nicotiana Tabacum*. I. A Report on the Results of Crossing certain Varieties,' *Univ. California Pub. Bot.*, V, No. 17, 1922, pp. 457–522. pp. 158, **176**

857. **Sirks, M. J.**, 'The Colour Factors of the Seed Coat in *Phaseolus vulgaris* L. and in *Ph. multiflorus* Willd.,' *Genetica*, 's Gravenhage, 1922, IV, pp. 97–138. pp. 159, 181

858. **Tammes, T.**, 'Genetic Analysis, Schemes of Co-operation and Multiple Allelomorphs of *Linum usitatissimum*,' *J. Genet.*, Cambridge, 1922, XII, pp. 19–46. pp. 158, **173**, 193

859. **Terasawa, Y.**, 'Vererbungsversuche über eine mosaik-

Page of text on which reference is made

farbige Sippe von Celosia cristata L.,' *Bot. Mag.*, Tokyo, 1922, XXXVI, pp. 45–83. — —

860. **Tjebbes, K.**, en **Kooiman, H. N.**, 'Erfelijkheidsonderzoekingen bij boonen. VII. Bloemkleur en zaadhuidkleur' (Blüthenfarbe und Samenhautfarbe), *Genetica*, 's Gravenhage, 1922, IV, pp. 447–456. pp. 159, 181

861. **Vavilov, N. I.**, 'The Law of Homologous Series in Variation,' *J. Genet.*, Cambridge, 1922, XII, pp. 47–89. p. 147

862. **1923. Anderson, E. G.**, and **Emerson, R. A.**, 'Pericarp Studies in Maize. I. The Inheritance of Pericarp Colors,' *Genetics*, Princeton, N. J., 1923, VIII, pp. 466–476. p. 159

863. **Gregory, R. P.**, **de Winton, D.**, and **Bateson, W.**, 'Genetics of *Primula sinensis*,' *J. Genet.*, Cambridge, 1923, XIII, pp. 219–253. pp. 159, **186**, 193

864. **Hagiwara, T.**, 'Genetic Studies of Flower-Colour in the Morning Glory,' *Bot. Mag.*, Tokyo, 1923, XXXVII, pp. 41–62, 71–84. pp. 158, **171**

865. **Hammarlund, C.**, 'Ueber einen Fall von Koppelung und freie Kombination bei Erbsen,' *Hereditas*, Lund, 1923, IV, pp. 235–238. pp. 159, 183

866. **Heuser, W.**, 'Beobachtungen über Farbvariationen der Samenschale von *Vicia Faba*,' *Zs. Pflanzenzüchtg*, Berlin, 1923, IX, pp. 178–184. pp. 159, 188

867. **Kajanus, B.**, 'Genetische Studien an Pisum,' *Zs. Pflanzenzüchtg*, Berlin, 1923, IX, pp. 1–22. p. 159

868. **Kakizuki, Y.**, 'Linked Inheritance of certain Characters in the Adzuki Bean,' *Genetics*, Princeton, N. J., 1923, VIII, pp. 168–177. — —

869. **Kappert, H.**, 'Ueber ein neues einfach mendelndes Merkmal bei der Erbse,' *Ber. D. bot. Ges.*, Berlin, 1923, XLI, pp. 43–47. pp. 159, 183

870. **Kristofferson, K. B.**, 'Crossings in Melanium-Violets,' *Hereditas*, Lund, 1923, IV, pp. 251–289. pp. 159, **189**

871. **Lindstrom, E. W.**, 'Genetical Research with Maize,' *Genetica*, 's Gravenhage, 1923, V, pp. 327–356. pp. 159, 190

872. **Miyake, K.**, and **Imai, Y.**, 'Genetic Studies in the Opium Poppy (*Papaver somniferum* L.). I. On the Flower Color,' *Bot. Mag.*, Tokyo, 1923, XXXVII, pp. 1–12. pp. 158, 179

873. **Punnett, R. C.**, 'Linkage in the Sweet Pea (*Lathyrus odoratus*),' *J. Genet.*, Cambridge, 1923, XIII, pp. 101–123. pp. 158, 173, 197

874. **Rasmuson, H.**, 'Ueber die Rübenpfropfungen von Edler und einige neue ähnliche Versuche,' *Hereditas*, Lund, 1923, IV, pp. 1–9. pp. 157, 165

875. **Tammes, T.**, 'Das genotypische Verhältnis zwischen

Page of text on which reference is made

dem wilden *Linum angustifolium* und dem Kulturlein, *Linum usitatissimum*,' *Genetica*, 's Gravenhage, 1923, v, pp. 61–76. p. 158

876. **Tedin, H.**, 'Eine mutmassliche Verlustmutation bei *Pisum*,' *Hereditas*, 1923, IV, pp. 33–43. p. 159

877. **1924. Becker, J.**, 'Schlussfolgerungen aus der Erscheinung der vegetativen Bastardspaltung,' *Zs. Pflanzenzüchtg*, Berlin, 1924, IX, pp. 189–215. — —

878. **Kajanus, B.**, 'Zur Genetik der Pisum-samen,' *Hereditas*, Lund, 1924, v, pp. 14–16. pp. 159, **183**

879. **Kristofferson, K. B.**, 'Colour Inheritance in the Seed Coat of *Phaseolus vulgaris*,' *Hereditas*, Lund, 1924, v, pp. 33–43. pp. 159, 181

INDEX

Numbers in heavy type denote papers in Bibliography.

Printed in the United States
By Bookmasters